Medical Device Cybersecurity for Engineers and Manufacturers

Second Edition

For a complete listing of titles in the
Artech House Computer Security Series,
turn to the back of this book.

Medical Device Cybersecurity for Engineers and Manufacturers

Second Edition

Axel Wirth

Christopher Gates

Jason Smith

ARTECH HOUSE

BOSTON | LONDON

artechhouse.com

Library of Congress Cataloging-in-Publication Data
A catalog record for this book is available from the U.S. Library of Congress.

British Library Cataloguing in Publication Data
A catalogue record for this book is available from the British Library.

Cover design by Joi Garron

ISBN 13: 978-1-63081-991-0

Artech House
685 Canton Street
Norwood, MA 02062

10 9 8 7 6 5 4 3 2 1

The authors would like to thank their spouses, Peggy Wirth, P.J. Gates, and Joyce Smith for their continued support during the composition of the second edition of this book. We appreciate your understanding and patience during the many extra evening and weekend hours when we, once again, uttered the phrase "I'm working on the book!"

Contents

Foreword to the First Edition

Whether we're in the doctor's office for a routine checkup, monitoring our heart rate with a wearable device on a morning jog, or having surgery in the operating room, we rarely think about the devices that we use for medical care being vulnerable to a cybersecurity attack. However, the benefits from rapid advances in the use and capabilities of connected health also come with potential risks.

According to the American Hospital Association (AHA), the United States is home to over 6,000 hospitals, ranging from small community hospitals with less than 50 beds to health systems with over 5,000 beds. In general, hospitals have 3–6 networked medical devices per bed. That means with a total of 931,203 staffed hospital beds across the United States, there are some 3–6 million connected medical devices at hospital bedsides alone, all of which must be protected against cyberattacks. In addition, health delivery organizations, including hospitals, depend on nonmedical devices such as infrastructure systems that keep elevators running, maintain temperature and air quality, and provide lighting and facilitate rapid remote communications for their continued operations. Thus, it is evident that patient safety and quality of care depend on cybersafety, yet we have a large risk exposure of an estimated 10–15 million total connected devices (including infrastructure and supporting devices) that, if compromised, could lead to patient harm or care delivery impact.

This fact is acutely on the minds of both healthcare providers who manage the devices and medical technology and health IT companies who manufacture them. Certain principles must be understood when we think about healthcare cybersecurity:

1. Because the threat landscape is constantly evolving, network and device security have difficulty keeping up;

2. Healthcare institutions do not have the time, money, capabilities, or resources to independently fix cyber vulnerabilities;
3. Patching updates and addressing vulnerabilities in the active medical device ecosystem can be more complicated than your average IT update because there may be a human, not an app or machine, connected to that device and system reboot may not be an option;
4. There are limits to the ability of government regulation to achieve the necessary balance of innovation, effectiveness, security, and privacy.

It is now recognized that these shared challenges are also a shared responsibility. In response, the public and private sectors are working together to address these healthcare cybersecurity challenges in multiple forward-looking ways. It can indeed be argued that working together is an imperative. Healthcare is considered one of the nation's 16 critical infrastructure sectors—like electricity, telecommunications, water, chemicals, transportation, and more. And because private industry owns and operates the vast majority of these critical infrastructures, which constitute essential public resources, it is their responsibility to serve the public good in addition to their commercial interests. That's why this shared responsibility is so important, from the engineers to the C-Suite, to work toward strengthening the security of our healthcare systems and the medical technology that has become indispensable for our nation's health.

The Health Sector Coordinating Council (HSCC)—a public-private partnership of health sector and government stakeholders dedicated to strengthening the nation's critical healthcare infrastructure against all hazards—convenes these interdependent stakeholders to improve the security and resiliency across the healthcare ecosystem. An HSCC task group addressed the medical device security issue head-on, working over 18 months to publish in January 2019 a best practices guide for medical technology companies—the Medical Device and Health I.T. Joint Security Plan (JSP). This is one of the many useful resources referenced in the book you're now reading—*Medical Device Cybersecurity for Engineers and Manufacturers.*

The JSP utilizes security by design principles throughout the product lifecycle of medical devices and health IT solutions. It encourages shared responsibility in the adoption of security-related standards, risk assessment methodologies, and vulnerability reporting requirements to improve information sharing between manufacturers and healthcare organizations. The JSP is a living document and will be updated as appropriate to adapt to the ever-changing threat environment for medical devices and health IT solutions.

More follow-on thought is now being given to how hospitals, device manufacturers, and government can coordinate to better communicate with patients about device vulnerabilities and security, and how we can deal with the challenging issue of aging medical technologies that have reached the end of supported life—whether because of security or operational efficiency—but are not easily replaced because of tremendous expense. Again, the shared challenge meets shared responsibility and collaborative solutions.

These and other critical issues are being addressed in sector-wide work streams, such as:

- An HSCC resource for health providers called the Health Industry Cybersecurity Practices (HICP);
- The FDA's September 10, 2019 Patient Engagement Advisory Committee, which sought to understand how best to communicate cybersecurity risks in health risk communications to patients (https://www.fda.gov/advisory-committees/patient-engagement-advisory-committee/september-10-2019-patient-engagement-advisory-committee-meeting-announcement-09102019-09102019);
- Efforts to define and operationalize the imperative for software bills of materials to help health systems understand which software components are in the devices and systems they purchase and hence how to manage associated risk (https://www.ntia.doc.gov/files/ntia/publications/ntia_healthcare_poc_report_final_draft_2019_0904.pdf);
- Expansion of the annual DefCon Biohacking Village Device Hacking Lab, where hackers, healthcare providers, and device manufacturers collaborate to identify vulnerabilities (https://www.villageb.io/device-lab);
- The International Medical Device Regulators Forum (IMDRF) cybersecurity working group, made up of industry and regulators and co-led by FDA and Health Canada, which is seeking to promote a globally harmonized approach to medical device cybersecurity via drafting of a cybersecurity guide (http://www.imdrf.org/consultations/cons-ppmdc.asp).

We see all these collaborations as signs of significant progress. In 2017, a health care cybersecurity task force of industry and government leaders diagnosed that healthcare cybersecurity is in critical condition. While we have had no reports to date that cyber incidents involving medical devices have led to direct patient harm, our collective action, premised on the recognition that patient safety depends on cybersafety, will go a long way toward maintaining public trust in the security, resiliency, and integrity of the life-saving and life-sustaining devices we depend on, and upgrading our healthcare cybersecurity diagnosis to stable.

Good cyber hygiene is found in the products and systems we design, manufacture, and maintain, and in the management of those devices and the clinical systems in which they operate. Active and continuous adoption and refinement of resources, like this guide, will help medical technology companies—their engineers, service departments, and executive management, as well as their health provider customers—step up to this shared responsibility.

Greg Garcia
Executive Director for the Health Sector
Coordinating Council's Cybersecurity Working Group;
Former Assistant Secretary of Cyber Security and Communications
for the U.S. Department of Homeland Security

Foreword to the Second Edition

The only thing optional about medical device security is whether you choose to read this book. While one could manually piece together the literature on medical device security and filter out marketing chaff posing as engineering, this book does that hard work for you. As the former acting director of medical device security for the U.S. Food and Drug Administration, I would routinely read new books and book chapters on medical device security engineering. Some were great resources, some were a little bit outdated, and, sadly, some were simply wrong. This book stands out as a great resource for anyone entering the field of medical device security or being tasked with the responsibilities of keeping a medical device secure against constantly evolving threats.

In 2008, only a handful of people were sounding the alarm about medical device security risks. Back then, I co-led a team that analyzed the security of implantable pacemakers and defibrillators, ultimately demonstrating how to induce ventricular fibrillation wirelessly by exploiting a design shortfall that did not protect against homemade radio transmitters that performed classic replay attacks on unauthenticated radio frequency (RF) traffic.

Since then, medical device security has exploded as a field and a marketplace. These authors bring valuable perspectives from their experiences in the field. Although we do not always agree on every topic (which should be expected, since this is a still-evolving, multifaceted field), I highly recommend this book to all persons with responsibility for medical device security, whether the role involves engineering, manufacturing, procurement, operation, maintenance, or regulation.

This book is also likely to open the reader's mind and stimulate dialogue because medical device security is a moving target. You can expect change in the years to come. Adversary methods change. Regulations change. New therapies, new diagnostics, and drug-device-data hybrids push the boundaries of technology. Then there is the coming paradigm shift of artificial intelligence (AI) on health care. If you are focused on protecting against one static threat or symptom (e.g., ransomware 1.0), you are likely to miss the forest for the trees and end up designing devices based on an ossified threat model for yesterday's problems. Medical devices must remain secure against both present and future threats.

Medical device cybersecurity is a complex socio-technical field of a dizzying array of stakeholders: patients, clinicians, engineers, medical device manufacturers, security researchers, academics, U.S. Food and Drug Administration officials, U.S. Congressional staffers, international regulators, and more. All these stakeholders, but especially engineers and manufacturers, can learn something new from this book. As the creator of the world's first graduate course in medical device security in 2013 at the University of Michigan (more recently, I have moved my laboratory, the Archimedes Center for Health Care and Medical Device Cybersecurity, and associated courses to Northeastern University in Boston, Massachusetts), I am so glad to have this book in my library.

The Archimedes Center continues to play a crucial role in advancing the field. Recently, we had our first event since the COVID pandemic in New Orleans, attended by many leaders from medical device manufacturing and healthcare delivery organizations. In addition to training on threat modeling and how to establish product security programs, major themes in 2024 at Archimedes included cybersecurity regulatory changes, advancements in AI and machine learning for medical device security, and collaborative approaches to cybersecurity.

The White House President's Council of Advisors on Science and Technology (PCAST) Working Group on Cyber-Physical Resilience has also played a crucial role in enhancing health care cybersecurity. This group made several recommendations to the president on strategies and policies to fortify the resilience of cyber-physical infrastructure. As a member of this working group, I contributed my expertise to help to shape national policies and promote best practices that ensure the security and integrity of healthcare infrastructure. Recent regulatory changes have further shaped the landscape significantly. The U.S. FDA issued draft guidance responding to Section 524B under the Consolidated Appropriations Act, and the finalization of the 2022 draft medical device security premarket guidance document has set new expectations for device security during regulatory review.

Advancements in AI and machine learning have introduced both new solutions but also new vulnerabilities with high expectations by ARPA-H and other sponsors of breakthrough technologies. It will be increasingly important that AI algorithms themselves are resilient and free from manipulation (confidentiality, integrity, availability).

High-profile cyber incidents continue to influence the field. For instance, a significant incident in 2024 involved a ransomware attack on Change Healthcare, which disrupted various healthcare services including billing for months within the United States and highlighted just how unwittingly dependent we are on shadow critical infrastructure that are single points of failure. Some small providers even had difficulty meeting payroll for clinicians because of the national outage. This incident prompted further regulatory scrutiny, Congressional review, and accelerated adoption of advanced security practices.

International collaboration and standards are also evolving. The International Medical Device Regulators Forum (IMDRF), the Association for the Advancement of Medical Instrumentation (AAMI), and ISO/IEC standards are leading to greater harmonization across borders, ensuring that medical devices are secure no matter where they are used.

Innovation in medical device technology continues to push boundaries. New technologies, such as telemedicine and the potential for adaptive remote therapy, introduce new security challenges in difficult to control home environments. Engineers must stay ahead of these innovations to ensure that they are secure by design.

Threat modeling, Software Bill of Materials (SBOM), penetration testing independent of the device design team, analysis of known vulnerabilities, software updates, postmarket management of vulnerabilities, exploits, and incidents—these topics are all motivating reasons to read the book and stay up to date on the topic.

I have lived through dramatic change in the field, especially how the community deals with vulnerability discovery and communication. When I presented my 2008 research to the manufacturer of the affected device, I ran into roadblocks (and other researchers reported similar experiences with other manufacturers). We literally had a meeting in a bathroom stall and whispered, all "off the record." Since then, we as an industry have made significant technical and social and cultural progress.

We are now light-years ahead and productively exchange security and vulnerability information within the community, openly. It is now much less common to vilify security researchers (or manufacturers), and more common to embrace and encourage interaction. Now I see a willingness to solve the problems identified by security researchers, evidenced by device manufacturers having formal "Coordinated Vulnerability Disclosure" processes, providing ways for researchers to submit newly discovered vulnerabilities through the manufacturer website, and communicate across the upstream and downstream supply chain via Information Sharing and Analysis Organizations (ISAOs). There is still quite some way to go, but we have improved significantly over the past 15 years.

Moreover, there are several influential and well-resourced groups contributing to the path forward. Organizations such as the Healthcare and Public Health Sector Coordinating Council (HSCC), the AAMI, and the Health Information Sharing and Analysis Center (H-ISAC) and MedISAO have contributions represented in this book. The Archimedes Center for Health Care and Medical Device Cybersecurity at Northeastern University continues to provide technical training and open workshops that drive the conversation for solving tomorrow's problems. AAMI produced consensus standards TIR57 (Principles for Medical Device Security–Risk Management) and TIR97 (Principles for Medical Device Security–Postmarket Risk Management for Device Manufacturers). In 2023, these AAMI standards were complemented by ANSI/AAMI SW96 (Standard for Medical Device Security – Security Risk Management for Device Manufacturers). HSCC created a number of helpful consensus documents such as the Joint Security Plan (JSP), currently being updated, Health Industry Cybersecurity – Managing Legacy Technology Security (HIC-MaLTS) for dealing with medical devices that outlast their threat models, and the Model Contract-Language for Medtech Cybersecurity (MC2) to assist with representing cybersecurity in procurement language. Greg Garcia's foreword to the first edition of this book goes into further detail on these topics.

Other examples of industry-wide activity include AdvaMed and other industry alliances that help to bring together parties and educate individual engineers and decision-makers, as well as the CyberMed Summit with real-world simulation of cyber events in a clinical environment. Events such as the BioHacking Village co-located with DEFCON also help to bring security vulnerability researchers and manufacturers together in productive ways.

At the time of this writing, the U.S. FDA had recently implemented its response to H.R. 2617 — Consolidated Appropriations Act (Omnibus Bill, December 2022, being the first federal laws requiring medical device security), specifically the "Cybersecurity in Medical Devices: Refuse to Accept Policy for Cyber Devices" (under Section 524B of the FD&C Act) and, in September 2023, finalized and released an updated version of the medical device cybersecurity premarket guidance.

I also want to suggest that persons outside the United States read this book. If a company plans to sell a medical device in the United States, I consider this book required reading. However, even other regions, for example, the EU under MDR/IVDR or Australia's TGA, have defined cybersecurity requirements for medical devices sold into their markets. Although details of the respective legal and regulatory approaches may vary, the way that an engineer would go about designing secure devices as covered in this book would not fundamentally change.

International regulators may differ in the specifics of their approach, but internationally evolving standards (see Chapter 3) are leading to harmonization on a technical and procedural level. See, for example, the FDA's recent clarification on their Cybersecurity Refuse to Accept (RTA) policy (now implemented and enforced via the FDA eSTAR program), which is firmly rooted in ISO/IEC 81001-5-1. Regulators throughout the world are participating in the International Medical Device Regulators Forum (IMDRF) to gather important perspectives and to align strategically.

As one of the first security researchers in medical device cybersecurity, I have been here since the beginning, and that gives me a unique vantage point. In my 15 years in the field, it has evolved and improved significantly. However, we also need to map out the path forward, and along the way, close the gaps that remain. I have known several of this book's authors and contributors for over a decade, having worked with them in standards bodies and other collaborative forums for driving consensus on technical matters, and I am confident that their current and future contributions will help to address these gaps and will help to lead us to a more secure, and therefore safer, future.

Kevin Fu
Archimedes Center for Health Care and Medical Device Cybersecurity
Northeastern University
Boston, Massachusetts
October 2024

1

Why Secure Medical Devices?

1.1 The Inspiration for This Book: The Original First Edition Introduction

We were originally inspired to write this book because of our long history of working with and for medical device manufacturers (MDMs) as well as with healthcare delivery organizations (HDOs), helping these organizations to produce and manage the cybersecurity posture of their medical device ecosystem. The medical device industry was among the first critical infrastructure industries to be targeted by cybersecurity regulation, which put MDMs in the odd position of being trailblazers despite cybersecurity not necessarily being their principal or traditional skill set. Secure device development has nothing in common with the practice of medicine; instead, it has everything to do with subtlety of design, quality of implementation, mature engineering practices, understanding of the threat landscape, and awareness of potential attack vectors. How to balance these concerns and accomplish security in the creation of a new medical device is the overriding topic of this book.

Approaches for incorporating secure development practices into the development lifecycle have not been taught in traditional education programs until very recently; nor have models or best practices specific to the medical device environment been available, although we have seen notable improvements between the first and second editions of this book. This lack of training for engineers, project managers, and senior leadership is a critical shortfall, making it very difficult for MDMs to produce proactively secured devices and preventing them from constructively engaging with their customers and regulators on this issue.

This book provides guidance to MDMs on how to implement a secure medical device lifecycle in a manner that is predictable, repeatable, trackable, and maintainable and produces artifacts needed for regulatory submission, thus improving the security standing of the individual medical devices as well as the larger device ecosystem. This book may also serve HDOs as an informative resource for understanding the security activities and support MDMs should perform, and how HDOs, MDMs, and other stakeholders (such as security researchers, regulators, security agencies, and even law enforcement) can collaborate to keep devices secure and patients safe.

There are many domestic and international guidances, practices, standards, and regulations that address security for medical devices (and IoT devices in general); however, they are inadequately aligned and do not provide sufficient details to actually implement such a program. In other words, they provide the what but not the how. This book helps to close that gap.

We are aware that this is a very dynamic topic and that much of the specific technical content that we address could change or even become obsolete in a few years. Notwithstanding, the principles guiding secure development practices presented in this book will remain viable and valuable even as technology evolves. Moreover, this book fulfills the need to identify a foundational position from which incremental improvements can be made toward maintaining and bettering the secure development lifecycle.

1.2 Why a Second Edition?

Section 1.1 was first written in early 2020, and, at the time of this writing in 2023, it is staggering how much has changed in medical device cybersecurity. Thankfully, the first edition of this book was very well received. It reached #1 on Amazon's engineering bestseller list, won two cybersecurity "book of the year" awards in 2020 and 2021, and is being used in various medical device cybersecurity training programs at the university and professional levels. However, our favorite accolades were the feedback, including constructive criticism, from the many people with whom we interacted who benefitted from the guidance that we and our fellow contributors tried to provide. We have especially loved to see readers hold up the book with a bunch of Post-it notes sticking out, marking specific sections with comments, questions, and pushback.

Our goal, audacious in scope, was to train our industry of medical device engineers and manufacturers on the core principles of securing a medical device. While the industry cannot be said to have matured across the board to the point of achieving that goal, it seems that this book, and the efforts it inspired or supported over the past 3 years, has helped to "move the ball down the field."

We write this new edition amid a seismic shift in the medical device cybersecurity legislative landscape. The most impactful change is that the U.S. Food and Drug Administration (FDA) now has the specific legal authority to directly enforce the security of medical devices [1], not just as a corollary to "safety and efficacy," and there is a planned cadence of annual updates to cybersecurity requirements [2]. These changes include a significant increase in the type and number of artifacts needed to prove that a device is cyber-secure and to provide such proof in the regulatory premarket submissions (e.g., FDA 501(k) or PMA). Under the recently issued Refuse to Accept (RTA) Policy, starting on October 1, 2023, FDA intended to reject device submissions that are deemed insufficient in their cybersecurity documentation [3]. Furthermore, FDA expects manufacturers to supply plans that outline how they will manage postmarket cyber risks, including vulnerability management, mitigation, and communication as well as a roster of periodic security testing to be performed for the life of the medical device (several decades, in some cases). This was further codified on September 23, 2023, when FDA released its final guidance document on "Cybersecurity in Medical Devices: Quality System Considerations and Content of Premarket Submissions" [4], and, no, "guidance" does not mean optional.

With this much change in play, and more expected soon, we can only guess how quickly sections of this book will be rendered out of date. Yet this edition is needed now to help manufacturers adapt to everything that has already happened in the past 3 years, plus future changes that can be anticipated.

In this new edition of our book, we have added new content, updated old content, and are grateful to have our efforts bolstered by more contributing authors who are all subject matter experts in their respective fields. Frequently, these are the people that we ourselves turn to for answers on difficult or niche topics when we encounter them in the course of our work with MDMs, and we are very pleased and honored to have their assistance here. Our contributors' names and biographies are available at the end of this book.

1.3 The Evolution of Cybersecurity in Healthcare

Over the past decades, our understanding of cybersecurity risks in the healthcare industry has been evolving. Healthcare has been moving from a compliance-driven industry to a more cybersecurity-aware one (compliance does not always equate with security). This is largely due to two factors:

1. Driven by government initiatives and the desire to reduce costs and improve efficiency, healthcare has increasingly digitalized (implemented more digital systems) and digitized (amassed more digital data).

This increasing connectivity and integration of systems and devices containing valuable information have not only created a broader attack surface and made health organizations a more attractive target of opportunity for malicious cyber actors, but has also increased dependence on the availability of these systems.

2. Cyber adversaries, driven by financial or political motivation, have become increasingly advanced and their attacks have become more sophisticated, targeted, stealthy, and, unfortunately, successful. Their impact on industries, governments, and individuals around the globe could have reached $8 trillion U.S. dollars in 2023 by some estimates [5].

Consequently, we are no longer thinking about cybersecurity risks merely in the context of our information systems and the data that they hold but now have a more comprehensive view, including the potential impact on patient health, care delivery, hospital operations, and even on public health systems at the national level.

Unfortunately, this learning has come from a number of painful experiences. Ransomware attacks have resulted in the payment of the demanded ransom [6] or have forced hospitals to go through a complex and costly recovery period [7]. For example, the WannaCry attack on the U.K. National Health Service (NHS) in May 2017 affected more than 1,000 IT systems and medical devices, resulting in the full or partial shutdown of 81 NHS Trust hospitals, the cancelation of 19,000 patient appointments, and a financial impact of an estimated £92 million [8]. This was followed by the NotPetya pseudo-ransomware (wiper) attack in June 2017, which impacted only a few healthcare delivery organizations, but impacted hosted documentation services, the availability of pharmaceuticals and vaccines, as well as global shipping and logistics companies [9]. Both attacks resulted in multibillion-dollar losses across the globe and across multiple industries. Yet we need to recognize that both WannaCry and NotPetya were imperfect malware containing design and execution flaws. In other words, in spite of the damage done, it could have been much worse.

1.4 The Unique Role of Medical Devices

As medical devices moved from the early days of beneficial uses of electricity [10] and radiation [11] to a tightly integrated system of systems with complex data flows between devices and between devices and hospital information technology (IT) systems, we became aware that these devices were not only more vulnerable and more difficult to protect, but also that the security compromise of any of these could result in patient harm or impact on care delivery in addition to the traditional security concerns around data confidentiality, integrity, and availability.

Industry stakeholders have been aware of the security risks of medical devices since security researchers demonstrated their ability to hack into an Implantable Cardiac Defibrillator (ICD) in 2008 [12]. Since then, we have seen individual researchers [13], hospitals [14], and government agencies [15] conduct their own additional research, all reaching the same conclusion: Our medical device ecosystem is highly vulnerable and we have historically failed to establish cybersecurity as a buying criterion and as a design objective. Regulators and leading MDMs have recognized the problem and are establishing a path forward.

It should also be understood that even though this book is generically using the term "medical devices," the lifecycle management and security best practices described can equally be applied to a wide range of operational technology (OT) devices that are used on hospital networks, not just the ones classified as medical devices by regulators. A cyber incident involving a hospital elevator impacts patient, caregiver, and equipment transportation; a change in temperature or humidity in the operating theater may force procedures to be delayed; shutting down a blood or organ refrigerator can have a serious impact on patients waiting for a transplant or infusion. Therefore, even though international regulators' guidances are focused on medical devices and their potential security risks to patient harm or care delivery, the practices outlined herein are equally beneficial and can be applied to other care-critical, albeit not regulated, device types and health technologies.

Clearly, cybersecurity for medical devices needs to be proactively addressed on the level of the individual device. However, since no device will ever be perfectly secure (and certainly not devices already in use), additional security measures should be applied on the level of the integrated system of devices, be it in the traditional, hospital-based care environment or in the evolving telehealth and home care settings. A system is only as secure as the sum of its parts and can only achieve a sufficient level of security if all stakeholders accept and perform their responsibilities. Notwithstanding, it has generally been accepted that the security posture of the device itself is the most critical link in the security chain. Regulators and purchasers alike are therefore eager to ensure that device manufacturers accept and perform their responsibility to provide secure designs for devices with maintainable security features [16]. The earlier in a device's development lifecycle security is applied, the more effective those measures will be, resulting in a better security posture at a lower cost over the life of a device.

Today's medical device ecosystem goes far beyond the device or its local network and includes electronic health record (EHR) and image management (PACS) systems, inventory and business automation systems, cloud-based value-added services, and access via mobile devices, home-based routers, and public networks. In order to provide sufficient security, each party needs to understand and accept their responsibility as well as provide their contribution.

In the most general sense, we need to fulfill the four main security tenets: Secure the device, manage the device, manage the ecosystem, and respond to incidents. This is detailed in Table 1.1 by outlining the respective security responsibilities of device manufacturers and device operators (e.g., the hospital).

1.5 Regulatory Environment

Over the past years, many international regulators have taken steps to provide a framework for and to establish their expectations on cybersecurity for medical devices. The U.S. FDA has been taking the lead with its recently finalized Premarket Guidance (2023) [17] as well as its Postmarket Guidance (2016) [18]. Other notable contributions have been and are being provided by the International Medical Device Regulators Forum (IMDRF) [19] and the European Union [20], in addition to a growing list of individual countries such as Canada, Japan, China, Australia, and France [21].

Table 1.1
The Four Security Tenets

Secure the Device	
Manufacturer	**Operator**
Software design best practices Hardened design Complementary endpoint security Cryptography-based information protection	Secure handling Secure media use Integration best practices Secure networking
Protect the Ecosystem	
Manufacturer	**Operator**
Secure remote access Strong authentication Security documentation Enablement and training	Network architecture Event monitoring Firewalls/gateways Enablement and training
Manage the Device	
Manufacturer	**Operator**
Lifecycle and vulnerability management Security verification and validation Vulnerability disclosure Software Bill of Materials (SBOM)	Procurement and contracting Asset management and maintenance Dependency management Risk assessment and management
Respond to Incidents	
Manufacturer	**Operator**
Threat monitoring Postmarket surveillance Customer support Regulatory reporting	Detect, respond, recover Forensic data capture Communication and decision-making External reporting as needed

The regulatory background will be further discussed in subsequent chapters; however, a commonly voiced objection has been that this is "guidance only" and therefore not a legally binding requirement. This is a common misunderstanding, but the fact is that many regulators, including the U.S. FDA, use the term "guidance" to provide clarification on current interpretation and implementation of existing regulations [22]. Adding to the potential confusion, nomenclature may have varying meanings depending on the region and publishing organization. For example, in 2019, a document published by the Medical Device Coordination Group (MDCG) of the European Commission titled "Guidance on Cybersecurity for Medical Devices" [23] sought to clarify the European Union's Medical Device Regulation (MDR) as it applied to cybersecurity, but this particular document, although it is called a "guidance," does not have the force of law [24].

Yet in no uncertain terms, regulators have stated that cybersecurity is not optional. The exact details on how to set up a secure lifecycle management process that results in sufficiently secure devices may be up to the manufacturer, but there is no ambiguity whatsoever over whether such a program is needed. Furthermore, recent government actions (including those mentioned above) have resulted in increased budgets, authority, and enforcement capabilities for regulators.

The medical device industry is highly regulated to ensure patient safety and reliable quality care delivery. Unfortunately, these regulations, combined with a formal design process and conservative decision-making, are just the opposite of what we are used to in information technology, where we want to be flexible and nimble so as to quickly respond to new cyber threats and risks. MDMs need to improve processes to be able to move fast in response to security challenges without compromising device safety or violating regulations. However, we also need to improve device design to provide a more secure state "out of the factory" so that fewer downstream field changes are needed, while also improving designs so that these fewer changes are easier and faster to deploy. This in itself is a significant challenge for manufacturers and device operators.

The challenge facing the medical device industry is to move to a more secure state without leaving the benefits of controlled processes and deliberate decision-making behind. We hope that this book will contribute to this needed change by providing the how-to of medical device security for device manufacturers and their engineers. (Guidance for healthcare providers will be discussed only on a high level, as detailed content for them has been provided elsewhere [25].) We will touch on security technologies and best practices to a certain extent, but will not offer a deep dive. Medical devices are too complex and vary too widely in the use case, risks, and technological capability to allow for a meaningful selection of security tools and technologies in this book. Instead, we will provide a detailed discussion of how to enable a security-capable organization and how to build a process framework to assure security-inclusive lifecycle management for the MDM process.

1.6 Looking Ahead

Over the past decade, much research has been published about the potential risks of failing to secure medical devices. We do not have the luxury of treating this challenge lightly. Security researchers have demonstrated over and over again the often-appalling security weaknesses of medical devices and how they could be exploited by an attacker. We should think of this problem not only in the context of a targeted attack with malicious intent:

- Since 2015, security researchers have repeatedly gathered evidence that medical devices are used as an entry point and beachhead for an attack on other targets [26].
- Even if the device is not the target at all, it may get caught up in an attack if the device happens to fit the threat profile, as happened during the WannaCry malware outbreak [27].
- In another nontargeted event, medical equipment in a Siberian hospital was shut down by ransomware in the midst of brain surgery. Even though the patient was reported unharmed, this was a close call [28].
- We know that even minutes of delay in care can have impact on emergency patient mortality rates [29] and that, in the aftermath of a cyberattack, hospitals saw additional patient deaths due to heart attacks [30]. Due to the care-critical nature of medical devices, any security incident affecting device and service availability will indirectly impact patient safety.

Before the incidents and studies above became available, it was often stated that we have "no reported case of direct patient harm due to a cyber incident." However, were we really able to assess the impact on patient outcomes from cybersecurity events? Did we look? Did we know where to look? Have we even had the right data and capabilities to detect such incidents? The answer to these questions, unfortunately, was no.

We need to resist the urge to allow this discussion to be driven by sensationalism, but we also need to proceed as we have begun, with continued urgency. We hope that this new edition will further contribute to a measured and reasonable approach that ultimately leads to a more secure, and therefore safer, healthcare ecosystem. In the end, this is all about preventing patient harm and preserving patient trust.

References

[1] FDA, "Cybersecurity in Medical Devices Frequently Asked Questions (FAQs)," March 29, 2023, https://www.fda.gov/medical-devices/digital-health-center-excellence/cybersecurity-medical-devices-frequently-asked-questions-faqs.

[2] Nelson, N., "The FDA's Medical Device Cybersecurity Overhaul Has Real Teeth, Experts Say," *Dark Reading*, March 31, 2023, https://www.darkreading.com/cloud/the-fda-medical-device-cybersecurity-overhaul-real-teeth.

[3] "Cybersecurity in Medical Devices: Refuse to Accept Policy for Cyber Devices and Related Systems Under Section 524B of the FD&C Act," FDA Guidance Documents, Final, March 2023, https://www.fda.gov/regulatory-information/search-fda-guidance-documents/cybersecurity-medical-devices-refuse-accept-policy-cyber-devices-and-related-systems-under-section.

[4] FDA, *Cybersecurity in Medical Devices: Quality System Considerations and Content of Premarket Submissions*, September 2023, https://www.fda.gov/regulatory-information/search-fda-guidance-documents/cybersecurity-medical-devices-quality-system-considerations-and-content-premarket-submissions.

[5] Morgan, S., "Cybercrime to Cost the World 8 Trillion Annually in 2023," *Cybercrime Magazine*, October 17, 2022, https://cybersecurityventures.com/cybercrime-to-cost-the-world-8-trillion-annually-in-2023/.

[6] Winton, R., "Hollywood Hospital Pays $17,000 in Bitcoin to Hackers; FBI Investigating," *Los Angeles Times*, February 18, 2016, https://www.latimes.com/business/technology/la-me-ln-hollywood-hospital-bitcoin-20160217-story.html.

[7] Davis, H. L., "ECMC Spent Nearly $10 Million Recovering from Massive Cyberattack," *The Buffalo News*, July 26, 2017, https://buffalonews.com/2017/07/26/cost-ecmc-ransomware-incident-near-10-million/.

[8] Smart, W., "Lessons Learned Review of the WannaCry Ransomware Cyber Attack," Department of Health & Social Care, NHS England, February 1, 2018, https://www.england.nhs.uk/wp-content/uploads/2018/02/lessons-learned-review-wannacry-ransomware-cyber-attack-cio-review.pdf.

[9] Greenberg, A., "The Untold Story of NotPetya, the Most Devastating Cyberattack in History," *Wired Magazine*, August 22, 2018, https://www.wired.com/story/notpetya-cyberattack-ukraine-russia-code-crashed-the-world/.

[10] Waller, A. D., "A Demonstration on Man of Electromotive Changes Accompanying the Heart's Beat," *The Journal of Physiology*, October 1, 1887, https://www.ncbi.nlm.nih.gov/pubmed/16991463/.

[11] Röntgen, W. C., *Über eine neue Art von Strahlen [On a new kind of rays]*, Würzburg, Germany: Stahel, 1896.

[12] Halperin, D., et al., "Pacemakers and Implantable Cardiac Defibrillators: Software Radio Attacks and Zero-Power Defenses," *2008 IEEE Symposium on Security and Privacy*, Oakland, CA, May 18–24, 2008, pp. 129–142, https://ieeexplore.ieee.org/document/4531149.

[13] "ICS Alert: Medical Devices Hard-Coded Passwords," *DHS CISA*, October 29, 2013, https://www.us-cert.gov/ics/alerts/ICS-ALERT-13-164-01.

[14] Bruemmer, D., and F. Hudson, "Medical Device Security in a Connected World," *11th AMC Conference on Securely Connecting Communities for Improved Health*, Chapel Hill, NC, June 22, 2015, https://nchica.org/wp-content/uploads/2015/06/Bruemmer-Hudson.pdf.

[15] "FDA and DHS Increase Coordination of Responses to Medical Device Cybersecurity Threats Under New Partnership; A Part of the Two Agencies' Broader Effort to Protect Patient Safety," *FDA News Releases*, October 16, 2018, https://www.fda.gov/news-events/press-announcements/fda-and-dhs-increase-coordination-responses-medical-device-cybersecurity-threats-under-new.

[16] Van Wagenen, J., "Medical Device Vulnerabilities Continue to Plague the Industry," *HealthTech Magazine*, December 11, 2018, https://healthtechmagazine.net/article/2018/12/medical-device-vulnerabilities-continue-plague-industry.

[17] FDA, "Cybersecurity in Medical Devices: Quality System Considerations and Content of Premarket Submissions, Guidance for Industry and Food and Drug Administration Staff," September 27, 2023, https://www.fda.gov/media/119933/download.

[18] "Postmarket Management of Cybersecurity in Medical Devices; Guidance for Industry and Food and Drug Administration Staff," FDA Guidance Documents, December 28, 2016, https://www.fda.gov/regulatory-information/search-fda-guidance-documents/postmarket-management-cybersecurity-medical-devices.

[19] "Principles and Practices for Medical Device Cybersecurity," *IMDRF*, December 2, 2019, http://www.imdrf.org/consultations/cons-ppmdc.asp.

[20] "MDCG 2019-16–Guidance on Cybersecurity for Medical Devices," MDCG European Commission, December 2019, https://ec.europa.eu/docsroom/documents/38941.

[21] Morgan, C., "Medical Device Cybersecurity Regulatory Publications," *Apraciti*, November 26, 2019, https://www.apraciti.com/blog/2019/11/25/global-regulatory-authority-publications-on-medical-device-cybersecurity.

[22] FDA, "Guidances," https://www.fda.gov/industry/fda-basics-industry/guidances.

[23] European Commission, "MDCG 2019-16 - Guidance on Cybersecurity for Medical Devices," June 2020, https://ec.europa.eu/docsroom/documents/41863.

[24] Biasin, E., and E. Kamenjasevic, "Cybersecurity of Medical Devices: Regulatory Challenges in the European Union," *The Future of Medical Device Regulation*, Cambridge, U.K.: Cambridge University Press, 2022, https://www.cambridge.org/core/books/future-of-medical-device-regulation/cybersecurity-of-medical-devices/AC01289C2DB05E44D0D98A9E66666562.

[25] Wirth, A., and S. Grimes, (eds.), *Medical Device Cybersecurity: A Guide for HTM Professionals*, Association for the Advancement of Medical Instrumentation (AAMI), May 2018, https://www.aami.org/productspublications/ProductDetail.aspx?ItemNumber=6489.

[26] TrapX Security, "Anatomy of an Attack – Medical Device Hijack (MEDJACK)," May 7, 2015, https://trapx.com/trapx-labs-report-anatomy-of-attack-medical-device-hijack-medjack/.

[27] Brewster, T., "Medical Devices Hit by Ransomware for the First Time in US Hospitals," *Forbes*, May 17, 2017, https://www.forbes.com/sites/thomasbrewster/2017/05/17/wannacry-ransomware-hit-real-medical-devices/#741c2b03425c.

[28] Pitalev, I., "Russian Hackers Switched Off Medical Equipment During Kid's Brain Surgery," *Russia Beyond*, July 6, 2018, https://www.rbth.com/lifestyle/328711-russian-hackers-switched-off.

[29] Jena, A., et al., "Delays in Emergency Care and Mortality During Major U.S. Marathons," *New England Journal of Medicine*, April 13, 2017, https://www.nejm.org/doi/full/10.1056/NEJMsa1614073.

[30] Holmes, A., "The Rise of Cyber Attacks and Data Breaches Against US Hospitals Has Been Linked to an Uptick in Heart Attack Deaths," *Business Insider*, November 13, 2019, https://www.businessinsider.com/cyber-attacks-hospitals-rise-in-heart-attack-deaths-study-2019-11.

2

Establishing a Cybersecurity Focus Within Medical Device Manufacturers

This book will present the key elements of a lifecycle-based approach for an MDM's device cybersecurity program. Good cybersecurity practices, especially in a regulated, quality-focused, and tightly controlled engineering and manufacturing environments, as we find it in the medical device industry, requires an approach that combines technology, process, organization, skill, communication, and culture.

We hope that this book will be useful to the individual reader across a spectrum of roles and will help manufacturers to develop or improve their cybersecurity programs. In this chapter, we will lay the groundwork that will help companies, whether they are at the beginning of their journey or have already taken steps to develop and refine their approach and organizational capability. We will provide an overview of the foundational steps: governance and organization, regulatory background, supply chain management, and engineering practices that need to be understood and embraced to be able to deliver secure, and therefore safe, medical devices to the market. These individual elements will then be analyzed in greater detail in the remainder of this book.

Although understanding the international regulatory environment is required for any MDM [1], we will define a lifecycle-based approach to cybersecurity that is largely independent of local regulations and can be applied no matter where your devices are developed, manufactured, and marketed. However, we will use the U.S. Food and Drug Administration's (FDA) cybersecurity guidances as the general reference point due to the FDA's leadership in medical device cybersecurity [2] as well as the fact that the United States is the single largest medical device market in the world [3] and will therefore likely be relevant for most manufacturers.

2.1 Governance and Organization

Cybersecurity for MDMs is not just about process, technology, and security know-how. It is as much a matter of strategy, leadership, and culture. No company is cyber-perfect. Some may have started earlier than others, but every company is on its own path and is following its own strategy. However, without leadership enabling an organization to execute and provide budgets and goals, the path to a better cyber future will be unnecessarily long (and costly) or fail altogether. "The engineers will take care of it" will not work here.

Healthcare providers and patients expect that the devices that they use are as secure as possible and will be secure for the foreseeable future. This is where governance comes in, as it provides not only vision and direction, but also enables and empowers an organization through structure, budgets, and culture. Governance is a framework of rules and practices, based on applicable laws and regulations as well as business objectives, by which leadership defines goals and objectives, establishes responsibilities (duties, privileges, and roles), ensures accountability, proper supervision and control, and information flow and monitoring of implementation and compliance. Governance should already exist as part of a company's business strategy and quality system, but expanding it to be inclusive of cybersecurity comes with its own challenges, including the admission that cybersecurity is a problem that needs to be solved for—or needs to be improved upon.

One unique challenge for medical device cybersecurity is its dynamic nature. Laws and regulations are evolving across the globe, care delivery and connectivity of medical devices are changing rapidly, new technologies regularly make their entrance in the medical device space, and, lastly, cyber-adversaries are becoming more sophisticated, capable, targeted, and malevolent. Therefore, no matter where a company is today, identifying an achievable path forward is critical. Measure where you are today (which includes the selection of the right metrics—further discussed later in this chapter), define goals relative to business and compliance objectives, provide stakeholders the authority, ability, and autonomy to achieve them, and continue to measure and adjust. "We will pass FDA approval" would not be a sufficient governance goal—but it would be an outcome of strong leadership and good governance that is:

- Based on market realities and regulatory expectations;
- Spans the entire organization and all products;

- Takes a risk-based approach to cybersecurity.

The larger healthcare industry is changing drastically with more devices being connected and workflows and data flows being increasingly integrated. Care is moving out of the hospital and devices are moving into patients' homes, not only for monitoring or care management purposes but increasingly for therapy such as home dialysis or home infusion [4]. Yet at the same time, the FDA has stated that [5]: "hospital networks are inherently hostile, therefore manufacturers are recommended to assume an adversary controls the network." If this what we must assume of a hospital network, an equally circumspect approach is at least as necessary in home healthcare settings.

As the larger healthcare industry continues to digitalize (adopting more digital and connected systems), its collective attack surface is growing, and as more health information is digitized (amassing more digital health information), it becomes a more attractive target for cyber criminals.

Although all of this is happening with the best intentions of improving efficiency, reducing costs, and improving diagnostic and treatment options, we need to realize that we also need to manage the downside (i.e., the exposure to cyber risk).

To illustrate the point, under the U.S. Health Insurance Portability and Accountability Act (HIPAA) Breach Notification Rule (45 CFR §§ 164.400-414), HIPAA-covered entities must provide breach notification to the affected individuals. Additionally, for breaches affecting more than 500 individuals, within 60 days they must provide notification to the affected patients, Health and Human Services (HHS), and, in certain circumstances, to the media [6]. Breach data is published by HHS on a web site [7] colloquially referred to as the "Wall of Shame." As shown in Figure 2.1, we can see a steady increase in the number of reported breaches since full-year reporting began in 2010.

The following specific trends can be observed:

- Continual year-over-year increase (except 2015) with an 10.5% CAGR 2010–2023, a more than threefold increase over 2010.
- "Hacking/IT Incidents" now account for 80% of all breaches (up from 5% in 2010) followed by a distant second "Unauthorized Access/Disclosure" at 17%.
- Even more remarkable is the fact that, since 2015, breach growth has been solely driven by "Hacking/IT Incidents."
- The number of very large (more than 1 million) breaches is now in the double digits at 10 (2021), 11 (2022), and 26 (2023).

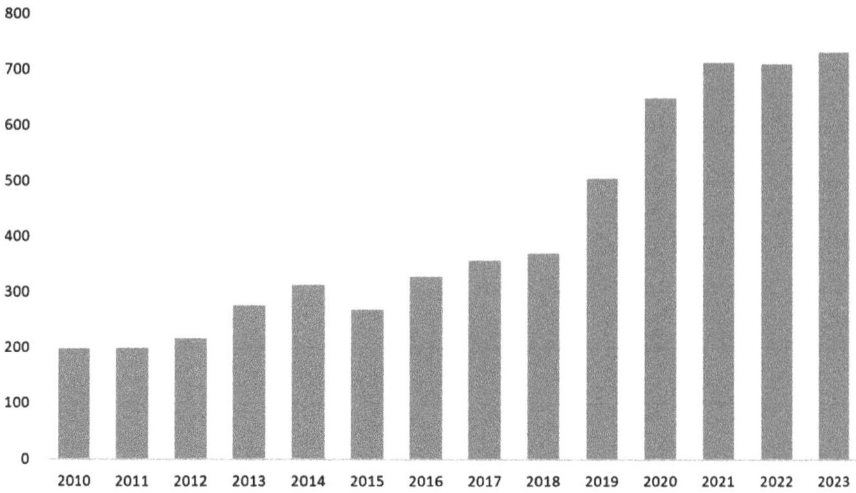

Figure 2.1 HHS "Wall of Shame" reported breaches affecting more than 500 individuals.

The data and its visualization in Figure 2.1 clearly show the steady increase and significant impact of ransomware and other protected health information (PHI)-breaching attacks on the healthcare industry.

As a result of the high exposure of the healthcare sector, and the potential significant consequences of a cybersecurity event, governments around the world and agencies responsible for security and safety in healthcare have started to establish new laws and regulations that address cyber risk on a national level, reduce the risk exposure of critical infrastructure, and, specifically in healthcare, improve patient safety and ensure the reliability of care delivery. Typically, these laws and regulations fall into three main categories:

1. Regulating care providers to reduce the vulnerability of the sector by protecting confidentiality, integrity, and availability of sensitive health information as well as ensuring safe and effective delivery of care. Examples are healthcare-specific laws such as HIPAA in the United States, Personal Information and Electronic Documents Act (PIPEDA) in Canada, and Personal Information Protection Act (PIPA) in Japan. Additional, nonsector specific laws and regulations may apply such as the European Union (EU) General Data Protection Regulation (GDPR) or California's Consumer Privacy Act (CCPA). These regulations define the rights of data subjects and the duties of data controllers and processors. No matter whether data is stored on hospital IT infrastructure, on personal health and wellness devices, or in the cloud, sensitive data at rest or in transit requires appropriate protection in alignment with applicable laws and regulations.

2. Regulating producers of health software, medical devices, and similar systems. Typically, these types of regulations deal with aspects of safety and effectiveness and require compliance with security processes, engineering practices, and testing to demonstrate that security requirements are met and security controls work as intended. These regulations are typically an extension of existing medical device regulations to address cybersecurity aspects. For example, U.S. FDA Premarket and Postmarket Guidances, EU MDR/IVDR (and specific security guidance under MDCG 2019-16), Health Canada, Singapore Health Sciences Authority, and Therapeutic Goods Administration (TGA) in Australia. However, regulators' focus is not limited to producing more secure devices; they also expect that manufacturers demonstrate how they plan to maintain devices' security posture throughout their useful life.
3. Using government buying power to steer the market, as, for example, demonstrated by U.S. Executive Order 14028 on Improving the Nation's Cybersecurity [8] and CISA's Secure Software Self-Attestation Common Form [9], each of which establishes requirements for software security as well as the supporting documentation to be provided.

2.2 Building a Security-Capable Organization

2.2.1 Implementation and Oversight

Once applicable laws and regulations are identified and understood and cybersecurity governance is established, an organization can start implementing (or improving) its product cybersecurity program. This includes the definition of goals, identification of metrics by which progress will be measured, and continual oversight and course correction as needed. As shown in Figure 2.2, the following are the key elements of such a program:

- *Governance and oversight:* Establishing cybersecurity governance is what will lead an organization to success. A governance team should be established to take on that responsibility. The exact makeup of that team will vary widely depending on an organization's size or whether it is a private or publicly traded company with a formal board of directors [10]. Either way, the right level of expertise is required, and external resources may be needed until sufficient maturity is reached. In addition to establishing the initial strategy, the governance team needs to stay informed on new risks, changing regulations and standards, and the experience of its own organization based on devices in the market.

Figure 2.2 Key elements of a product cybersecurity program.

- *Leadership and culture:* There should be no question that cybersecurity is an identified strategic mission. Establishing the right culture across all levels of an organization, and even extending externally to include partners and suppliers, will provide the necessary buy-in and will lead to successful execution. There should be no doubt about unwavering adherence to the defined strategic security initiatives.
- *Cross-organizational adherence:* No organization is homogeneous. Especially in the medical device industry, many larger manufacturers have grown through acquisition and different entities may be at different levels of security maturity. The key is to identify security capabilities across the organization, develop a plan to bring everybody to the same level (while recognizing differences between products and use cases), and implement the capabilities to do so (e.g., through a corporate security division such as a Security Center of Excellence). The end goal is to have complete buy-in and adherence to the same strict standards across business segments and products.
- *Goals and roadmap:* It would be easy to bring an organization to a standstill with the attempt to move from no security to a full, mature program out of the gate. A more sensible and economically feasible approach is to identify a set of goals that will be achieved over time as the organization matures. These goals can still be aggressive, but, with top leadership support, should be attainable and should follow a well-planned stepwise approach.

- *Measures and metrics:* With goals comes the need to track how an organization is doing in the implementation. No plan is perfect, and no execution will be impeccable. Therefore, identifying the right maturity metrics (further discussed later in this chapter) and continually measuring progress relative to it enables an organization to adjust plans or improve on execution. Cybersecurity will never be "done"; therefore, continual measurement of progress should be a routine process and should be on the agenda of every board or executive meeting.
- *Budgets:* Define the business case for cybersecurity (see Section 2.3) and translate needs into budgets. The challenge with this is that cybersecurity program implementation has specific costs (staffing, tools, training, consultancy), but the benefits are more difficult to measure. Typically, the hypothetical future cost includes potential loss of business, delays in the market introduction, regulatory fines, lawsuits, cost of recalls and corrective action, damage to reputation, and loss of staff. As Dr. David Brailer, former National Coordinator for Health IT, once stated [11]: "Infrastructure is a hard sell: Concentrated cost with diffuse benefit." This is equally true for cybersecurity.
- *Organizational structure:* The overall implementation of a security-capable organization will vary widely. With small manufacturers, this will come down to people having to wear multiple hats. Larger organizations will be able to afford dedicated cybersecurity expertise at varying levels. The largest manufacturers will even be able to establish corporate security functions and teams that can lead strategically, help with expertise where needed, and become the public face in all matters of security. However, no matter the size or actual implementation, it is key to avoid conflict of interest. Persons making critical security decisions (e.g., whether a new product is sufficiently secure to be released to market) should not have direct economic interests that could be impacted by that decision.
- *Capability:* Review current competency and capability and define the need for additional staff, additional training, or complementation through external resources. Cybersecurity hiring is tough and finding the right people for medical device security roles is even more difficult. Therefore, an approach of capability development augmented with external expertise will most likely be the practical path for most organizations.
- *Training:* As part of capability development, make resources available to your teams (based on individual needs). Training courses [12] and train-

ing material, such as this book, have now become available and should be utilized.

- *Motivation:* For the purpose of motivation, continual improvement, and communication, members of the security teams should be encouraged to participate in internal security threats (e.g., internally organized hackathons) and external security events (e.g., RSA, DEFCON, Biohacking Village) as well as industry-wide sharing events (note that before participating in an open-registration or public security testing event, careful consideration should be given to the possible exposure of intellectual property (IP) that could be exposed in the course of the event).
- *Maintenance:* Recognize the changing nature of security and remain proactive by monitoring changes in technology, market needs, threat landscape, regulations, and standards. Establish means and metrics that allow the organization to identify gaps, measure maturity, and provide vision (roadmap) and a path forward. Update governance as needed and ensure communication across organizational stakeholders.
- *Communication:* Develop the right communication model(s) for internal and external, regular and crisis, and technical and regulatory communication. Cybersecurity topics can be sensitive, but, in general, open communication is viewed as a sign of security maturity. Constructive engagement with industry organizations and regulators will go a long way. Security is a team sport; the more we share, the better we will be prepared.

2.2.2 A Lifecycle Approach to Cybersecurity

We will discuss the details of a lifecycle-based approach to security engineering in later chapters; however, a general overview of the secure lifecycle development process is warranted here. From a governance perspective, it is critical that the mandate for formal premarket and postmarket processes is clearly defined and prescribed. Without a plan, there is no direction, and with no direction, the probability of reaching the desired target on time and at cost is unlikely.

(A quick note on terminology: The FDA expects management of the Total Product Lifecycle (TPLC). This is accomplished, in part, through the tenets and principles of the Secure Development Lifecycle (SDLC), which, in turn, may be implemented via a Secure Product Development Framework (SPDF) and a Secure Software Development Framework (SSDF)).

Considerations for building an appropriate lifecycle framework should include:

- A collection of security policies, plans, and processes (PPPs) aligned with the organization's quality system: Develop a set of SDLC PPPs for managing the TPLC, stating with governance and covering all aspects of the development process through market release and postmarket management. Where needed, identify touch points with nonsecurity PPPs; for example, the security risk management process should be linked with the safety risk management process for security risks with safety implications and for safety risks that may impact security (discussed in more detail in AAMI TIR 57). Separate security PPPs are not required in all cases; depending on organization and product, security aspects may be covered as a security extension or section to an existing nonsecurity PPP.
- Premarket cybersecurity management lifecycle: Specifically establish a set of PPPs that govern all security activities and ensure documentation of activities performed, from initial product concept to market release. Ensure and document traceability of security requirements and identified risks to final product verification and validation (V&V) to demonstrate that requirements are met and risks are mitigated and reduced to an acceptable level.
- Postmarket management and continual improvements: Establish PPPs that define how security risks for fielded products will continually be monitored, managed, and mitigated, including cybersecurity signal monitoring, vulnerability management and communication, and feedback into the organization to allow for continual improvements. Most regulators (including the FDA) expect MDMs to have postmarket management plans in place prior to market release.

2.2.3 Design and Implementation of Best Practices and Principles

Although executive decision-makers should not be involved in the day-to-day workings of how security is achieved (via tools and practices), it is still their responsibility that the appropriate decisions are made and that implementation is enabled through budgets, staffing, training, as well as measures and gap assessment. Specifically, the following should be on everybody's checklist:

- *Tools:* Implement the right set of tools to enable efficient and reliable execution of security processes. This includes tools to support secure coding practices, security scanning and testing, risk and vulnerability management, and the generation of regulatory and customer-facing security artifacts such as Software Bill of Materials (SBOM).

- *Practices:* Identify and implement security engineering best practices such as secure coding standards and hardened design and follow security engineering principles such as:
  - *Security and privacy by design:* A set of focused security and privacy practices, including threat models, security risk analysis, privacy assessments, security testing, and vulnerability management activities.
  - *Minimization of the attack surface:* Reduce, as far as reasonable and practical, the entire area of a system or network that could be exposed to cyberattacks. The attack surface should be minimized so as to minimize the opportunity for an attacker without impeding the device's functionality or usability.
  - *Data minimization:* Collection of personal identifiable information (PII) as well as other sensitive data (e.g., IP or PHI) should be minimized as required by the use case. Ensure that data collection is adequate, relevant, and limited to what is necessary in relation to the purposes for which they are acquired and processed.
  - *Minimize data retention:* Reducing data retention time reduces legal and security risks of exposure due to storing too much information or storing it for an unnecessarily long time. Data retention management features should be designed into the device, including the ability to wipe all sensitive data when the device is decommissioned or changes owners.
  - *Role-based access controls (RBAC)*—Robust device access controls complement user access controls to ensure that devices cannot access the networks with the proper authorization. A zero trust system must monitor the devices attempting to access the network to minimize its attack surface.
  - *Cryptographic protection of sensitive data at rest and in transit:* Encryption of data on systems or devices as well as encryption of data exchanged between systems to ensure confidentiality and to preserve integrity and authenticity (cryptography will be further discussed in Chapter 12).
  - *Protection of cryptographic material:* Protecting keys from exposure or compromise. Within a cryptographic design, keys are the most important asset; a compromised key can give access to systems or data or allows a malicious user to perform the same actions as any authenticated user.
  - *Implementation of trust zones:* Defining areas of a system where the connections within the zone are treated as at the same level of trust.

The different zones within a system provide hardware and software-based levels of isolation for the purpose of protection.

- *Design for zero trust:* Users, devices, and connections are designed so they are never trusted by default, even if they are connected to a permissioned network such as a corporate LAN and even if they previously had been authenticated. The zero-trust principle applies to the design of the device itself (especially for more complex systems of systems) but also includes the requirement to be able to operate in a zero-trust network environment.
- *Default to the highest level of security:* When delivered to a user, by default, the device should be secure, and it should be up to the user to make conscious decisions about reducing the level of their security. For example, the use of hardcoded, shared, or default passwords would be a clear violation of this principle.
- *Least privilege access:* The concept and practice of restricting access rights for users, accounts, or resource permissions and granting each user or entity, based on their need (i.e., respective role and function), the minimum necessary access permissions. Also known as principle of least privilege (PoLP), principle of minimal privilege (PoMP), or the principle of least authority (PoLA).
- *Defense in depth:* Layering of security risk control measures so to protect valuable data and information If one control fails, another control steps up as an additional barrier against an attack.
- *Security event detection and recovery features:* Identification, logging, and alerting to security events as well as specific features that aid with recovery from successful attacks.
- *Failsafe mechanisms:* Ensure that if software fails, it fails securely, that is, when in failure mode, the security risk remains acceptable and/or does not introduce new security risks. Ideally, a device should maintain the required minimal functionality even when compromised.
- *Security maintenance and update:* Provisioning of updates or patches to devices in the field to address newly identified security risks. To provide for timely delivery of patches and to enable efficient update of fielded devices (with or without the support of local personnel), over-the-air (OTA) update capability is desired, which implies security features like code signing to prevent compromised code (e.g., computer viruses) from being installed.

- *Key activities and documents* produced during design and implementation, derived from [5, 13]:

- Security Risk Management Plan should:
  - Address how software components will be monitored for disclosure of vulnerabilities and how these identified vulnerabilities are evaluated for security risk in the medical device.
  - Address how postmarket surveillance of all utilized third-party software components is performed.
  - Address how any security-related information collected by a fielded medical device is reported back to the manufacturer and its subsequent evaluation.
  - Reference an associated Coordinated Vulnerability Disclosure policy and procedure.
  - Address how software updates are performed on fielded medical devices.
  - Address how the software update process is also utilized to perform a cadence of periodic patches to fielded devices.
  - If available from previously fielded medical devices, then the following metrics should also be included:
    - Percentage of identified vulnerabilities that are updated or patched.
    - Time period between first knowledge of a vulnerability to the general availability and release of the update or remediation.
    - Time periods required to reach percentage milestones during an update or patch operation, such as 50%, 60%, 70% (may require data collection from fielded deices, directly or through customer reporting).
  - Should address which secure product development framework was adopted and the justification for that selection.
- Threat modeling, including justifications for the methods utilized, covering:
  - Supply chain;
  - Device design and architecture;
  - Production (including transfer to manufacturing);
  - Deployment;
  - Interfaces;
  - Patches and updates;
  - Decommissioning features and process.
- Security architecture views:
  - Device hardware;

    - Configuration and other devices in the system;
    - Expected healthcare delivery organization (HDO) operating environment;
    - Communications/networking;
    - Connected servers;
    - Global system;
    - Risk of multipatient harm;
    - Updateability;
    - Security use cases;
    - Sequence diagrams of security-related communications.
- Supply chain vendor assessments;
- Management of third-party software components:
    - Licensing and contracting;
    - Level of support;
    - End of support dates;
    - Risk assessment of all known vulnerabilities;
    - Method of identifying vulnerabilities;
    - Compensating controls utilized;
    - SBOM.
- Risk assessment of all known software vulnerabilities;
- Testing of effectiveness of all utilized security mitigations;
- Vulnerability testing;
- Security Risk Management Report:
    - Summary of all security activities, including identified vulnerabilities;
    - Summary of all utilized mitigations;
    - Summary of any remaining known vulnerabilities and dispositioning of their respective risks;
    - Tracing from security design inputs to implementations and testing of mitigations.
- Manufacturer Disclosure Statement of Medical Device Security (MDS^2), defining the key security properties of a product.

Note that these practices are not limited to the product itself but should also be inclusive of the larger ecosystem such as the production environment, remote support, and the target clinical and technical integration environment (including integration with enterprise security systems).

2.2.4 Security Testing

An insecurely designed product cannot be made acceptable through testing, but a well-designed product can be improved through testing and the testing outcome can act as evidence that the product is secure and will operate as intended. During the various lifecycle stages, security requirements are defined, a secure design and architecture has been developed and assessed using threat modeling, and secure implementation and integration practices have been followed. Correspondingly, testing and assessment activities mirror the design and development activities so to establish with confidence that the security objectives of each phase have been met.

During the individual design and implementation phases, requirements, risks, and controls (or acceptance of residual risk) are documented. The various testing activities should establish traceability back to risks and controls and document evidence of successful mitigation and implementation.

Once the finished product is assembled, it can be tested in its completeness and the final V&V activities demonstrate that the product meets its security requirements, that implemented security controls work as intended, and that it is fit for market approval. However, it would be a folly to expect the final V&V testing to be successful if no other testing (and resolution of deviations) has been performed at previous lifecycle stages. If only the finished product were tested, the number of findings would likely doom any pending market release.

The relationship between security verification testing and security validation testing is illustrated in Figure 2.3.

Testing activities may include:

- Definition of testing procedures, documentation requirements, and metrics as well as plans to support continuous improvement.
- Specification, identification, and qualification of testing tools.
- Incoming inspection and testing of third-party code as part of supply chain risk management activities.
- Code review for adherence to identified practices.
- Specific testing activities throughout development and integration phases (Malformed inputs system testing, Code Review (CR), Static Application Security Testing (SAST), Dynamic Application Security Testing (DAST)).
- Use of automated testing tools (e.g., fuzz testing).
- Verification testing and documentation to establish traceability to identified requirements, risks, and controls.

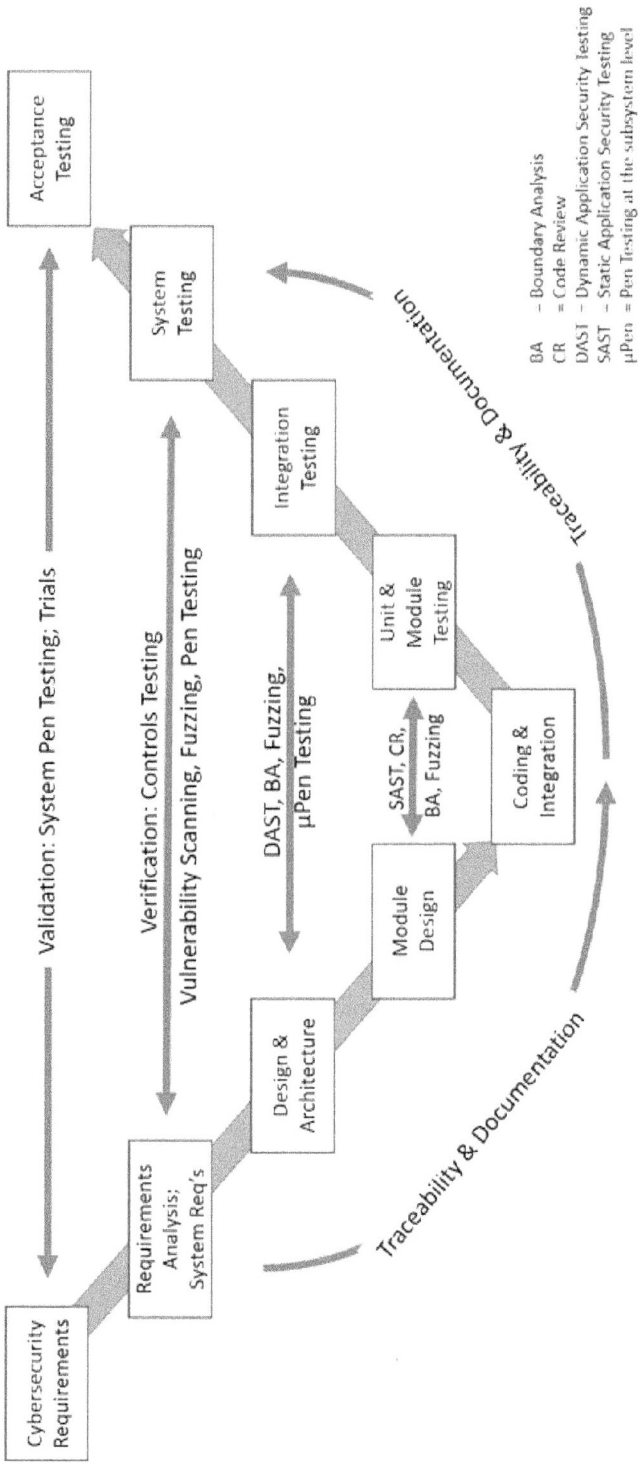

Figure 2.3 Cybersecurity traceability across the lifecycle.

- Testing and documentation activities to support requirements of specific customers (e.g., U.S. Department of Defense (DoD)).
- Validation of security mitigations' effectiveness and documentation in a traceability matrix.
- Penetration testing (often called simply "pen testing"). Note that pen testing should be performed by an entity that has no business interests associated with the product. Due to this unique requirement, contracting external resources may be advisable.

2.2.5 Postmarket Management

Once released to market, MDMs are required to continually monitor cybersecurity signals and perform postmarket management and vigilance activities. This is codified in various regulations by national agencies through, for example, FDA Postmarket Management, or for the EU, under MDCG 2019-16. Typical postmarket management activities include:

- Establish a Coordinated Vulnerability Disclosure (CVD) and response plan to allow for the submission of newly discovered vulnerabilities through external entities (e.g., researchers) as well as plans and processes to manage the overall postmarket activities.
- Monitoring of product supply chain (open source, commercial, contracted, and embedded software/firmware) for new vulnerabilities, reported exploits, and changes in support and end-of-life (EOL) status.
- Monitoring of fielded devices and identification and mitigation of new risks based on discovery through security event detection and logging, customer reports, security researcher reports, and returned devices. MDMs should also establish continual testing cadence to account for changes in threats, vulnerabilities, and testing capability.
- Establish communication and response plans to enable appropriate action based on identified risk level, especially risks that could affect patient safety.
- Communication of identified risks to customers, information sharing organizations, and, if needed, to regulators.
- Release of mitigations, support of deployment if needed. Update of product security documentation as needed (e.g., SBOM and MDS[2]).

Public and private entities have adopted the practice of Coordinated Vulnerability Disclosure (CVD) to forestall future breaches and prevent their spread. As part of response planning, CVD allows a controlled, confidential process for the submission of newly discovered vulnerabilities through external entities (e.g., security researchers) to a software manufacturer. As such, it helps to guide the overall process of managing response activities, boosting the efficiency of communication and coordination while maintaining appropriate boundaries around sensitive information, including personal information (PI) and IP. CVD programs include a path for reporting potential safety risks and connecting those reports and associated response activities to existing regulatory reporting, corrective action, or recall processes. Industry organizations, such as ISACs or ISAOs, can be helpful in supporting CVD activities.

These policies and practices represent a significant change from the early days of medical device security research, where researchers reported having difficulties communicating their findings to manufacturers. Over the past few years, several leading medical device manufacturers have designed and implemented CVD policies, and as a result have been globally recognized as industry leaders with fully developed and operationally mature processes behind their policy. Various government and public-private partnerships with interests securing critical infrastructure have also provided significant leadership in this area. Notable examples include Philips [14], Beckton Dickinson [15], Medtronic [16], Johnson & Johnson [17] as well as government agencies such as the CERT [18], HHS [19], and CISA [20] in the United States, and ENISA [21] in Europe. When public media attention is drawn to security incidents, it is these organizations that are often singled out as the ones prepared for difficult security issues and capable of addressing them.

2.2.6 Internal Training for Non-Cybersecurity Engineers and Managers

Cybersecurity is for everyone. In today's connected world, all of us would do well to understand the basics of good cybersecurity habits and practices. Our actions are likely to have implications not just for ourselves, our families, and the organizations for which we work, but could also impact the lives of thousands of people we may never meet, especially if we work in a critical infrastructure industry like medical device manufacturing.

MDMs employ people in many roles, and it can be easy for anyone who does not work directly on cybersecurity to reduce cybersecurity to an afterthought or frame it as somebody else's problem. This attitude can be common for nontechnical staff, management, and business leaders as well as technical contributors. Do electrical engineers consider security when laying out a design for a new printed circuit board (PCB)? Are mechanical engineers accounting for security when selecting hardware? Are quality assurance (QA) personnel prepared to account for cybersecurity risks in their safety evaluations? Do the executive assistant and the office manager understand why it is a bad idea to use a universal serial bus (USB) thumb drive to transfer a file to the office printer as backup process for an unresponsive network? Does the comptroller know how to recognize a possible phishing email, or what to do with a suspicious embedded link?

Quantifying the human element in root-cause analyses of cybersecurity incidents is complex and yields varying results, but virtually all studies agree that human error is a major contributor. Attackers are frequently able to gain access to their targets, not because of a flaw in code or hardware, but because a person did something that he or she should not have, or failed to do something that he or she should have. Estimates for cyberattack success rates due to human error range from 20% to as high as 95% [22]. It is therefore imperative that all critical infrastructure industries, including MDMs, undertake a program of regular training of all personnel. Training should be mandatory for staff at all levels, contractors, vendors, certain supply chain personnel, and, if appropriate, maintenance, building managers, and others who have access to the physical spaces where the MDM's work is performed.

From the perspective of the SDLC, training for non-cybersecurity staff should be role-based and scope-based on the physical and logical proximity of the trainee's function and work activities to medical devices. In other words, the engineer designing the circuit boards that go into the product needs a different type of training than the comptroller. However, if the office manager's responsibilities require connecting to the same network accessed by the quality manager for research and development (R&D), both may need similar or the same training with respect to that network.

For some roles and responsibilities, specifically, those furthest removed from actual medical device development, production, and support, and with no access to the people, processes, and equipment used for those purposes, training in traditional IT cybersecurity hygiene may be sufficient. However, all personnel whose roles or responsibilities include access to the people, processes, or equipment used in medical device development, production, or support should be regularly trained in at least the basics of operational technology (OT) and medical device cybersecurity, plus any role-based or function-specific training as it pertains to their role in developing, manufacturing, or supporting a medical device throughout its useful life.

2.3 Communicating Cybersecurity Needs, Costs, and Risks to Senior Leadership

Leadership buy-in is critical to ensure success in building and maintaining a medical device's cybersecurity. Alas, communicating about product cybersecurity needs, costs, and risks to senior leadership is often an exercise in frustration for all involved. Cybersecurity is often perceived as a cost rather than a value-add, and therefore resources always seem to fall short of needs and risks may go unaddressed until something bad happens. However, as internet connectivity becomes ubiquitous and vulnerabilities increase in number and pervasiveness, executive decisions are needed more often and more rapidly. Once a medical device has reached the postmarket phase of its lifecycle, managing newly discovered and evolving risks requires a quick response and sound decisions. These decisions can impact the whole organization. Cybersecurity teams need trust, support and resources, and fast decisions from leadership.

First, start with what you can agree on. Begin a conversation with a leader by getting on the same page. In this case, consider starting by acknowledging that the world has changed, and therefore the old approach to product security will have to change. Historically, many leaders have not had a need to understand or execute cybersecurity requirements for medical device development, so cybersecurity investment is indeed a change. Until relatively recently, cybersecurity was perceived as an optional or esoteric side-effort rather than an integrated function through the development lifecycle. Now, to meet the (newly clarified) cybersecurity requirements, medical device executives need to accept that there will be changes to how products are developed. Changing the way that products are developed requires a change to how things have traditionally been done. Building a product or a program with sufficient and appropriate security requires new skills, different timelines, different budgets, and new tools. An initial conversation starting with acknowledgement of this new reality goes a long way toward getting leadership buy-in on cybersecurity needs.

Second, be clear on the goal for this communication. Know what you are seeking. Do you need a decision to be made? An investment? Are you only trying to ensure that you have communicated the risks that you perceive, but you do not have an ask? Clarify what is needed before approaching leadership. Consider multiple options. Consider whether you are ready to ask for a decision, or rather you are establishing the beginning of a conversation without a specific objective in mind.

Third, avoid common pitfalls. From our experience, these are common causes for failed communications with leadership about cybersecurity needs:

1. Lack of common understanding of cybersecurity risks;
2. Lack of context for the risk and its tradeoffs;

3. Talking to the wrong leader.

We describe the common pitfalls and solutions next.

2.3.1 Lack of Common Understanding of Cybersecurity Risks

Cybersecurity risk is complex, requiring technical understanding that is in short supply. Anyone paying attention to the state of the market is very aware of the large shortfall of cybersecurity experts [23], yet sometimes we seem to communicate as if we expect leadership to quickly grasp the security scenarios before us and understand the risks and all the implications thereof in real time, as vulnerabilities and exploits are exposed. We seem to forget that people outside of cybersecurity do not have a common understanding of the risks that come from cybersecurity weaknesses. It is only safe to assume that our colleagues have the awareness that could be heard on the evening news (sound bites about breaches and ransomware).

One solution to this is training and continual education for leadership, training that is short and to the point of what expectations should be for cybersecurity responsibilities, reporting, and decision-making inside of their organization. Even the leadership's awareness of changes to legal obligations may not be up to date, including such sea changes as the recent change to the Food, Drug, and Cosmetic Act, which gives a mandate to the FDA to ensure that new medical devices are secure [24].

Barring organizational training for leaders, individual experts will have to carefully articulate cybersecurity concerns and needs by creating a common understanding within individual conversations. This usually requires translation, translating a particular risk or need out of cybersecurity terms and into business terms.

Humans tend to underestimate the likelihood of rare or distant risks. For example, when faced with a decision about risk, especially new types of risk that have not yet occurred in our personal experience, humans tend to underestimate the probability that it will happen to them [25]. This is called "optimism bias," a cognitive bias that causes people to believe that they are personally less likely to experience bad events, even when they are aware of others experiencing the same events. Humans systematically and predictably make risk communication mistakes due to a long list of well-established biases [26]. Optimism bias directly impedes our efforts to identify and mitigate risks, thus preventing bad things from happening. In the case of medical devices, the bad things include hospital networks shutting down or patients getting harmed, usually through the lack of availability of the device's clinical function [27].

Cybersecurity risks have broader impact than other forms of postmarket risk. One attack can bring widespread failure. In the past, when patients have been harmed by medical device failures, it was usually only small numbers of patients who experienced specific, individual failure events over an extended time period, so leadership is not accustomed to addressing cybersecurity scenarios that can rapidly escalate and could impact large numbers of patients.

Communicate vivid and real examples to reduce normal human biases. To counteract psychological blind spots, cybersecurity advocates can use the same psychological biases to our advantage. For example, we can make a rare risk seem vivid by sharing case studies and examples of recent events in the context of the associated business risks:

- Organizational loss of reputation;
- Negative impacts to stock prices;
- Regulatory actions (e.g., mandatory recalls, submission delays, additional oversight);
- Regulatory fines levied or lawsuits filed;
- Loss of the organization's IP;
- Published accounts of care disruption or even patient harm.

When past examples of risk occurrence are scarce, descriptions of hypotheticals can be useful. Plausible hypothetical newspaper headlines may do the trick. Describe what has changed about the environment (e.g., ransomware, FDA actions, interoperability, cloud connections) and then draw the connection to the cybersecurity risk that results from the change. This will help the leader answer the internal question "I have never needed this before in my more than 20-year successful career. Why is it different now?"

Working to communicate in a way that decision-makers can understand is difficult but important. Infamously, engineers at one NASA subcontractor knew that the O-rings on the Space Shuttle *Challenger* might not maintain a seal in freezing temperatures. The engineers attempted to share their knowledge of the risk, but their middle managers, reluctant to be the bearers of bad news (and its attendant implications for cost and schedule impacts), softened the engineers' report as they passed the message up the management chain and across to the prime contractor and again to NASA itself, with the result that true extent of the risk was never effectively communicated to decision-makers and senior leadership went forward with the launch anyway [28]. This event has marked history as a risk communication failure, resulting in preventable loss of life. Failure to effectively communicate cybersecurity risk may result in the same.

2.3.2 Lack of Context for the Risk and Its Tradeoffs

Cybersecurity risk is just one of many risks and benefits that must be considered when making decisions about medical devices. When communicating with leadership, we are either asking for something or we are communicating risk to inform their decision-making. Cybersecurity risks are inputs to decision-making rather than decisions in and of themselves. Decisions may be to increase budgets, change timelines, change software architecture, discontinue use of components or code from previous product generations, establish new programs, reduce expected profits, invest in new solutions, or any number of other business decisions. None of these are absolute; rather, each requires a tradeoff. If we present cybersecurity needs as absolute, we risk managers and leaders ignoring us because we are not embedding our knowledge in a realistic decision context.

When communicating cybersecurity needs or risks, follow these steps:

1. Frame the need or risk as a business decision or input. What is the problem? What is the desired outcome? Is the problem that there are cybersecurity gaps? (Answer: No, because there will always be cybersecurity gaps.) Is the problem that there are foreseeable, preventable risks to the product's success? (Answer: Yes, that frames the problem as a business decision.)
2. Present information relative to business goals, with pros and cons. Cybersecurity engineers are always going to find more gaps and more ways to improve. Executives are accustomed to being asked for more (more time, budget, scope) and must hold boundaries to ensure that the product gets out the door. Within this context, they are accustomed to accepting or rejecting proposals based on risk and they usually have more knowledge about the total decision than others. Present cybersecurity flaws or needs relative to other benchmarks or data to help to contextualize the recommendation. Include related costs and schedule impact. Keep requests tangible and properly framed. An example of a bad approach is: "We need asymmetric cryptography." Instead, consider: "In the past decade, our products were not designed to be sold separately. Now that they are, we have to build a secure means to establish trust between our unbundled parts."
3. Use regulatory risk as a proxy for communicating patient risk. Cyber advocates should reference compliance effectively. Regulatory burdens are a proxy for safety risks. FDA and other global regulatory bodies exist in part to assess risk to patients, so regulatory actions (e.g., hold letters, marketing denials) can be treated as a proxy for patient risk. Stakeholders are reticent to appear as if their decisions are made pri-

marily to comply, rather than to meet the direct principle of patient safety. This is a false dichotomy. Cybersecurity advocates would do well to embrace their regulatory oversight, as regulatory actions serve as concrete, tangible examples of risk to businesses directly resulting from patient safety risks.

4. Present options. Leaders make decisions from a selection of options. When we present one solution, their options are "yes" or "no" (or "I'll consider it," which often becomes "no" by default). Instead of presenting one solution to which they could say no, present three options, each of which would be a good path forward. For example, instead of "we need to invest in asymmetric cryptography now," offer: "Here are three paths forward, and here are the pros and cons of each path." Articulate the pros and cons in terms of dollars, time, scope, and brand.

If we do not have enough information to look at the whole picture, we should collaborate with someone who does. Doggedly pursuing a specific decision without considering the whole decision context is likely to be received as demanding, impractical, and misaligned with business objectives. This is not to say that we should subjugate our security principles to business metrics. Rather, for our principles to be sufficiently heard and accounted for, we would do well to consider the whole picture.

2.3.3 Talking to the Wrong Leader, That Is, Not the Decision-Maker

Unfortunately, even the most well-presented risk will go unanswered if presented to the wrong decision-maker. A project manager on a defined budget and a deadline to get software released in 3 months will immediately reject a proposal to design and implement a new public key infrastructure and encryption design. So do not ask them to. People usually respond to issues of risk with their own decision model and the responsibilities of their current job. A long-term, budget-busting decision needs to be made at a higher level, but the cybersecurity engineer who understands the need and risk may not have a relationship with the right decision-maker (e.g., a vice president (VP) of R&D). In that case, the communication needs to be in terms of the project manager's needs and decision authority. Instead of presenting a project manager with the choice of either expensive encryption controls or delaying the project, present the project manager with the decision to communicate the risk upwards (i.e., stay silent about risk or raise the risk). A good project manager will always want to raise a material risk to their project.

To achieve buy-in for a cybersecurity effort that spans multiple products or product lines, the communication needs to be made up, down, and sideways through an organization. That takes time and effort. Consider creating a presentation (one that does not necessarily require you to personally present in order to be understood) for systemic change, using the approaches described in this chapter, for circulating to the right decision-makers at the appropriate levels in your organization.

We often avoid being the alarmist, justifiably fearing real consequences for our jobs. It is all too common to find businesses where the unwritten rules of the organization include observable phenomena such as, "Bringing problems to leadership is career-limiting." This may or may not actually be the case, but whether real or perceived the reality is that in these situations, leadership is being painted a pretty picture of minimal business risk and are therefore blind to real risk that could be mitigated before breaches occur.

In organizations where it really is the case that leadership does not want to hear any negative issues, there may be little that any one individual can do to make a difference. Speaking "truth to power" (clearly, tactfully, patiently, persistently, with integrity) can be difficult. Advocating for cybersecurity may make you unpopular and may increase the odds that you become the scapegoat when the worst should happen. Consider whether the organization has the will to change. Remember that good-faith advocacy is acting in the best interests of the organization and its customers.

We believe that most organizational conflicts about cybersecurity in medical devices can be alleviated. However, systemic negligence can and does occur. As in any job, know your personal principles and rights. If you are concerned, have an exit strategy that allows you to preserve and protect your personal integrity.

Communication with leadership is one of the most critical elements of medical device cybersecurity. Without leadership support, there is little that one individual can perform that will improve the organization's security hygiene. Although not always immediately successful, persistent, well-crafted communication up the chain of command is essential for gaining such support. Take the time to educate your leadership and convey in simple terms the risks, rewards, timelines, and costs as they relate to cybersecurity. Frame security issues in business terms; present options; get creative about overcoming cognitive biases surrounding risk and "bad news"; find ways to reach the right decision-makers at the right levels of the organization; be clear, patient, and persistent. These approaches will help you succeed as an advocate for cybersecurity.

2.4 Security and Lifecycle Management Overview

A lifecycle-based approach to medical device cybersecurity is actually the combination of four distinct lifecycles and the interaction between them (see Figure 2.4). This may be described variously as an aspect of TPLC, SDLC, and/or

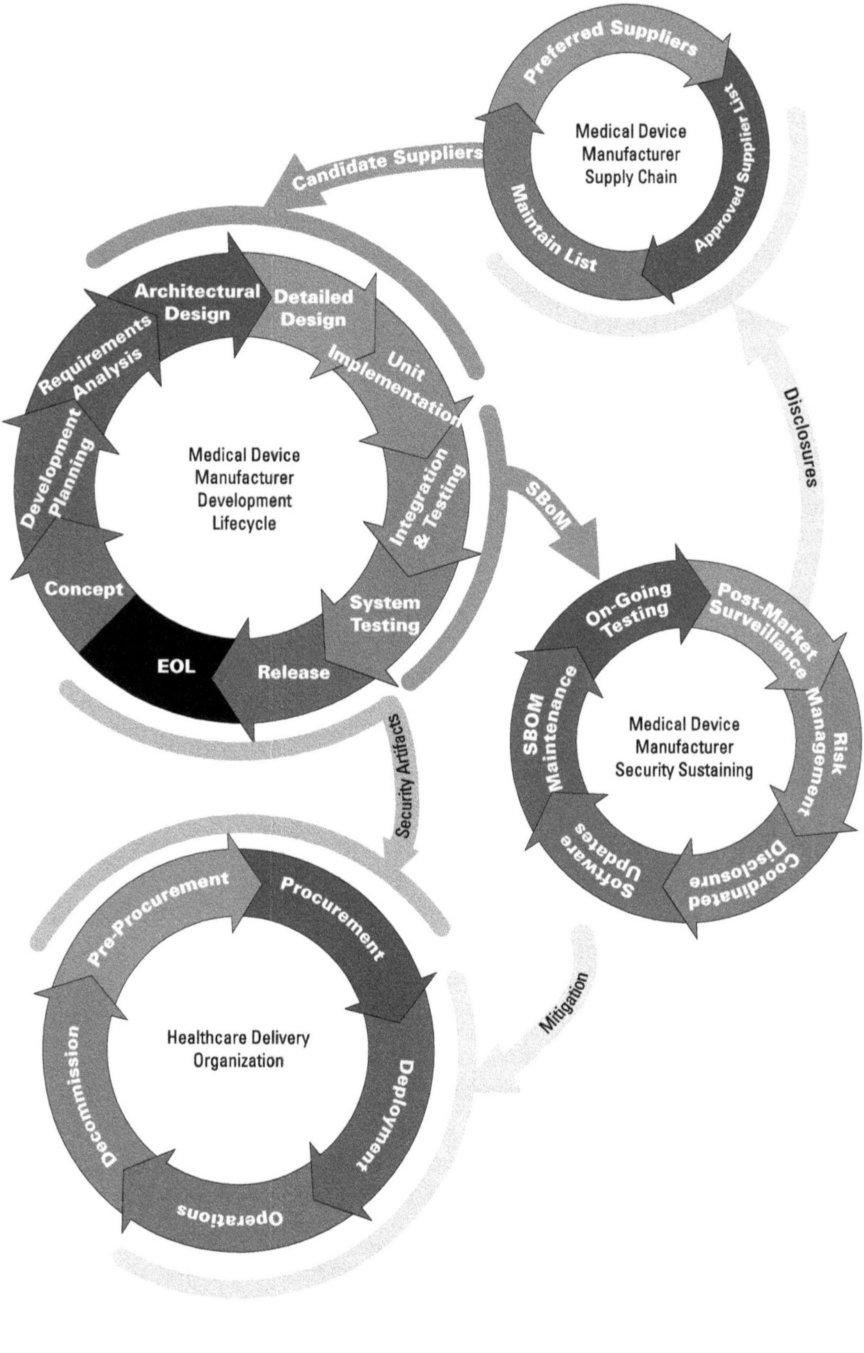

Figure 2.4 Medical device cybersecurity lifecycle components.

Secure Product Development Framework (SPDF). There will be shades of variance between what these terms encompass, some of which is discussed in later sections, but all share the same basic approach to achieving Secure by Design status.

Besides managing the complexities and intricacies of each of the individual cycles, a challenge for most MDMs is to understand and interface with the components that are not under their direct control, yet are required to function in harmony to provide the desired outcome. In other words, in addition to the internal processes it also becomes an issue of the interfaces to other stakeholders (i.e., the flow of information and coordination with their respective processes). This may require contractual agreements, process definitions, or organizational memberships (as, for example, in an Information Sharing and Analysis Organization (ISAO)). Most importantly, it requires cultural change that embraces a willingness to share and communicate.

As with any model, the lifecycles depicted in Figure 2.4 are an abstraction that illustrates the most general case. It cannot reflect all the details and variations that are required in the real world. However, it does provide a usable reference that can be followed and can be used for explanatory purposes, as it is in this book.

The four key lifecycles we need to be aware of and manage, discussed in further detail through the remainder of this book, are:

1. *Manufacturer development lifecycle:* From a regulatory perspective, the development lifecycle encompasses what is typically referred to as premarket activities (i.e., the steps required prior to a product being approved for and introduced to the market). Starting with the inception of a new product or version, to design steps, to manufacturing transfer, postmarket management, and eventually end-of-life, it is the manufacturer's leadership that is ultimately responsible for enabling an organization to successfully deliver secure products to the market.

   The typical steps for a security-enabled development lifecycle are not necessarily different than in traditional software engineering; however, cybersecurity requires some unique considerations, for example, the need to make certain security review and testing activities a continual aspect of engineering activities, rather than a distinct step at the end. Furthermore, cybersecurity can be very dynamic and manufacturers are required to respond to newly discovered vulnerabilities, evolving threats, or customer-reported incidents no matter whether a product is already in the market or if it is still in development.

   In today's complex and global business world, not all these phases may be conducted by the manufacturer. To complement internal talent, capability, and capacity, certain development, testing, or manufacturing tasks may be conducted under an outsourcing arrangement. However, even though the manufacturer may not conduct all of these steps, it still needs to control them because it is the manufacturer and distributor of the finished product who ultimately has regulatory and legal responsibility.

We usually consider these activities as circular, rather than linear, processes with a beginning and an end. The conception and development of the next version or next generation of a product will typically be already well on its way when the current version reaches the end of its commercial life. Furthermore, lessons learned from the previous product or version should be incorporated in the planning and requirements development for the next one, as discussed in more detail in Section 2.8.

2. *Supply chain lifecycle:* Modern medical devices are complex and so are their technology ecosystem. Today's medical device manufacturers commonly contract out part of their design, utilize commercial off-the-shelf (COTS) components, including software components, in addition to open-source software. It is essential that all components, and the vendors supplying them, are properly managed to ensure that any third-party component is managed to the same standard as software and components designed and built in-house. This does not necessarily mean that the supplier itself complies with FDA regulations and has appropriate supporting processes in place, although some do. However, the supplied software will need to be integrated, validated, and documented under the device manufacturer's quality system and applicable regulations.
3. *HDO lifecycle:* This covers the typical activities required on the HDO side from security-inclusive pre-procurement planning, execution of the procurement process, onboarding and deployment of devices, operation and maintenance, and, ultimately, the end of useful clinical life and proper decommissioning (discussed in more detail in Section 2.10). The security objectives and requirements for the clinical lifecycle are to reduce (to an acceptable level) the risks associated with patient safety, information privacy, care delivery disruption, and business.
4. *Manufacturer security sustaining lifecycle:* In the postmarket phase, the manufacturer is expected to undertake a set of activities that, collectively, are intended to ensure the ongoing security posture of a device and to enable response to what the FDA calls "cybersecurity signals." To summarize, these activities include:
    - Periodic cybersecurity testing;

- Periodic reassessment of the attack surface and known risks;
- Security incident response;
- Periodic patches to the fielded devices;
- Updates to the fielded devices as needed;
- SBOM surveillance.

Note that these are just generalized process steps, which could vary based on device design and risk specifics. For example, a design change may be indicated by a hardware-based vulnerability, possibly leading to hardware changes and/or an update to hardware-residing microcode, or, although it is common that a newly identified risk is related to a newly discovered vulnerability or threat, there could also be the case that a change in the device's intended use requires an updated risk assessment because a risk previously defined as acceptable may no longer be so in the new use case. These activities are further discussed in subsequent chapters.

2.4.1 Coordination Between the Four Lifecycles

These four lifecycles, although managed independently by different stakeholders, have a relationship at certain touchpoints that serve specific functions to ensure proper management of security requirements and, ultimately, delivery of a proactively secured device:

- *Supplier management:* The MDM's management of its supply chain is a continual, ongoing process that requires careful selection of preferred suppliers, enforcement and maintenance of an Approved Supplier List (ASL), and management of the supplier and supplied components. Approved supplier candidates may include both custom and off-the shelf software manufacturers that are proposed to be phased into the medical device product during the design, implementation, and V&V.
- *Sales process:* As part of the sales process and product delivery, the MDM typically provides a set of security artifacts to the HDO, including but not limited to:
    - Hazard analysis, mitigations, and design considerations (including list of cybersecurity risks considered and cybersecurity controls applied);
    - Traceability matrix that links cybersecurity controls to the respective risks;
    - A plan for postmarket management of software updates and patches;
    - Documentation of software integrity controls (i.e., ensuring that the software remains free of compromise);

- Instructions and specifications for cybersecurity controls (e.g., anti-malware software or firewall management), as well as for secure deployment, handling, and integration. These artifacts include, but are not limited to, documentation of security risk assessment, device software composition (SBOM), and service security properties (MDS^2) [29], as well as assurance of software security (e.g., evidence of testing or certification). Agreement on the supporting documentation may be formalized in form of a contract between the parties and should enable the HDO to securely deploy and maintain the device during its useful clinical life.
- *MDM postmarket:* This is the most connected of the lifecycles and requires coordination between the MDM, any third-party supplier to the MDM (via disclosure of any issues that may be discovered as part of postmarket surveillance), and the HDO (via disclosure and mitigation of any newly discovered vulnerabilities or threats). A critical reference point for many of these activities is the device's SBOM, which enables surveillance relative to the device's software components as well as management, disclosure, and mitigation of any issues uncovered.

This section reviewed the four main lifecycles that govern the steps and processes that enable us to manage the regulatory, engineering, operational, and maintenance processes. This high-level overview was focused on developing a general understanding of the lifecycle-based approach and how they apply and connect to individual stakeholders. This will be discussed in much more detail in the remainder of this book.

2.5 Organizational Roles and Responsibilities

Obviously, there is no single approach to defining the right organizational model for a cyber-capable manufacturer, nor is there a set of well-defined roles and responsibilities that would work for every manufacturer, every device type, and every target market. However, we can make some general assumptions about certain functions that need to be provided that then will map to respective roles and responsibilities. Individual organization's needs will determine if all of these roles are required, whether they should be dedicated or if they can be fulfilled part-time, if they can be extensions of existing roles or should be created as new and dedicates positions, or if they can even be contracted to a third party.

This section will provide some insight and suggestions for building a device manufacturer's cybersecurity organization. As such we will focus on product security, with limited discussion of traditional information technology (IT) and IT security roles as applicable to product security and internal infrastructure and communication needs. For example, a manufacturer's support organization may take on specific security-related tasks in their relationship with their customers, and for that they may require remote access that leads to certain manufacturer-internal IT needs like virtual private network (VPN) access, strong authentication, and similar.

In addition to touching on the role of the individual and his or her roles and responsibilities, we will also discuss the role and benefits of working with external stakeholders (such as information-sharing organizations) that play key parts in solving the medical device cybersecurity challenge.

2.5.1 Roles and Responsibilities: Overview and Rationale

Recognizing the complexity of today's medical devices and medical device ecosystem, as well as the unique cybersecurity challenges facing the industry, it is clear that well-defined roles to be filled by skilled and experienced staff are of utmost importance. Furthermore, the ability to effectively and constructively communicate between internal and external stakeholders, within the scope of the defined roles, is a prerequisite for success.

The understanding and proper execution of cybersecurity responsibilities goes "up the chain of command," starting with the individual engineer, to engineering management, regulatory and legal, and all the way up to executive leadership and its responsibility to establish the right cybersecurity culture and governance, and ultimately bearing the responsibility for risk acceptance decisions. This complex distribution of roles requires not only the clear identification of responsibilities, but also continual education on medical device cybersecurity, as we are facing a constantly evolving field with new cyber threats and new regulatory requirements emerging continually.

Building a cybersecurity-capable organization can be quite a challenge, however. Based on conversations that we have had with manufacturers that did so, we would assume that in many cases, new security positions will have to be filled with new hires in order to obtain the right skill set and shoulder the additional workload. To add to the challenge, the required skill sets are currently not easy to be found in software engineers and other required roles, although we would hope that this will improve in the future.

It is essential to establish clearly defined roles and responsibilities, ensuring that stakeholders can meet what is required of them as well as allowing for the integration of individual functions within the larger ecosystem. A common approach to accomplish and communicate this is the so-called RACI matrix, which helps identify and define activities for which stakeholders are "Responsible, Accountable, Consulted, and Informed," as shown in Table 2.1.

Table 2.1
Example of a High-Level RACI Matrix

	Role											
Responsibility	Executive Leadership	Product Cybersecurity Executive	Regulatory and Legal	IT and IT Security	Sales and Sales Operations	Product Management	Engineering Management	Architecture/Systems Eng.	Security Engineering	Software Quality	Sustaining (Security) Eng.	Service (Maintenance and Support)
Establish Security Governance and Budgets	R/A	C	C	I	I	I	I	I	I	I	I	I
Identify Regulatory Objectives	I	C	R/A	I	C	C	C	I	I	I	I	I
Ensure Staffing and Training Levels	I	R/A	I	I	C	C	C	I	I	I	I	I
Security Requirements	—	C	C	I	I	R/A	C	C	C	C	I	I
Risk Management	I	C	C	I	C	C	C	C	R/A	I	I	I
Develop Security Architecture	I	I	I	I	I	C	C	R/A	C	C	I	I
Secure Code Design	—	—	I	—	—	I	I	C	R/A	C	I	I
Code Review	—	—	—	—	—	I	I	C	C	R/A	I	I
V&V	—	I	I	—	—	I	I	C	C	R/A	I	I
Transfer to Production	—	I	I	—	—	I	I	C	C	I	R/A	I
Postmarket Vulnerability Management	I	C	C	—	I	C	I	C	I	I	C	R/A
Customer Communication	C	R/A	C	—	C	C	C	C	C	I	I	I
Security Maintenance	—	I	I	—	I	R/A	C	C	C	I	I	C
Incident Response	I	C	C	—	C	C	C	C	C	I	I	R/A

R: Responsible, A: Accountable, C: Consulting, and I: Informed.

The manufacturer-internal roles span the range from the responsibility for enabling the organization to execute (typically senior management responsibilities like budgets, staffing, enablement), to defining the right security requirements (driven by regulations and shifting market needs) and implementation (software design best practices and applicable security technologies), to providing a proactively secured product to the market, together with the appropriate regulatory approval and documentation artifacts that are useful to the customer.

Furthermore, large or mid-sized manufacturers may benefit from separating the roles into two categories: cybersecurity responsibilities of dedicated roles as part of a central organization and cybersecurity responsibilities (dedicated or not) as part of existing engineering or similar roles. In a sense, this would be a horizontal overlay with general, mainly strategic responsibilities that are differentiated from vertical and more tactical roles that focus on a specific product or part of the process. The advantage would be the separation of responsibilities and avoidance of conflicting objectives, whereas the disadvantage could be challenges in alignment of goals and communication as well as limitations in quickly responding to change.

In that sense, and recognizing that each organization is unique and will likely find its own approach on how to organize around cybersecurity, here some general guidance on how to approach and establish the respective responsibilities:

- *Executive leadership:* Although executive leadership typically does not own responsibility for specific products, it does own responsibility for the company's general priorities and direction. It needs to have strong influencing skills, the ear of company management, and some degree of authority so to constructively influence direction. It is executive leadership's role to establish a culture of security and to support and enable the business case for security. This should also include defining cybersecurity governance, providing strategic direction, defining risk tolerance and business cases, establishing the right organizational structure, and funding the execution through budgets and staffing. Leadership's involvement in security matters should be continuous in order to understand, and correct as needed, gaps and respond to developments that may require attention (e.g., security incidents reported by customers or newly evolving regulations).
- *Product cybersecurity executive role:* Some, especially larger, manufacturers may find it beneficial to create a dedicated strategic security role, for example, a Chief Product Security Officer, who, as a member of the executive team, will be responsible for the overall program and implementation as well as the management of the company's security strategy. Especially with large organizations that need to align their security objectives and approach across multiple product lines, legal entities, and geographic locations, this type of role is essential.

Furthermore, such a senior executive will also provide an important external-facing function by engaging with industry, regulators, and customers. In other words, he or she is the public face of the company on all cybersecurity matters as it relates to the company's products as well as its public image and the representation on cybersecurity topics.

- *Regulatory and legal:* Cybersecurity is a developing topic that, over the past few years, has been added to the responsibilities of regulatory/quality/legal roles. Understanding the evolving regulatory requirements and providing guidance to the organization (leadership and engineering) are critical as we are still far from a settled regulatory landscape and final (stable) requirements.

  In essence, new cybersecurity tasks for regulatory and legal departments include understanding developing international regulations, inclusion of cybersecurity foundational requirements and processes in the software quality management system, establishing cyber-relevant complaint management and regulatory reporting, managing supply chain including the adoption of applicable legal requirements and their inclusion in the company's contract framework, and review of customer contracts with regard to security issues in those contracts.

  Further tasks may be required in addition, such as support of regulatory filings and approvals, response to customer inquiries, review of contracts, and supporting the preparation (and, if needed, delivery) of communication to customers, regulators, and the public in response to a cyber event as well as helping management (and other roles) to stay educated about the evolving regulatory landscape.

  Larger companies may benefit from a dedicated cybersecurity standards specialist, a person or team focused on and knowledgeable in the evolving cybersecurity standards and responsible for translating these standards into policies and procedures as well as defining and managing internal practices for implementation across various functions including engineering and services.

- *IT and IT security:* Manufacturer-internal IT/IT security's involvement in product security is typically peripheral. However, there are some manufacturers that have established direct or dotted line reporting of product security into IT security, presumably to leverage expertise.

  Also, there are some areas where IT/IT security involvement typically is required:

  - Ensure security and integrity of the manufacturing environment so that neither the manufactured devices nor the manufacturing equipment (IT gear, industrial control systems (ICS), robots) are at risk of a security compromise. Any security incident impacting the manufacturing environment can result in compromised devices or device software or compromise of manufacturing quality, either of which may result in costly remediation and recalls. For example, according to the 2020 Health Information Management and Systems Society (HIMSS) Cybersecurity Surveys [30], 7% of respondents have reported receiving "hardware or software infected with malware 'off the shelf' (e.g., pre-loaded malicious software)."

- Ensure security and integrity of the service (maintenance and support) IT infrastructure, including providing tools for secure remote access to customer-based devices through tools like VPN or 2-factor authentication; assuring security of field personnel IT equipment (laptops, data carriers); and protecting manufacturer IT infrastructure (local and hosted) that holds customer data.
- Provide and maintain a secure environment for engineering software tools and data (e.g., documentation, source code) to ensure confidentiality (e.g., protection of IP), integrity (e.g., protect source code repositories), and availability (e.g., backup and recovery).
- As needed, support IT integration and review and audit third-party partners and suppliers.

Note that if service activities involve access to PHI or if PHI is transferred to the manufacturer's IT infrastructure as part of a service provided, under the U.S. HIPAA, the manufacturer will most likely be considered a Business Associate (BA), defined as any third party that receives, transmits, maintains, or creates PHI on behalf of a HIPAA-covered entity. This would then require contracting (Business Associate Agreement (BAA)), HIPAA-compliant processes and IT infrastructure for transmission and storage of PHI, and HIPAA training of employees who interact with or have access to PHI. To determine whether an MDM would need to comply with HIPAA, we advise seeking appropriate legal counsel [31].

- *Sales and sales operations:* As the people who face the customer, sales staff needs to be educated on cybersecurity in general, the company's overall security strategy and value statements, and the security particulars of the individual products. Furthermore, although sales may not be responsible for providing security content and answering security-related technical questions, they are certainly the communication channel for providing security information to the customer and for responding to their inquiries, be it through a formal process like RFI/P, contract, or MDS^2, or through an informal process of general customer questions.

This is a new responsibility for most sales organizations and, in many cases, may require some retraining, as many sales people, for too many years, have learned to brush off security questions or have used incorrectly constructed security arguments to justify a shortcoming (e.g., unavailability of a patch). However, this is an area where solid and honest commercial support will be recognized by customers as a differentiator and will earn trust that goes a long way.

What may complicate this particular area of organizational and cultural change is the complexity of many sales organizations that may have to serve different countries (and need to deliver different regional security messages), or in other cases where the sales organization may be one level removed (e.g., when a manufacturer uses a reseller in certain regional markets).

Larger organizations may create a dedicated back-office role or team, with close ties to product and technical security functions, to respond to customer inquiries, fill out security questionnaires, maintain MDS^{2}s, and so forth.

- *Product management:* As the typical "owner" of a product, product management's main responsibilities with regard to cybersecurity are to analyze and understand the evolving international regulatory requirements as well as changing customer needs. These can be aligned with the company's business objectives (e.g., in which global markets to sell the product), resulting in the development and management of product security requirements specifications and roadmaps as input to the engineering process. Product management also plays a vital role in supporting regulatory filings, internal education, customer communication, and postmarket management.
- *Engineering management:* The main role of engineering management is one of enablement to ensure that the MDM's engineering organization can successfully execute on delivering secure devices as per defined requirements. This is a combination of budget management, organizational structure, resource planning and staffing, providing the right tools and complementary services, and enabling education on the topic. Furthermore, engineering management should ensure compliance with corporate governance and product specifications and therefore plays a key role in internal stakeholder communication and support of key process steps like manufacturing transfer. Lastly, it is typically engineering management's role to triage security issues and help to develop an action plan to address the engineering part of the problem, be it a customer-

reported incident, a newly discovered vulnerability, or an internal issue during product development.

One important role of engineering management is the maintenance of cybersecurity information, be it the source code itself or the documentation artifacts created for regulatory filing, internal documentation, or artifacts that are developed to share with customers. Although the documentation requirements, formats, and methods may be defined by others (e.g., by a quality organization), developing and managing the processes and systems that produce and collect product security information are of critical importance and part of the engineering organization's responsibility. Artifacts that need to be managed, some for internal purposes and some for external purposes, can include risk analysis, threat modeling, security controls, MDS^2 and supporting data, SBOM, traceability matrix, and information on vulnerabilities and mitigation (e.g., updates and patches).

- *Security systems engineering:* The art of systems engineering consists of achieving an optimum balance between all elements in the system. This can be especially difficult when each element has different attributes with different competing priorities; compounding this, changes to any single attribute may change another element's attributes, causing a ripple effect throughout the design, prompting systematic reassessment. This delicate refactoring and releveling can be performed over and over again in the effort to seek the "correct" balance in the system, where "correct" itself may be a changing goal. Today's regulatory and customer expectations mean that the balance of priorities familiar to traditional systems engineering must also include cybersecurity considerations.

  A hypothetical example of competing priorities in a development project is an implantable medical device that needs to physically fit in a sealed titanium enclosure with battery space extremely limited; yet the implantable device needs to have a battery life measured in years, due to surgery being required to gain physical access to the battery (with all the attendant risks accompanying that surgery). To achieve this long battery life, the central processing unit (CPU) needs to be in a low power mode for as much time as possible, occasionally coming up to full power to perform its principal activity. When, for example, using a hardware-accelerated cryptographic engine, utilizing cryptography only adds microseconds to the full power operation; yet trying to perform these same operations purely in software would consume hundreds of milliseconds of additional full-power operation, obviously taking a significant toll on the battery life, leading to possibly not meeting one of the device's core objectives, total useful life.

The placement of security mitigations in the design has to be purposely performed to address specific weaknesses and risks, and the electronic hardware needs to be included in this so that features such as hardware-accelerated cryptography will be present in the design. Otherwise, the design may require a bigger battery and bigger case, or else the designer must accept a shorter battery life leading to more frequent surgical replacement.

Once this balance has been achieved, it is the responsibility of the security systems engineer to formalize this balanced system through the use of Interface Control Documents (ICD). This document describes the relationships that hold all the elements of the system together, including communications, functionality, timing, and security mitigations. It is a highly technical document containing sequence diagrams, data flows, details of packet contents, and more. The ICD is the ultimate authority on system interfaces and all activities passing back and forth between them. It should be written at a level that enables an average staff software engineer to implement the specified design.

When defining the topology of the system, the final, "complete" system should be stipulated, even if the device's rollout is in a phased deliverable approach where the initial version of the system may not contain all elements that will be supported in future releases or in various configurations of the system. A good example of this is to support firmware updating via a communications medium only, even where there are plans to roll out additional functionality over that communications medium in future versions or if there are plans to provide more efficient updating mechanisms as installed volumes increase, such as Firmware Over-the-Air (FOTA) update, further discussed in Chapter 8.

The "system-of-systems" elements are not necessarily defined by the device's physical boundaries, but by logical grouping at a given level of abstraction and potentially spanning across network and data storage layers (e.g., cloud storage). Furthermore, in that system-of-systems, the individual components may have different levels of regulatory controls. An example would be that commonly with a Picture Archiving and Communication System (PACS), the diagnostic viewing software is regulated as a Software as a Medical Device (SaMD), but the actual workstation, its operating system, backend servers, and storage may not be regulated as medical devices and, in certain circumstances, may not be regulated at all. With that, another part of the activities to be performed by a security systems engineer is to identify within a system, where regulatory boundaries need to be mapped onto logical and/or physical separation of these systems. This would be true for the applicability of regulatory security requirements across system layers and components but also to correctly understand and implement corresponding protective measures how privacy controls need to be implemented.

- *Security architecture:* Whether security architecture is a separate role or part of systems engineering is typically dependent on the company (e.g., size), system architecture (e.g., complexity), and use case (e.g., patient risk). In either case, a separate and dedicated security architecture effort early on in the design phase is critical to ensure that the right security decisions are made and that security objectives are met without undue compromise.

  This includes applying security engineering best practices to the architecture itself (device internal as well as target integration architecture), support of the selection of complementary security technologies and tools, definition of the right security controls and metrics, and monitoring design deviations relative to the requirements during the development phase. Furthermore, security architecture and systems engineering typically play an important role in regulatory filings and the creation of documentation and artifacts.
- *Security engineering:* Again, several factors determine whether security engineering is a separate role or a function being provided by security-capable software or system engineers (if available). Either way, this function is critical to ensure a secure final product that is supported by secure development activities that follow established software design best practices and standards. This includes the development itself as well as support of testing, regulatory filing, documentation development, manufacturing transfer, and, on an as-needed basis, postmarket issues. Also, typically as a task conducted in cooperation with security architecture, this role is a critical competency required to support security analytics tasks, such as security risk assessment or threat modeling.
- *Software quality:* In the context of cybersecurity, quality assurance's responsibility is similar to traditional QA. This includes testing, V&V activities, and producing supporting documentation prior to and as part of regulatory filing and production transfer. However, what is unique for the security processes is that these reviews and testing ideally should happen much earlier in the engineering process and is, in many cases, a continual or iterative process (for example, at critical component selection or integration steps), rather than a distinct activity that happens

at the end of the development phase. Another difference is that, during the course of the secure development lifecycle, cybersecurity issues that potentially affect patient safety and efficacy will be referred into the software QA process for risk-based evaluation and determination of potential impacts.

Because cybersecurity is a unique problem area requiring a unique skill set and the use of specialized tools, organizations may benefit from a dedicated Cybersecurity Tester (or group) that is knowledgeable on various testing techniques including automated and manual, along with interpretation and disposition of results. Typical processes, ideally supported by tools, applied to ensure secure and quality software are code review, fuzz testing, pen testing, and vulnerability scanning (further discussed in Chapter 6).

Out of similar considerations, the function of Vulnerability Management may require a dedicated resource that has the right skill set and is dedicated to managing the intake, tracking, assessing, and facilitating response to newly identified vulnerabilities.

Certain roles, including, for example, individuals or teams performing pen testing, need to be sufficiently independent or even an independent external entity so to avoid conflicts of interest or perception bias.

- *Sustaining (security) engineering:* This is a common continuing support function that typically takes on maintenance engineering responsibilities after production transfer. This includes supporting the initial transfer, ensuring that product and documentation requirements are met as well as the resolutions of any issues that may occur during this process, and managing and synchronizing future updates and issues that may manifest during manufacturing.
- *Service (maintenance and support):* Every manufacturer has some form of maintenance, support, or service function. In the context of cybersecurity, this can be broken down into two main sets of responsibilities:
    1. Support customers with any security issues that may arise (e.g., incident response; see Chapter 9) and help with the management of security issues such as deployment of updates and communication of issues.
    2. Ensure that any service or maintenance activities do not pose a security risk to the customer environment. This includes, but is not limited to, secure remote access (VPN, two-factor authentication) and preventing on-site transfer of malware from service equipment (e.g., via service laptop, USB flash drive, or data carrier) into the customer environment.

Other, general responsibilities for service and maintenance include the escalation of any customer security issues as well as compliance with local regulations (e.g., HIPAA, EU General Data Protection Regulation (GDPR)) and requirements with regards to handling of customer equipment and/or data.

The above is a somewhat generic view of the commonly required roles. As stated before, every manufacturer, device, use case, or market have different requirements and require a different approach and capabilities. These roles may be developed as a new and dedicated function but also could be an overlay to or matrixed into already existing roles, or a mix of the two approaches. Also, every manufacturer (and individual employee) starts at a different level of security capability and a careful plan should be developed to form a sensible path forward. Although cybersecurity is critically important, moving towards a cyber-capable organization cannot be rushed, nor can it stop the business for an extended time. Understanding an organization's current state and development of a step-by-step plan to reach the desired final state are essential and require, as discussed at the beginning of this chapter, proper planning and support by leadership.

Also, there are certain tasks that would span across several of these roles and that are discussed in more detail elsewhere in this book. For example, supplier and supply chain risk assessment and management, response to customer questionnaires and contracts, vulnerability management and disclosure, or incident response. Because these can occur and impact any (or multiple) of the phases in the development lifecycle, a cross-functional approach is typically required.

2.5.2 Role-Specific Training and Education

Each of the roles reviewed above requires appropriate security training and education. In today's medical device world with devices and customers facing increasingly sophisticated cyber threats, nobody should be exempt from keeping their cybersecurity knowledge up-to-date. Obviously, the need for continual education would be very different across the roles discussed above, be they technical or nontechnical, engineering, or administrative. For example, an executive decision-maker or regulatory staff would need quite a different perspective than an engineer who writes code.

In cybersecurity, education is a continuous process as security technology and best practices, security research, the threat landscape, and regulatory requirements continuously evolve. Security education should be a well-planned and sufficiently funded aspect of any organization. This can happen through participation in conferences, providing access to literature and online resources, or educational courses and certifications provided by security industry organizations or software platform providers.

The following is a list of organizations, several of which provide conferences and events, that we found useful to support general and medical device-specific cybersecurity education needs:

- *H-ISAC:* Webinars, working groups, topic-focused meetings and events, annual Spring and Fall Summits, international summits;
- *MedISAO:* Manufacturer-focused cybersecurity information and webinars;
- *HIMSS:* Online resources, webinars, local chapter events, as well as HIMSS Annual Conference;
- *AAMI:* Online resources, webinars, as well as annual AAMI eXchange;
- *AdvaMed:* Online resources, webinars, MedTech Conferences;
- *Archimedes Center:* Conferences, virtual learning events, training programs;
- *Q1 Productions:* Various events focused on select topics (regulatory, security engineering);
- *MedTech Intelligence:* Events and webinars;
- *TT LifeSciences:* regular EU and U.S. conferences on various life sciences topics, including medical device software engineering and cybersecurity;
- *Software platform vendor-provided resources:* Almost every provider of commercial software solutions has published hardening guidelines that should be followed. Especially critical system components like platform (operating system), databases, or software tools for access, authorization, and authentication should never be ignored.

Organizations that provide security education and certifications:

- *SANS Institute:* Training courses and certifications on various security subspecialties (e.g., cyber and network defenses, penetration testing, incident response, digital forensics, auditing).
- *ISC2:* Security training and certification, including HCISPP (Healthcare Information Security and Privacy Practitioner) and CISSP (Certified Information Systems Security Professional) certifications.

- *Center for Internet Security (CIS):* Critical Security Controls (CSC) cybersecurity tools, best practices, and education, including hardening guidelines, benchmarks, and metrics.
- *Information Systems Audit and Control Association (ISACA):* Professional association focusing on IT governance as well as providing some well-recognized certifications like Certified Information Systems Auditor (CISA), Certified Information Security Manager (CISM), and Certified in Risk and Information Systems Control (CRISC).
- *CompTIA:* Foundational and advanced IT and security training and certifications, including Network+ and Security+ certifications for beginners.
- *Open Web Application Security Project (OWASP):* Articles, methodologies, documentation, tools, and technologies in the field of web application security.
- *National Association of Corporate Directors (NACD):* Board-level and director-level resources on cybersecurity oversight and responsibility.
- Velentium's online virtual training for medical device cybersecurity [32].

Government and partnerships:

- *National Institute for Standard and Technology (NIST):* Wide range of publications on software, IT, and security topics.
- *National Cybersecurity Center of Excellence (NCCoE):* Security technology as well as medical device-specific security use case examples.
- *Department of Homeland Security (DHS):* Security resources, including general and medical device security advisories (ICS-CERT) and security education (NICCS).
- *Health Sector Coordinating Council (HSCC) Cybersecurity Working Group (CWG):* Established as a public-private partnership and recognized by the U.S. Department of Health and Human Services as the critical infrastructure industry partner tasked with coordinating strategic, policy, and operational approaches to prepare for, respond to, and recover from significant cyber and physical threats to maintain the ability of the sector to deliver critical assets and services to the public.
- *European Union Agency for Cybersecurity (ENISA):* General and topic-specific cybersecurity resources.
- *European Telecommunications Standards Institute (ETSI):* General and topic-specific cybersecurity resources.

Literature and online resources:

- NACD Director's Handbook on Cyber-Risk Oversight;
- Health Care Industry Cybersecurity Task Force: Report on Improving Cybersecurity in the Health Care Industry;
- Healthcare and Public Health Sector Coordinating Council (HSCC): Health Industry Cybersecurity Practices: Managing Threats and Protecting Patients;
- Healthcare and Public Health Sector Coordinating Council (HSCC): Joint Security Plan version 2 (JSP2);
- CyBok, The Cyber Security Body of Knowledge (v1.1, free download), University of Bristol in cooperation with several other leading European and U.S. academic institutions;
- CIS Security Benchmark Mapping Guidance;
- OWASP Secure Medical Device Deployment Standard and many other resources related to cybersecurity practices and metrics;
- IEEE Computer Society: Building Code for Medical Device Software Security.

There are many more resources than can be listed here, so it is worthwhile to watch the developing security educational space. Also, we should not undervalue the benefits of internal learning. Many organizations have at least a few people that were early adopters and have been dealing with cybersecurity for a number of years. These employees have developed valuable skills, especially product-specific and market-specific knowledge that would be difficult to find in public courses or material. It is strongly advised that organizations leverage these experienced people and make internal teaching/learning part of their cybersecurity journey.

It is a generally accepted best practice to make a cybersecurity skills assessment (as appropriate for the role) part of annual personnel assessments or performance reviews, as well as to include specific cybersecurity goals and activities in staff members' annual development goals and training plan.

2.6 Regular Review of Security Maturity

Whether an MDM is just setting out on the journey towards becoming a cyber-capable organization or already has taken steps in the right direction, understanding one's current security maturity, defining the desired future state, understanding external drivers and changes, and measuring progress along the path are of critical importance. For that purpose, security maturity models can be applied, enabling an organization to assess, plan, and adjust.

In general, a maturity model is a tool that allows an organization to understand its effectiveness at achieving a particular goal, in this case, cybersecurity capabilities (engineering, manufacturing, and communication processes) and outcomes (the delivered end product of the desired level of security). A good security maturity model gives an organization's leadership a reliable measure on its current capabilities, desired future state, and tangible actions to take to move up the maturity curve. It aligns business goals with day-to-day operations and enables measuring progress.

Any organization, its employees, and applied practices require review and adjustment on a regular basis. Processes may be found to have shortcomings that can be improved upon, or the environmental conditions may change due to new regulatory requirements, changes in market needs, or newly discovered cyber risks. It is essential that an organization stays vigilant by observing these changes, identifying gaps, and developing and progressing along a path forward.

Assessing security maturity requires a regular "so, how are we doing?" exercise with input provided from internal and external stakeholders. It also requires that an organization has identified the data sources that are consulted for this purpose, assigned the individuals or roles responsible for regularly collecting that data, implements regular assessment milestones, and defines the participants that manage the maturity review process or provide input to it.

Improving an organization's security capabilities and maturity is both a tactical step-by-step exercise and a long-term strategy ("where do we want to be in X years?"). Did the organization embrace security practices and demonstrate security culture, or is it an overlay that is forced onto its processes and employees?

Typical events that provide grounding and instigation for a maturity review are the major handoff steps during the lifecycle, as well as both key events and predetermined review intervals taking place after product has been released and has some market experience "under its belt." What worked, where did we fail? Can we improve efficiency, reduce errors, or improve the product? These are typical questions that should be asked.

A security maturity review should be a complete, end-to-end undertaking. Is the finished product sufficiently secure and did it meet regulatory and customer requirements? Does it deploy and operate as intended and is it capable to withstand the security challenges of its normal operating environment?

Maturity reviews should always include both an assessment of the success of the device's design (does it meet the defined security requirements?) as well as a review of the quality of the processes that were applied (did their use result in a sufficiently secure product?). Furthermore, prior to launching the next development cycle (new product or update), one should assess if any of the environmental variables have changed (target markets, use cases, regulations, threat landscape), as they likely have. This then should result in an update of processes, training and skill set, design requirements, and technology used.

Security maturity reviews are commonplace in traditional IT security. Models exist that can be applied to or adopted for assessing security maturity of a security engineering organization and the applied lifecycle processes. Some examples for security maturity models are: The Systems Security Engineering Capability Maturity Model (SSE-CMM) [33] (security engineering and software design), NIST's Program Review for Information Security Assistance (PRISMA) (federal, critical infrastructure) [34], Department of Homeland Security's Cybersecurity Capability Maturity Model White Paper (process and capability) [35], or European Union Agency for Cybersecurity: ENISA CSIRT Maturity Assessment Model (organization, human, tools, processes) [36].

There is a saying in cybersecurity that "you can't protect what you can't see." The analogous statement for security maturity would be: "you can't improve what you don't assess." Security maturity models fill this gap. Lastly, as with any plan, encounters with reality need to be considered (e.g., security events or regulatory rejection) and maturity models and plans need to be updated from time to time.

2.6.1 Research on Maturity Models

Under sponsorship by the U.S. FDA, the Medical Device Innovation Consortium (MDIC) researched the use of maturity models with the goal to document the use of maturity models in other industries. Maturity models help organizations to assess their current operational state and to identify and implement improvements. FDA's objective for this 2015 study was to identify possible quality system maturity models that can guide organizations in regulated industries, like healthcare, and establish an approach of continual improvement.

The study analyzed a total of 22 models across 7 industries and conducted detailed research on the 5 most promising models through interviews of 210 professionals from 130 companies and across industries. The Capability Maturity Model Integration (CMMI) model was the most common model (76%), followed by the Test Maturity Model Integration (TMMi) (16%) as a derivative of CMMI specific for testing processes, with a remaining 8% using a variety of other models [37].

In a subsequent study published in 2022, MDIC surveyed the cybersecurity maturity of MDMs based on criteria derived from the Healthcare Sector Coordinating Councils' (HSCC) Joint Security Plan (JSP). As such, this approach can also be used by MDMs to define and assess their own cybersecurity practices [38]. The study was based on 44 questions across 4 categories that were derived from the JSP and scoring each answer using the CMMI Maturity

Level 1-5 scale. The average combined score was 1.51, corresponding to CMMI characteristics between "Initiated" and "Managed," with no company exceeding a score of 2.5.

2.6.2 Building Security in Maturity Model (BSIMM)

BSIMM [39] originated as a concept in 2008 based with the goal to build a security maturity model from actual data gathered from a select number (initially 9) of leading software firms. By 2022, BSIMM had grown to 130 company members, represented through 3,350 software security experts. It is a closed and proprietary scoring system that is performed by an outside firm. Also, it is intended to be performed infrequently, usually on an annual basis. This is best used to compare an organization to peers in the industry.

As such, BSIMM has been developed and matured to become a widely accepted software security framework based on an assessment of categories of identified activities, helping software developing organizations to plan, implement, and measure their security initiatives. As a data-driven evaluation tool, it can be applied to support planning of resources, schedule, investments, and priorities.

Annually, BSIMM publishes a report that analyzes hundreds of assessments from industry verticals to be used as a benchmark, allowing organizations to continually tune and improve their practices. The identified leading application security practices enables organizations to assess and improve their own programs, supported by a community of security professionals that provide collaboration and exchange on techniques and practices.

The four main Software Security Framework Domains contain three practices each:

- Governance;
- Strategy and metrics;
- Compliance and policy;
- Training;
- Intelligence;
- Attack models;
- Security features and design;
- Standards and requirements;
- SSDL touchpoints;
- Architecture analysis;
- Code review;

- Security testing;
- Deployment;
- Penetration testing;
- Software environment;
- Configuration management and vulnerability management.

Within the above framework and under the 12 BSIMM practices a total of 125 individual activities are assessed, allowing each organization to apply a consistent metric but also allows comparison against peers in similar industries.

2.6.3 Software Assurance Maturity Model (SAMM)

SAMM [40] originated in the same company as BSIMM; however, they take very different approaches to rating security. SAMM is a prescriptive model, an open framework that is simple to use, and fully defined. The target auditor can be anyone in the organization and assessments can be performed as often as necessary. The assessment consists of less than 300 questions in a freely downloadable Excel or Google sheet.

SAMM is designed for organizations to be able to:

- Analyze their current software security practices;
- Build a custom security program in defined iterations;
- Monitors progressive improvements in secure practices;
- Define the target maturity for each organization.

It is oriented around five business functions:

- Governance;
- Design;
- Implementation;
- Verification;
- Operations.

SAMM provides specific, actionable activities for continual improvement of the organization's security posture. It also has excellent reporting capability to display the current status and progress made over a defined series of phased improvements. Further information and resources are available from OWASP at https://owaspsamm.org/resources.

2.6.4 CMMI

The CMMI [41] is a process model aiming to improve organizational performance. It is structured around 5 Maturity Levels and 3 Capability Levels that can be applied to a wide range of processes, including product development or services. As such, it can equally be applied to assess an organization's cybersecurity capability, processes, and maturity. After more than 20 years, CMMI has matured and is generally well accepted and can be adopted across industries, company sizes, and regions.

The CMMI Maturity Levels characteristics are:

- Level 1: Initial—Processes unpredictable, poorly controlled, reactive.
- Level 2: Managed—Processes characterized for projects and are often reactive.
- Level 3: Defined—Processes characterized for the organization and are proactive.
- Level 4: Quantitatively Managed—Processes measured and controlled.
- Level 5: Optimizing—Focus on process improvement.

2.6.5 TMMi

The TMMi maturity model [42], based on established industry testing best practices, provides a reference model and assessment framework to help organizations identify their testing strengths, weaknesses (risks), and best practices and will provide an understanding of the established level of testing maturity and guidance for improvements. Users of TMMi have demonstrated benefits in effectiveness/product quality and test efficiency and productivity. Organization complying with TMMi model criteria can be formally assessed and certified.

Although TMMi can stand on its own, it is complementary to CMMI as it largely follows the same maturity structure but provides more granularity and support specifically for software and system testing.

TMMi Maturity Levels are defined as follows:

- Level 1—Initial.
- Level 2—Managed: established and managed but may vary from project to project.
- Level 3—Defined: all projects are following the same standards and procedures.

- Level 4—Measured: measurement of activities and outcomes are applied, advanced reviews performed.
- Level 5—Optimization: all activities and outcomes are assessed and optimized; continuous improvement toward defect prevention and optimized quality.

2.6.6 Cybersecurity Capability Maturity Model

The Cybersecurity Capability Maturity Model (C2M2) [43] was developed by the Pacific Northwest National Laboratory (Battelle) under contract for the U.S. Department of Energy. It is specifically suitable for organizations that operate and need to secure IT and/or OT network environments and is being used by organizations across many sectors and industries to assess and improve their cybersecurity programs and strengthen their resilience. The majority of users are in the energy sector (39%) and the IT sector (36%), with some degree of adoption in government (7%), finance (5%), communications (4%), and other industries.

C2M2 provides cybersecurity practices to manage IT and OT assets as well as the environments in which they operate. As such, it can be used to strengthen an organization's cybersecurity posture, to continuously evaluate and benchmark their cybersecurity capabilities, to prioritize actions and budgets, and to share knowledge and practices. It is performed as a self-assessment and can be used to develop new cybersecurity programs or to measure and improve an existing cybersecurity program.

C2M2 provides a consist way to assess cybersecurity program maturity and to communicate findings and improvements across an organization and across roles from executive decision-makers to managers in charge of resources. The most recent version, 2.1, is in alignment with U.S. national strategic guidance on cybersecurity and is aligned with internationally recognized standards and practices, including the NIST Cybersecurity Framework Version 1.1.

C2M2 is organized in 10 domains, each with a defined purpose, that incorporate a set of individual cybersecurity practices:

1. Asset, Change, and Configuration Management (ASSET);
2. Threat and Vulnerability Management (THREAT);
3. Risk Management (RISK);
4. Identity and Access Management (ACCESS);
5. Situational Awareness (SITUATION);
6. Event and Incident Response, Continuity of Operations (RESPONSE);

7. Third-Party Risk Management (THIRD-PARTIES);
8. Workforce Management (WORKFORCE);
9. Cybersecurity Architecture (ARCHITECTURE);
10. Cybersecurity Program Management (PROGRAM).

Within the practice assessment, and independently of each domain, four maturity levels are defined (summarized):

- MIL0: Practices are not performed.
- MIL1: Initial practices are performed but may be ad hoc.
- MIL2: Practices documented and adequately resourced; generally, more advanced than MIL1.
- MIL3: Guided by policies; responsibilities are assigned to personnel with adequate skills and knowledge; effectiveness tracked; generally, more advanced than MIL2.

With C2M2 being focused on operators of IT/OT infrastructure, it is more suitable for medical device manufacturers and their production and service/support organizations as well as any IT infrastructure provided to support an engineering organization or to host customer data.

2.7 Conclusion

We reviewed and discussed many aspects related to organizational readiness and aspects of this chapter, ranging from executive responsibility and communication to the discussion of individual responsibilities and capabilities. Every organization's maturity level is different, and everyone will need to find their own path. A large MDM with many individual and relatively independent business units will take an entirely different approach to cybersecurity than a small startup company with just a few resources that need to be shared across responsibilities.

There is a simple piece of business advice one of us was given many years ago: "There is only one way to run a business: Make a plan and then follow it." It sounds obvious, but it is certainly true for maturing an organization's cybersecurity capabilities. Identify where you are and where you need to be, define sensible steps, educate and train, and measure. Start at the top (with governance), augment your weaknesses, and continue reading this book.

References

[1] Morgan, C., "Medical Device Cybersecurity Regulatory Publications," *Apraciti*, November 26, 2019, https://www.apraciti.com/blog/2019/11/25/global-regulatory-authority-publications-on-medical-device-cybersecurity.

[2] U.S. FDA, "Cybersecurity," January 23, 2020, https://www.fda.gov/medical-devices/digital-health/cybersecurity.

[3] KHIDI, "Status of the Global Medical Device Market," 2021, https://www.khidi.or.kr/board?menuId=MENU01253.

[4] NIST NCCOE, "Securing Telehealth Remote Patient Monitoring Ecosystem," https://www.nccoe.nist.gov/healthcare/securing-telehealth-remote-patient-monitoring-ecosystem.

[5] FDA, "Cybersecurity in Medical Devices: Quality System Considerations and Content of Premarket Submissions," *FDA Guidance Documents*, September 2023, https://www.fda.gov/regulatory-information/search-fda-guidance-documents/cybersecurity-medical-devices-quality-system-considerations-and-content-premarket-submissions.

[6] HHS, "Breach Notification Rule," July 26, 2013, https://www.hhs.gov/hipaa/for-professionals/breach-notification/index.html.

[7] HHS, "Breach Portal: Notice to the Secretary of HHS Breach of Unsecured Protected Health Information," June 2023, https://ocrportal.hhs.gov/ocr/breach/breach_report.jsf.

[8] The White House, "Executive Order on Improving the Nation's Cybersecurity," *Executive Orders*, May 12, 2021, https://www.whitehouse.gov/briefing-room/presidential-actions/2021/05/12/executive-order-on-improving-the-nations-cybersecurity/.

[9] CISA, "Secure Software Self-Attestation Common Form," April 27, 2023, https://www.cisa.gov/resources-tools/resources/secure-software-self-attestation-common-form.

[10] Douville, S., and E. Svetcov, "Cybersecurity Board Director Candidate: Is That Board Interview Worth Your Time?" *Medium*, September 7, 2022, https://sherridouville.medium.com/cybersecurity-board-director-candidate-is-that-board-interview-worth-your-time-9af770466605.

[11] "Abstracts of the EPSM-ABEC 2008 Conference," *Australas. Phys. Eng. Sci. Med.*, Vol. 31, No. 4, 2008, p. 415, https://link.springer.com/content/pdf/10.1007/BF03178607.pdf.

[12] https://www.velentium.com/cybersecurity-training.

[13] AAMI TIR 57.

[14] https://www.usa.philips.com/a-w/security/coordinated-vulnerability-disclosure.html/.

[15] https://www.bd.com/documents/brochures/across-bu/BD_BD-Coordinated-Vulnerability-Disclosure-Program_BR_EN.pdf.

[16] https://global.medtronic.com/xg-en/product-security/coordinated-disclosure-process.html.

[17] https://www.productsecurity.jnj.com/.

[18] https://resources.sei.cmu.edu/library/asset-view.cfm?assetid=503330.

[19] https://www.hhs.gov/vulnerability-disclosure-policy/index.html.

[20] https://www.cisa.gov/coordinated-vulnerability-disclosure-process.

[21] https://www.enisa.europa.eu/news/enisa-news/coordinated-vulnerability-disclosure-policies-in-the-eu.

[22] Moustafa, A. A., A. Bello, and A. Maurushat, "The Role of User Behaviour in Improving Cyber Security Management," *Frontiers in Psychology*, June 18, 2021, https://www.frontiersin.org/articles/10.3389/fpsyg.2021.561011/full.

[23] ISC^2, "(ISC)^2 Research Reveals the Cybersecurity Profession Needs to Grow by 3.4 Million People to Close Global Workforce Gap," October 20, 2022, https://www.prnewswire.com/news-releases/isc-research-reveals-the-cybersecurity-profession-needs-to-grow-by-3-4-million-people-to-close-global-workforce-gap-301654730.html.

[24] Wirth, A., and N. Schwartz, "Omnibus Act Impact on Medical Device Cybersecurity," *Medium*, January 10, 2023, https://medium.com/medcrypt/omnibus-act-impact-on-medical-device-cybersecurity-55f7266e5fff.

[25] Alnifie, K. M., and C. Kim, "Appraising the Manifestation of Optimism Bias and Its Impact on Human Perception of Cyber Security: A Meta Analysis," *Journal of Information Security*, January 2023, https://www.researchgate.net/publication/368693379_Appraising_the_Manifestation_of_Optimism_Bias_and_Its_Impact_on_Human_Perception_of_Cyber_Security_A_Meta_Analysis.

[26] Kahneman, D., *Thinking, Fast and Slow*, New York: Farrar, Straus and Giroux, 2013.

[27] McKeon, J., "How Healthcare Cyberattacks Can Impact Patient Safety, Care Delivery," *Health IT Security*, September 8, 2022, https://healthitsecurity.com/news/how-healthcare-cyberattacks-can-impact-patient-safety-care-delivery.

[28] Winsor, D. A., "Communication Failures Contributing to the Challenger Accident: An Example for Technical Communication," in K. J. Harty, (ed.), *Strategies for Business and Technical Writing*, 6th ed., Boston, MA: Pearson Education, 2008.

[29] NEMA, "Manufacturer Disclosure Statement for Medical Device Security," October 8, 2019, https://www.nema.org/Standards/view/Manufacturer-Disclosure-Statement-for-Medical-Device-Security.

[30] HIMSS, "2020 HIMSS Cybersecurity Survey," Chicago, IL, 2020, p. 16, https://www.himss.org/resources/2020-himss-healthcare-cybersecurity-survey.

[31] Rose, R., "How Does HIPAA and the HITECH Act Impact Medical Device and Pharma Companies?" *Becker's Hospital Review*, January 11, 2013, https://www.beckershospitalreview.com/healthcare-information-technology/how-does-hipaa-and-the-hitech-act-impact-medical-device-and-pharma-companies.html.

[32] https://www.velentium.com/cybersecurity-training.

[33] Systems Security Engineering Capability Maturity Model (SSE-CMM), https://www.sse-cmm.org/model.htm.

[34] National Institute of Standards and Technology (NIST), "Program Review for Information Security Assistance Project," March 29, 2018, https://csrc.nist.gov/Projects/Program-Review-for-Information-Security-Assistance.

[35] Department of Homeland Security, "Cybersecurity Capability Maturity Model," August 4, 2014, https://niccs.us-cert.gov/sites/default/files/Capability%20Maturity%20Model%20White%20Paper.pdf.

[36] European Union Agency for Cybersecurity (ENISA), "Challenges for National CSIRTs in Europe in 2016: Study on CSIRT Maturity," April 30, 2019, https://www.enisa.europa.eu/publications/study-on-csirt-maturity.

[37] MDIC, "MDIC Case for Quality Program Maturity Model Research Report," June 2015, https://mdic.org/wp-content/uploads/2015/06/MDIC-Deloitte_Maturity-Model-Research-Report_June-2015.pdf.

[38] MDIC, "MDIC Medical Device Cybersecurity Maturity: Industry Benchmarking Report 2022," January 20, 2023, https://mdic.org/resource/cybersecurity-benchmarking-report/.

[39] Synopsys, "Building Security In Maturity Model (BSIMM)," https://www.synopsys.com/software-integrity/software-security-services/bsimm-maturity-model.html.

[40] OWASP, "Project SAMM," https://owasp.org/www-project-samm/.

[41] CMMI Institute, https://www.cmmiinstitute.com/ & Software Engineering Institute at Carnegie Mellon University, https://www.sei.cmu.edu/.

[42] TMMi Foundation, https://www.tmmi.org/.

[43] DoE, "C2M2 Version 2.1," June 2022, https://c2m2.doe.gov/.

3

Global Regulations, Standards, and Guidances

3.1 The Distinction Between Laws, Regulations, Standards, and Guidance

The medical device industry is a regulated industry, and the primary goal of global regulatory authorities is to ensure the safety and effectiveness of medical devices for patients and users. The basic structure against which global regulatory authorities gauge a manufacturer's compliance and their products' safety and efficacy is the Quality Management System (QMS). A QMS includes policies, plans, procedures, and work instructions that demonstrate compliance with regulatory requirements and result in records that demonstrate compliance when the QMS is executed, that is, a product is developed following the QMS framework. Cybersecurity work within a QMS extends throughout the product lifecycle, from policy-setting through design controls and manufacturing and into postmarket management, with a specific focus on full lifecycle security risk management throughout.

As it relates to the cybersecurity of connected medical devices, there are two primary regulatory goals that have some degree of overlap. Medical devices should be sufficiently protected against intentional attacks or unintentional misuses that could result in: (1) physical harm to a patient, environment, or user; or (2) the compromise of sensitive information such as protected health information (PHI), account data and credentials, or intellectual property (IP).

What are the cybersecurity documents used to build a QMS and ensure safety, efficacy, and compliance with regulatory expectations? Where are regulatory expectations documented?

3.1.1 Laws

Laws are passed by legislative bodies within governments (e.g., the U.S. Congress, U.S. state legislatures, or national parliaments). Laws define rules of conduct or behavior. Two laws in the United States driving medical device requirements (including cybersecurity) are the federal Food, Drug, and Cosmetic Act (FD&C Act) and the Health Insurance Portability and Accountability Act (HIPAA). The FD&C Act was most recently amended to strengthen cybersecurity requirements (including a requirement for SBOMs in premarketing applications for cyber devices) in the Consolidated Appropriations Act of 2023 (H.R. 2617), §3305.

3.1.2 Regulations

Regulations are written by executive departments and administrative agencies defining details regarding how laws will be enforced. In practice, regulations have the force of law. For example, Title 21 of the U.S. Code of Federal Regulations, Part 820 (Quality System Regulation) is the medical device regulation in the United States enforced by the U.S. FDA. Similarly, Regulation (EU) 2017/745 (Medical Device Regulation) provides enforceable medical device regulatory requirements for the European Union.

3.1.3 Standards

A standard is neither a law nor a regulation and does not inherently have the force of law. Standards are developed and published by industry representatives, experts, and other interested parties and may exist either at the national or international level. Standards are published by recognized Standards Development Organizations (SDOs) [e.g., Association for the Advancement of Medical Instrumentation (AAMI), International Organization for Standardization (ISO), International Electrotechnical Commission (IEC), and Institute for Electrical and Electronics Engineers (IEEE)] and establish a common understanding across one or more industries for a given set of concepts or practices. Such standards are known as "consensus standards." Within the medical device industry, standards attempt to drive the industry to an agreed level of safety and quality.

Although the legal methods vary nationally, from time to time a regulatory authority officially recognizes a particular consensus standard as meeting the requirements of a law or regulation. Such standards become known as "regulatory standards."

For example, in the European Union (EU), one of the European Standards Organizations (ESOs) [i.e., European Committee for Standardization (CEN), European Committee for Electrotechnical Standardization (CENELEC), and European Telecommunications Standards Institute (ETSI)] authors or adapts a consensus standard to ensure that it complies with a European directive or regulation. These standards are then published in the *Official Journal of the European Union*, given the prefix "EN," and classified as "harmonized." The European Commission maintains a list of harmonized standards for medical devices [1].

In the United States, the FDA publishes a list of recognized consensus standards, which qualifies them as [2]: "appropriate for manufacturers of products to declare conformance to meet relevant requirements in the FD&C Act, including premarket submission requirements."

3.1.4 Guidance (or Guidelines)

A guidance is material published by a regulatory authority (or organization recognized by regulatory authorities) that is supplemental to a law or regulation and provides clarification regarding best and acceptable practices when complying with regulatory requirements. In most countries, a guidance is published by the regulatory authority (e.g., U.S. FDA, Health Canada, Australia's TGA). In the EU, guidance for compliance with the EU Medical Device Regulation (EU MDR) and In Vitro Device Regulation (EU IVDR) is published by the Medical Device Consortium Group (MDCG). On an international level, the International Medical Device Regulators Forum (IMDRF) brings together 11 national regulatory authorities to publish guidances recognized to varying degrees around the world.

Although a guidance is not considered legally binding, and a company cannot "violate" a guidance, most regulatory authorities (including notified bodies) hold tightly to principles and practices described in guidances when evaluating product submissions and performing audits and inspections. It is not uncommon for guidances to be quoted as the rationale behind deficiencies in product submissions, notified body audit findings, and observations in regulatory inspections.

3.1.5 The Unenforceable

This chapter will not focus on the myriads of related white papers, discussion papers, books, articles, and other publications available, even if written by regulatory authorities. Although these publications represent important and valuable thinking on the topic of medical device cybersecurity, they do not represent regulatory requirements and would be both extraneous to and misleading in consideration of such requirements.

3.2 Who Is Writing and Enforcing Cybersecurity Regulatory Requirements?

As with other elements of medical device QMSs, cybersecurity regulatory requirements are published by a variety of organizations. Regulatory requirements are enforced globally by both national regulatory competent authorities and for-profit notified bodies.

Due to the high-profile nature of cybersecurity issues, some requirements are established by higher-level government entities and are scoped to include cybersecurity in its broadest sense. Executive Orders released by the U.S. President and the General Data Protection Regulation (GDPR) written by the European Commission and passed by the European Parliament are two examples.

3.2.1 Government Regulatory Bodies and Competent Authorities (A Sampling)

Government regulators and competent authorities are responsible for the implementation and enforcement of national laws. One means to accomplish this is through the establishment of regulations. To codify a set of best and acceptable practices and further clarify regulations, these bodies will publish and/or recognize applicable guidance and standards.

3.2.1.1 The U.S. White House

In May 2021, U.S. President Joseph Biden issued executive order 14028, which, among other things, mandates that federal agencies and their contractors improve their cybersecurity profile through the use of SBOMs, Zero Trust Architecture, better cloud platform security, and vulnerability disclosure programs and otherwise follow guidelines from the National Institute of Standards and Technology (NIST).

3.2.1.2 The U.S. Food and Drug Administration (FDA)

The FDA operates as an agency under the Department of Health and Human Services (HHS) and is responsible for the regulatory oversight of food, drugs, cosmetics, medical devices, and veterinary medicine. As will be outlined later in this section, the FDA authors regulations and guidance (two directly related to cybersecurity) to elaborate and clarify the FD&C Act. The Agency also participates with numerous U.S.-based and international bodies in authoring cybersecurity standards and guidances.

3.2.1.3 The U.S. HHS Office for Civil Rights (OCR)

The OCR within HHS is responsible for implementing and enforcing HIPAA, which requires the privacy and security of PHI transmitted by U.S.-based "covered entities" (health plans, health care clearinghouses, and healthcare providers, as well as their business associates).

3.2.1.4 The European Commission

The European Commission (EC) functions as the executive branch of the EU, publishing and enforcing regulations such as the EU Medical Device Regulation (EU MDR) and the General Data Protection Regulation (GDPR).

3.2.1.5 Health Canada

In addition to its responsibility for Canada's general Medical Devices Regulations (SOR/98-282), Health Canada also publishes guidance in support of the regulations, including one centered on medical device cybersecurity.

3.2.1.6 Therapeutic Goods Administration (TGA), Australia

The TGA is similarly responsible for Australia's Therapeutic Goods (Medical Devices) Regulations 2002 and for authoring related guidance, including its own medical device cybersecurity guidance.

3.2.2 Other Government Actors (A Sampling)

3.2.2.1 The Medical Device Coordination Group (MDCG)

The MDCG was created under the EU MDR/IVDR and is responsible, among other things, for the oversight of Notified Bodies and the development of guidance documents supporting the EU MDR, including one specific to medical device cybersecurity (MDCG 2019-16).

3.2.2.2 Notified Bodies

Although not government actors in a literal sense, Notified Bodies assess conformity to regulatory requirements on behalf of competent authorities and certify MDMs to standards. Notified Bodies do not publish regulations, standards, or guidance, but they do regularly publish documents outlining how they intend to assess conformity, which is sometimes a de facto exercise of regulatory enforcement discretion.

3.2.3 Standards Development Organizations

Standards Development Organizations (SDOs) establish both standards and guidance (e.g., Technical Information Reports (TIRs)). For documents published by SDOs to have the weight of regulatory requirements, they must be harmonized or recognized by a national regulatory or competent authority. Some SDOs also offer certification and/or testing services related to the standards that they publish. The following is a list of SDOs that have written, or are currently writing, standards and/or guidance applicable to medical device cybersecurity.

- The International Organization for Standardization (ISO);
- The International Electrotechnical Commission (IEC);
- Underwriters' Laboratories (UL);
- The Association for the Advancement of Medical Instrumentation (AAMI) (note: the FDA has representation on many AAMI standards working groups);
- The American National Standards Institute (ANSI) (note: the FDA has representation on ANSI and is involved in some standards work through this organization);
- The Institute of Electrical and Electronics Engineers (IEEE);
- National Electrical Manufacturers Association (NEMA).

3.2.4 Other Cybersecurity Guidance-Writing Stakeholders

3.2.4.1 The International Medical Device Regulators Forum (IMDRF)

The successor to the now-defunct Global Harmonization Task Force (GHTF), the IMDRF started in 2011 and consists of regulatory authority representatives from Australia, Brazil, Canada, China, the EU, Japan, Russia, Singapore, South Korea, the United Kingdom, and the United States. The World Health Organization (WHO) and Argentina's National Administration of Drugs, Food, and Medical Devices are Official Observers.

IMDRF's guidance documents are recognized by many global regulatory authorities as sufficient to establish compliance to their own national regulatory requirements.

3.2.4.2 The National Institute of Standards and Technology

The National Institute of Standards and Technology (NIST) operates as a nonregulatory agency under the U.S. Department of Commerce. This agency has published a cybersecurity framework and numerous technical guidance documents referenced by the FDA and various other global healthcare and medical device cybersecurity guidances. As an aside, the NIST also manages the National Vulnerability Database (NVD).

3.2.4.3 The Healthcare and Public Sector Coordinating Councils

The Healthcare and Public Sector Coordinating Councils (HSCC) advises the HHS and the FDA on threats to the U.S. healthcare infrastructure. The HSCC has published two documents on cybersecurity practices that have achieved unofficial near-guidance status with the FDA and, as such, will be included later in this section.

3.2.4.4 Massachusetts Institute of Technology Research and Engineering

The Massachusetts Institute of Technology Research and Engineering (MITRE) is a federally funded research and development organization with several areas of focus, including cybersecurity. MITRE has completed projects funded by FDA and in collaboration with other groups (e.g., MDIC) to address cybersecurity issues applicable to connected medical devices. Similar to HSCC, some of their work has the backing of FDA to the degree that it would be irresponsible to ignore it in a reference list of regulatory guidance.

3.2.4.5 The Medical Device Innovation Consortium

The Medical Device Innovation Consortium (MDIC) is an industry group focused in several areas of regulatory science including cybersecurity. They frequently collaborate with the FDA, other industry groups, SDOs, and research organizations in the development of guidance, standards, and other tools.

3.3 Road to the Current State of Cybersecurity Regulatory Literature

The FDA's guidance regarding medical device software dates to its original "Design Control Guidance for Medical Device Manufacturers" and the "General Principles of Software Validation," which were both published in 1997. A year later, the FDA published its "Guidance for the Content of Premarket Submissions for Software." The "General Principles of Software Validation" specifically calls for software security provisions in software requirements, software design, and software testing.

In 2001, ANSI and AAMI published a landmark standard labeled ANSI/AAMI SW68:2001 (Medical Device Software – Software Life Cycle Processes), which also pointed to security requirements during software development planning, requirements definition, and verification. Shortly thereafter, the FDA revised its "General Principles of Software Validation," but it was SW68 that became the basis for a global approach to software development.

First, the FDA aligned a revision of the "Guidance for the Content of Premarket Submissions for Software" with SW68 in 2005. However, in parallel, the IEC and ISO established a Joint Working Group to develop an international standard based on SW68. This standard became known ANSI/AAMI/IEC 62304:2006 (Medical Device Software – Software Life Cycle Processes) and is the core software lifecycle model still in use today around the world for software in or as a medical device.

Despite the inclusion of software security elements in two FDA guidance documents and the IEC 62304 standard, the medical device industry paid scant attention to cybersecurity in any meaningful way until two catalytic events caused an awakening to the risks of vulnerable devices. In 2008, a paper at the IEEE Symposium on Security and Privacy by nine academic researchers was published revealing that implantable cardiac defibrillators (ICDs) could be hacked by "software radio attacks" [3]. One of those researchers was Kevin Fu, who would later become the FDA's first Acting Director of Medical Device Cybersecurity for a 1-year term in 2021. Three years later, at the 2011 Black Hat Technical Security Conference in Las Vegas, a diabetic man named Jerome Radcliffe took the stage and showed the audience how he could hack his own insulin pump from up to a half-mile away [4]. As a result of these highly publicized vulnerabilities, the Government Accountability Office published a report titled [5]: "Medical Devices: FDA Should Expand Its Consideration of Information Security for Certain Types of Devices."

The FDA began engaging in discussions with researchers, MDMs, and hospitals to promote shared responsibility, cooperation, and communication and in 2014 released its first cybersecurity-specific guidance, "Content of Premarket Submissions for Management of Cybersecurity in Medical Devices."

Over the next several years, the FDA continued to provide new and revised guidance regarding medical device cybersecurity. Other countries and industry groups have since added numerous standards and guidance, while regulators struggle to keep pace with the constantly moving medical device cybersecurity freight train.

In 2017, the European Union went beyond standards and guidance and included requirements for information security directly within the text of Regulation (EU) 2017/745, the EU Medical Device Regulation. Other countries have followed suit since.

As the global medical device cybersecurity regulatory and best-practices literature (including this book) have evolved, publishers have aligned on several common threads:

- Inclusion of cybersecurity in QMS, governance, engineering, and lifecycle management processes;
- Establishment of cybersecurity risk management through threat modeling and vulnerability assessment, with appropriate links to safety risk management;

- Establishment of an SBOM to enable supply chain vulnerability and postmarket cybersecurity risk management;
- Cybersecurity-specific requirements, architecture, and design;
- Cybersecurity engineering best practices and principles;
- Cybersecurity-specific design verification;
- Postmarket cyber risk surveillance and management;
- Coordinated Vulnerability Disclosure (CVD);
- Cybersecurity information-sharing;
- Support capability for planned and structured incident response (i.e., incidents impacting the manufacturers IT, OT, or hosting infrastructure or support of customers experiencing a medical device incident, discussed in more detail in Chapter 9).

Although repeatedly addressed in guidance, the linkage between cybersecurity risk and safety risk was codified in a standard in the 2019 revision of ANSI/AAMI/ISO 14971:2019 (Medical Devices – Application of Risk Management to Medical Devices), which is the primary standard for addressing safety risk management in medical devices, and elaborated in its accompanying guidance (AAMI/ISO/TIR24971:2020 (Medical Devices – Guidance on the Application of ISO 14971)). This revision altered the definition of "harm" by removing a single word… "physical." Harm, as defined by 14971, was no longer restricted to physical injury (at least as perceived by industry up to that point) but now clearly included data security [6]: "The process described in ISO 14971 can be applied to hazards and risks associated with the medical device. Risks related to data and systems security are specifically mentioned in the scope, to avoid any misunderstanding that a separate process would be needed to manage security risks related to medical devices. This does not preclude the possibility of developing specific standards, in which specific methods and requirements are provided for the assessment and control of security risks. Such standards can be used in conjunction with ISO 14971…"

3.4 A Cybersecurity Regulations, Standards, and Guidance Reference

This section lists regulations, standards, and guidance either directly related to cybersecurity or indirectly related due to its mandate for adequate design controls and links to downstream security standards or guidance. The scope of the list is regulatory literature recognized, harmonized, or referenced by one or more of the Medical Device Single Audit Program Countries (MDSAP) (Australia, Brazil, Canada, Japan, United States) plus the European Union, which is currently an MDSAP observer (MDSAP observers participate in MDSAP Working Group activities but do not make use of MDSAP in its regulatory audit or inspection scheme).

3.4.1 General Medical Device Regulations and Standards

3.4.1.1 Australia

- Therapeutic Goods (Medical Devices) Regulation 2002, most recently amended October 20, 2022, includes specific cybersecurity language.

3.4.1.2 Brazil

- RDC No. 665 of 30 March 2022 (Provides for the Good Manufacturing Practices for Medical Products and In Vitro Diagnostic Products), includes language requiring design control.

3.4.1.3 Canada

- SOR/98-282 (Medical Devices Regulations), most recently amended September 27, 2022, does not include specific product realization or design control language but requires certification to ISO 13485.0

3.4.1.4 European Union

- Regulation (EU) 2017/745 (Medical Device Regulation) and Regulation (EU) 2017/746 (In Vitro Diagnostics Regulation), both published April 5, 2017, include language requiring information security for SiMD, SaMD, and IT systems.
- Regulation (EU) 2016/679 (General Data Protection Regulation), published April 27, 2016, note that, although not specific to medical devices, the impact on cybersecurity within the medical device industry, as it relates to EU residents, is profound.
- NIS 2 Directive (Directive (EU) 2022/2555), published December 2022, legislative act to achieve a high common level of cybersecurity across the EU.
- EU Cybersecurity Act: Introduces an EU-wide cybersecurity certification framework for ICT products, services, and processes. A proposed amendment will enable the future adoption of European certification schemes for "managed security services" (e.g., incident response, penetration testing, security audits, and consultancy).

3.4.1.5 Japan

- MHLW MO169 (The Ministerial Ordinance on Standards for Manufacturing Control and Quality Control for Medical Devices and In-Vitro Diagnostics), revised in 2021 to align with the 2016 revision of ISO 13485, includes product realization requirements consistent with ISO 13485:2016.

3.4.1.6 United States

- 21 CFR Part 820 (Quality System Regulation), most recently revised November 9, 2022; note that 21 CFR Part 820 has been revised to incorporate by reference ISO 13485:2016 under the new name Quality Management System Regulation (QMSR). (The QMSR will become effective February 2, 2026). Includes language requiring design controls.

3.4.1.7 International

- ISO 13485:2016 (Medical Devices – Quality Management Systems – Requirements for Regulatory Purposes), most recently revised March 1, 2016, includes product realization requirements for medical devices and security requirements for records.

3.4.2 Medical Device General Design and Development Guidance and Standards

3.4.2.1 United States

- FDA Design Control Guidance for Medical Device Manufacturers, published in 1997, Elaborates on general design control requirements.

3.4.2.2 International

- IEC 60601-1 (Medical Electrical Equipment – Part 1: General Requirements for Basic Safety and Essential Performance), most recently revised in 2020, addresses information security as a source of safety risk and as an element of software processes, References IEC 62304 as the standard for the software development lifecycle.
- IMDRF/GRRP WG/N47 Final:2018 (Essential Principles of Safety and Performance of Medical Devices and IVD Medical Devices), published October 31, 2018, References IEC 62304 and specifically calls out the need for cybersecurity protections for SiMD, SaMD, and IT systems.

3.4.3 Medical Device General Software Design and Development Standards That Contain Cybersecurity Elements

3.4.3.1 International

- IEC 62304:2006/AMD1:2015 (Medical Device Software – Software Life Cycle Processes), most recently revised in 2015, includes software security requirements during software requirements definition.
- IEC 82304-1:2016 (Health Software – Part 1: General Requirements for Product Safety), published in 2016, incorporates software security throughout the software-only health software lifecycle.

3.4.4 Medical Device General Software Design and Development Guidances That Contain Cybersecurity Elements

3.4.4.1 Australia

- Consultation: Regulation of software, including Software as a Medical Device (SaMD), published February 2019, recommended changes to the Therapeutic Goods (Medical Devices) Regulation 2002, references IMDRF software guidance documents.
- Note that compliance with IEC 62304:2006 is evidence of compliance to Australia's Essential Principles for MDMs.

3.4.4.2 Canada

- SaMD: Definition and Classification, published October 3, 2019, references IMDRF software guidance documents.
- Note that Canada recognizes IEC 62304:2006 as an acceptable approach for software development.

3.4.4.3 European Union

- Note that MDCG 2019-11 (Guidance on Qualification and Classification of Software in Regulation (EU) 2017/745 – MDR and Regulation (EU) 2017/746 – IVDR) is an important software guidance in the EU and references IMDRF software classification schemes, but makes no mention of cybersecurity by inclusion or reference. However, the EU has harmonized to IEC 62304:2006.

3.4.4.4 United States (FDA)

- General Principles of Software Validation, most recently revised January 11, 2002, incorporates software security throughout the software development lifecycle.

- Content of Premarket Submissions for Device Software Functions, published June 14, 2023, supersedes the Guidance for the Content of Premarket Submissions for Software Contained in Medical Devices (2005), which made no mention of cybersecurity by inclusion or reference, incorporates cybersecurity through reference to FDA's premarket and postmarket cybersecurity guidance documents.

3.4.4.5 International SDOs

- AAMI TIR45: 2012 (R2018) (Technical Information Report Guidance on the use of AGILE practices in the development of medical device software), most recently revised on August 20, 2012, References IEC 62304:2006.
- IMDRF/SaMD WG/N23 FINAL: 2015 (Software as a Medical Device (SaMD): Application of Quality Management System), published on October 2, 2015, incorporates software security throughout the software development lifecycle.

3.4.5 Medical Device (and Broader Healthcare) Cybersecurity Standards

- ISO 27799:2016 (Health Informatics—Information Security Management in Health Using ISO/IEC 27002), most recently revised in July 2016, ISO/TC 215 is chaired by ANSI, which includes U.S. FDA representation.
- UL 2900-2-1:2018 (Software Cybersecurity for Network-Connectable Products, Part 2-1: Particular Requirements for Network Connectable Components of Healthcare and Wellness Systems), published September 1, 2017, application of UL 2900-1-1 to medical devices, accessories to medical devices and medical device data systems.
- IEEE Std 11073-40101-2020 (Health Informatics—Device Interoperability Part 40101: Foundational—Cybersecurity—Processes for Vulnerability Assessment), published on September 24, 2020, applies to personal health devices and point-of-care devices, and defines vulnerability assessment process.
- IEEE Std 11073-40102-2020 (Health Informatics—Device Interoperability Part 40102: Foundational—Cybersecurity—Capabilities for Mitigation), published on September 24, 2020, applies to personal health devices and point-of-care devices, and defines cybersecurity protection measures for incorporation during software design and development.

- IEC 81001-5-1:2021 (Software as a Medical Device (SaMD): Application of Quality Management System), published in 2021, maps cybersecurity activities directly to IEC 62304:2006.

3.4.6 Medical Device Cybersecurity Guidance

3.4.6.1 Australia

- Medical device cyber security guidance for industry, published in July 2019, notes harmonization with medical device cybersecurity guidance from the U.S. FDA, EU MDCG, and IMDRF.

3.4.6.2 Canada

- Premarket Requirements for Medical Device Cybersecurity, published on June 17, 2019, incorporates full lifecycle cybersecurity guidance, references ISO 13485:2016, IEC 62304:2006/AMD1:2015, AAMI TIR57:2016, UL 2900-1:2017, UL 2900-2-1:2018, IEC 8001-1:2010, FDA's Content of Premarket Submissions for Management of Cybersecurity in Medical Devices, FDA's Market Management of Cybersecurity in Medical Devices, ISO 14971, NIST's Framework for Improving Critical Infrastructure Cybersecurity, and NIST's Guide for Conducting Risk Assessments.

3.4.6.3 European Union

- MDCG 2019-16 (Guidance on Cybersecurity for Medical Devices), published in December 2019, references IMDRF/CYBER WG/N 60 (DRAFT Principles and Practices for Medical Device Cybersecurity).

3.4.6.4 Japan

- PSEHB/MDED Notification No. 0724-1 (Guidance on Ensuring Medical Device Cybersecurity), published on July 24, 2018, references IEC 80001-1:2010, IEC TR 80001-2-2:2012, IEC TR 80001-2-8:2016, and NIST SP800-53.

3.4.6.5 United States (FDA)

- Cybersecurity for Networked Medical Devices Containing Off-the-Shelf (OTS) Software, published on January 14, 2005, structured in a question-and-answer format.
- Postmarket Management of Cybersecurity in Medical Devices, published on December 28, 2016, recommends incorporation of NIST's Framework for Improving Critical Infrastructure Cybersecurity, refer-

ences as "recognized consensus standards" ISO/IEC 30111:2013, ISO/IEC 29147:2014, and ANSI/AAMI/ISO 14971:2007/(R)2010.

- Cybersecurity in Medical Devices: Refuse to Accept Policy for Cyber Devices and Related Systems Under Section 524B of the FD&C Act, published on March 30, 2023, establishes four prerequisites for review of premarket product submissions: (1) list of "postmarket cybersecurity vulnerabilities and exploits" and coordinated vulnerability disclosure procedures; (2) procedures for ensuring premarket and postmarket cybersecurity (including procedures related to ongoing cybersecurity patches; (3) SBOM; (4) other requirements defined in subsequent regulations.
- Cybersecurity in Medical Devices: Quality System Considerations and Content of Premarket Submissions, published on September 27, 2023, Supersedes Content of Premarket Submissions for Management of Cybersecurity in Medical Devices published on October 2, 2014, establishes a Secure Product Development Framework (SPDF), references as "recognized consensus standards" IEC 81001-5-1, ANSI/ISA 62443-4-1 (equivalent to ISO/IEC 62443-4-1), ISO 14971, AAMI TIR57; ANSI/AAMI SW91, ANSI/UL 2900, IEC TR 80001-2-2, IEC TR 80001-2-8, and IEC TR 80001-2-9, further references NIST's Framework for Improving Critical Infrastructure Cybersecurity, the Healthcare & Public Health Sector Coordinating Council's Medical Device and Health IT Joint Security Plan, the MDIC/MITRE Playbook for Threat Modeling Medical Devices, the Common Vulnerability Scoring System (CVSS) Version 3.0, the Cybersecurity and Infrastructure Agency (CISA) Known Exploited Vulnerabilities Catalog (KEV), National Telecommunications and Information Administration's (NTIA's) Framing Software Component Transparency: Establishing a Common Software Bill of Materials; National Electrical Manufacturers Association (NEMA) Manufacturer Disclosure Statement for Medical Device Security (MDS^2); NIST FIPS 140-3 (Cryptographic Module), and NIST FIPS 140-2 (Suite B).
- Note that the FDA has also published several white papers and discussion papers regarding medical device cybersecurity including *Best Practices for Communicating Cybersecurity Vulnerabilities to Patients* and *Strengthening Cybersecurity Practices Associated with Servicing of Medical Devices: Challenges and Opportunities.*

3.4.6.6 International SDOs and Industry Stakeholders

- IEC/TR 80001-2-2:2012 (Application of Risk Management for IT-Networks Incorporating Medical Devices—Part 2-2: Guidance for the

Communication of Medical Device Security Needs, Risks and Controls), published in July 2012; TS/SC 62A includes ANSI representation.

- AAMI/IEC TIR80001-2-8:2016 (Application of Risk Management for IT Networks Incorporating Medical Devices—Part 2-8: Application Guidance—Guidance on Standards for Establishing the Security Capabilities Identified in IEC 80001-2-2), published on November 18, 2016, and the working group includes U.S. FDA representation.
- IEC TR 80001-2-9:2017 (Application of Risk Management for IT-Networks Incorporating Medical Devices - Part 2-9: Application Guidance—Guidance for Use of Security Assurance Cases to Demonstrate Confidence in IEC TR 80001-2-2 Security Capabilities), published on January 30, 2017. The working group includes U.S. FDA representation.
- AAMI TIR57:2016 (Principles for Medical Device Security – Risk Management), published on June 5, 2016. The working group included U.S. FDA and Department of Veteran Affairs representation.
- HSCC Medical Device and Health IT Joint Security Plan version 2 (JSP2), originally published in January 2019 and updated in March 2024. The HSCC Joint Cybersecurity Working Group includes U.S. FDA representation.
- AAMI TIR97:2019 (Principles for Medical Device Security – Postmarket Risk Management for Device Manufacturers), published on September 27, 2019. The working group was cochaired by a U.S. FDA representative.
- ANSI/NEMA HN 1-2019 (American National Standard – Manufacturer Disclosure Statement for Medical Device Security), published in 2019 and maps to IEC TR 80001-2-2:2012, ISO/IEC 27002:2013, and NIST 800-53 r4.
- IMDRF/CYBER WG/N60 FINAL:2020 (Principles and Practices for Medical Device Cybersecurity), published on March 18, 2020. As noted above, IMDRF member countries include regulatory representatives from 11 countries plus Argentina and the WHO as observers.
- NIST Special Publication 1800-24 (Security Picture Archiving and Communication System (PACS): Cybersecurity for the Healthcare Sector), published in December 2020 and provides detailed technical guidance for PACS cybersecurity.
- Playbook for Threat Modeling Medical Devices (MITRE/MDIC), published on November 30, 2021, published under contract with the U.S. FDA.

- Medical Device Cybersecurity: Regional Incident Preparedness and Response Playbook (MITRE), published in November 2022 and published under contract with the U.S. FDA.
- ANSI/AAMI SW96:2023 (Standard for Medical Device Security—Security Risk Management for Device Manufacturers), published January 13, 2023, and uses an ISO 14971 medical device risk management framework to establish a cybersecurity risk management process in alignment with AAMI TIR57 and AAMI TIR97.
- IMDRF/CYBER WG/N70 FINAL:2023 (Principles and Practices for the Cybersecurity of Legacy Medical Devices), published on April 11, 2023, and leverages other IMDRF guidance, AAMI TIR57, AAMI TIR97, numerous IEC and ISO standards, and national cybersecurity guidance documents to map new product development activities to legacy medical devices.
- IMDRF/CYBER WG/N73 FINAL:2023 (Principles and Practices for Software Bill of Materials for Medical Device Cybersecurity), published on April 13, 2023, and leverages numerous NTIA SBOM publications to establish this international SBOM guidance. (Note that the NTIA has produced several SBOM-related documents that culminated in the publication of The Minimum Elements for a Software Bill of Materials (SBOM) in support of a U.S. White House Executive Order (14028) on May 12, 2021.)

3.4.7 General Cybersecurity Standards and Guidance

Note that there are a multitude of detailed technical cybersecurity guidance documents from the NIST (e.g., SP 800 series), IEEE, and others not included in this list. The focus of this section is general standards and guidance that drive software development lifecycle processes incorporating effective cybersecurity activities.

- NIST Special Publication 800-115 (Technical Guide to Information Security Testing and Assessment), published in September 2008.
- NIST Special Publication 800-66 (An Introductory Resource Guide for Implementing the Health Insurance Portability and Accountability Act (HIPAA) Security Rule), published in October 2008.
- NIST Special Publication 800-144 (Guidelines on Security and Privacy in Public Cloud Computing), published in December 2011.

- NIST Special Publication 800-30 (Guide for Conducting Risk Assessments), published in September 2012.
- ISO/IEC 27000:2018 (Information Technology — Security Techniques — Information Security Management Systems — Overview and Vocabulary), most recently revised in February 2018.
- ISO/IEC 29147:2018 (Information Technology—Security Techniques—Vulnerability Disclosure), most recently revised in October 2018.
- NIST Special Publication 800-37 (Risk Management Framework for Information Systems and Organizations: A System Life Cycle Approach for Security and Privacy), most recently revised in December 2018.
- ISO/IEC 30111:2019, most recently revised in October 2019.
- NIST Special Publication 800-128 (Guide for Security-Focused Configuration Management of Information Systems), published in October 2019.
- NIST Special Publication 800-207 (Zero Trust Architecture), published in August 2020.
- NIST Special Publication 800-53, most recently revised in September 2020.
- NIST Special Publication 800-160, Volume 2 (Developing Cyber-Resilient Systems: A Systems Security Engineering Approach), published in December 2021.
- NIST Special Publication 800-218 (Secure Software Development Framework (SSDF) Version 1.1: Recommendations for Mitigating the Risk of Software Vulnerabilities), published in February 2022.
- ISO/IEC 27002:2022 (Information Security, Cybersecurity and Privacy Protection — Information Security Controls), most recently revised in March 2022.
- NIST Special Publication 800-161 (Cybersecurity Supply Chain Risk Management Practices for Systems and Organizations), published in May 2022.
- ISO/IEC 27001:2022 (Information Security, Cybersecurity and Privacy Protection — Information Security Management Systems — Requirements), most recently revised in October 2022.
- NIST Cybersecurity Framework (CSF) 2.0, published on February 26, 2024, establishes a core framework consisting of six framework functions (Identify, Protect, Detect, Respond, Recover, and Govern),

with categories and subcategories under each function to aid in implementation.

3.4.8 Industrial Control Cybersecurity Standards and Guidance

Industrial control system cybersecurity standards and guidance are not specific to the medical device industry, and compliance is not required by any regulatory authority. However, these documents are useful for establishing cybersecurity processes and practices for systems used in MDM and for IT environments constituting the ecosystem in which medical devices reside.

- IEC TS 62443-1-1:2009 (Industrial Communication Networks - Network and System Security - Part 1-1: Terminology, Concepts and Models), published on July 30, 2009.
- IEC TR 62443-3-1:2009 (Industrial Communication Networks - Network and System Security - Part 3-1: Security Technologies for Industrial Automation and Control Systems), published on July 30, 2009.
- IEC 62443-2-1:2010 (Industrial Communication Networks - Network and System Security - Part 2-1: Establishing an Industrial Automation and Control System Security Program), published on November 10, 2010.
- ISO/IEC 62443-3-3:2013 (Industrial Communication Networks - Network and System Security - Part 3-3: System Security Requirements and Security Levels), published on August 7, 2013.
- NIST Special Publication 800-82 (Guide to Industrial Control Systems (ICS) Security), most recently revised in June 2015.
- ISO/IEC 62443-4-1:2018 (Security for Industrial Automation and Control Systems - Part 4-1: Secure Product Development Lifecycle Requirements), published on January 15, 2018.
- ISO/IEC 15408-2:2022 (Information Security, Cybersecurity and Privacy Protection — Evaluation Criteria for IT Security — Part 2: Security Functional Components), most recently revised in August 2022.
- ISO/IEC 15408-3:2022 (Information Security, Cybersecurity and Privacy Protection — Evaluation Criteria for IT Security — Part 3: Security Assurance Components), most recently revised in August 2022.
- ISO/IEC 15408-4:2022 (Information Security, Cybersecurity and Privacy Protection — Evaluation Criteria for IT Security — Part 4: Framework for the Specification of Evaluation Methods and Activities), most recently revised in August 2022.

- ISO/IEC 15408-5:2022 (Information Security, Cybersecurity and Privacy Protection — Evaluation Criteria for IT Security — Part 5: Predefined Packages of Security Requirements), most recently revised in August 2022.
- ISO/IEC 15408-1:2022 (Information Security, Cybersecurity and Privacy Protection—Evaluation Criteria for IT Security—Part 1: Introduction and General Model), most recently revised in October 2022.

3.5 What Does the Future Hold?

The regulatory landscape for medical device cybersecurity is one of ever-shifting sand. There is risk in publishing any forward-looking view on this topic in book format. New work may have begun or been completed during the publishing process, so the following must be viewed as a snapshot in time and portions may be outdated by the time this book goes to press.

3.5.1 Guidance: Medical Device Stand-Alone Software, Including Apps (including IVDMDs) (U.K.)

After Brexit, the MHRA in the United Kingdom continues to revise its general SaMD guidance and makes reference to the U.K.'s Royal College of Physicians publication *Using Apps in Clinical Practice Guidance.* The guidance is primarily concerned with device classification and contains only rudimentary design and development requirements.

3.5.2 Ongoing Cybersecurity Work in Industry SDOs

Virtually every SDO (e.g., AAMI, MDIC, ISO, IEC, IMDRF, ANSI) has cybersecurity committees and working groups. These teams continue to develop new standards and guidance and revise documents in the existing regulatory library. As seen in the above reference list, many of these organizations collaborate in their standards work, which speaks to continued efforts at harmonizing regulatory expectations for medical device cybersecurity globally.

3.5.3 Selling to the Federal Government

If your intended marketplace includes the federal government, such as the U.S. military or the Veterans Administration (further discussed in Chapter 4), then additional considerations are needed. These include FIPS-140 certified cryptographic libraries and CISA's "Secure Software Development Attestation Form," which takes its authority from Executive Order 14028 and applies to any vendor selling to the U.S. federal government.

This form is a vendor's declaration of conformance to some of the requirements found in NIST SP 800-218A "Secure Software Development Framework" (more commonly referred to as the "SSDF"). The SSDF has 19 categories which include a total of 42 requirements. The new attestation form only references 4 of those categories and enforces only 29 requirements. This is a self-attestation form; however, FedRAMP-approved third-party auditors can be used without having to sign the form.

3.5.4 The Link to Healthcare Delivery

In addition to the above referenced efforts specific to medical device manufacturers, we must note that hospitals and other healthcare delivery organizations (HDOs) and caregivers are experiencing pressure to address cybersecurity from both an operational and a privacy perspective; see for example HHS' most recently published [7]: "HPH Cybersecurity Performance Goals." Although voluntary at present, they have started to raise the bar for cybersecurity expectations across the healthcare industry.

Therefore, security considerations continue to be integrated into purchasing decisions, vendor approval processes, and contract language. The work of groups such as the HSCC will continue to strengthen the bond of cybersecurity accountability between HDOs and MDMs, as evidenced in their March 2022 publication "Model Contract-language for Medtech Cybersecurity (MC2)." ENISA in Europe similarly published "ENISA Procurement Guidelines for Cybersecurity in Hospitals" in 2020.

However, securing medical devices solely or primarily on the level of the HDO network is a less effective and more expensive way of implementing security. The FDA has stated that MDMs should consider hospital networks as "hostile environments." A shift to better secured devices is, therefore, inevitable (think "zero-trust architecture") in that it reduces the dependence on security external to the device and provides a better security posture at a lower total lifetime cost.

The threshold of cybersecurity acceptability within medical device operating environments will likely continue to rise. The ability to demonstrate conformity to standards and regulatory expectations can be key to proving that a medical device has been designed and is managed in a way that security is sufficiently addressed. Increasingly, HDOs are appreciating that their networks are only as secure as their weakest link, so they have been more selective in their review and approval of medical devices that cannot provide evidence of a robust security profile. As the industry continues to mature, all stakeholders, including regulators, hospitals, and MDMs, are likely to raise the bar on security. This proactive approach is necessary to stay ahead of the threat actors, who are also evolving and maturing.

References

[1] European Commission, "Regulation (EU) 2017/745 on Medical Devices - Summary List as pdf Document," March 8, 2024, https://ec.europa.eu/docsroom/documents/48579.

[2] FDA, "Appropriate Use of Voluntary Consensus Standards in Premarket Submissions for Medical Devices; Guidance for Industry and Food and Drug Administration Staff," September 2018, https://www.fda.gov/regulatory-information/search-fda-guidance-documents/appropriate-use-voluntary-consensus-standards-premarket-submissions-medical-devices.

[3] Halperin, D.,, et al., "Pacemakers and Implantable Cardiac Defibrillators: Software Radio Attacks and Zero-Power Defenses," *Proceedings of the 2008 IEEE Symposium on Security and Privacy*, 2008.

[4] Ngak, C., "Black Hat Hacker Can Remotely Attack Insulin Pumps and Kill People," *CBS News*, August 29, 2011, https://www.cbsnews.com/news/black-hat-hacker-can-remotely-attack-insulin-pumps-and-kill-people/.

[5] Government Accountability Office (GAO), "FDA Should Expand Its Consideration of Information Security for Certain Types of Devices," August 31, 2012, https://www.gao.gov/products/GAO-12-816.

[6] ANSI/AAMI/ISO 14971:2019, Annex A.

[7] U.S. Department of Health and Human Services, "HPH Cybersecurity Performance Goals," December 2023, https://hphcyber.hhs.gov/performance-goals.html.

4

Use Environments

Healthcare is anything but homogenous. The complexity of the human body and its ailments is reflected in the complexity of the healthcare delivery system, along with added layers of administrative burden. Recent advances in technology have allowed us to expand the spaces and places where healthcare delivery takes place from a hospital-centric approach to a patient-centric one. The COVID-19 pandemic has accelerated this further, and healthcare now increasingly happens where the patient is [1]. In addition, we have populations with unique healthcare service needs, which has led (in the United States) to the development of the Department of Defense (DoD), Veterans Affairs (VA), and Indian Health Service (IHS) systems.

This chapter will provide an overview, albeit only scratching the surface, of different care delivery scenarios and systems and how cybersecurity considerations can vary based on the device's use environments. These considerations pose a set of unique requirements for which device manufacturers must account when designing secure devices and systems.

4.1 Healthcare Facilities

When the previous edition of this book was published, COVID-19 was just beginning to wreak havoc. On March 13, 2020, it was declared a national emergency in the United States. While we are still trying to grasp the pandemic's full impact on society and on healthcare, we know that it has been massive.

The move to remote healthcare had already begun, but no one could have anticipated the kind of accelerant that COVID-19 would provide to transforming the healthcare sector—from realizing advantages through telehealth, which had been resisted due to the cost of change, to the disadvantages and harms that would be introduced into healthcare from the supply chain risk, staffing shortages, to a massive shift to remote work with almost no preparation for the technology needed or for the people working remotely. COVID-19 gave us all a crash course in "digital health," a phrase bandied about in 2020, ill-defined and even less well understood. We are only now beginning to understand what digital health means, and, from a cybersecurity perspective, it is dangerous and complex and requires a level of constructive collaboration, integration, and interaction that has never been required before in healthcare, in information technology (IT), and certainly not in health IT.

The emergency is over, for now. However, the term "healthcare facilities" will never have the same meaning that it did before COVID-19. Care can be delivered by providers (to any patient, anywhere, anytime) who may be nowhere near the patient. Your home may be your hospital now, thanks to Best Buy [2]. Clinics may be at your corner drug store or at the nearest Walmart, or your doctor may see you from his or her office while you are traveling. An electrocardiogram (EKG) can be done using a smartphone while you are on vacation and your smartwatch may be collecting, managing, and connecting your key health and fitness data directly into the electronic medical record (EMR) that your providers are using, 24-7. The days of the patient-touching technology in healthcare environments are far from over, but they may not necessarily take place in the traditional healthcare setting. It is increasingly likely that the healthcare environment will be everywhere. No longer are medical devices and their associated electronic medical records or electronic health records (EMR/EHR) our primary connection to physicians, nurses, and other caregivers. The medical technologies used in patient care now run from traditional medical devices and EMR/EHR, to patient-worn devices and smart clothing, therapeutic gaming, 3-dimensional (3D) printing of radiology scans, laboratory equipment, and even "smart home" devices such as speakers, security cameras, and lighting. Even the definition of "lab" turns out to be flexible. Remember that labs during COVID-19 may have been massive drive-through facilities in parking lots at hospitals or other facilities where tens of thousands of noses were swabbed each day. The need for effective cybersecurity and privacy to ensure medical device functionality, reliability, and safety has become more important, more complex, and much more difficult with the increasing use of wireless and connected devices that consumers want. More of these devices can now be ordered or prescribed by licensed professionals.

Yet cybersecurity incidents have left medical devices and hospital networks inoperable, disrupting the delivery of patient care across healthcare facilities. Such incidents may lead to patient harm through delays and/or errors in diagnoses and/or treatment interventions. All stakeholders within healthcare must share the accountability for medical device cybersecurity. To share that responsibility effectively, all stakeholders must have a better understanding of their role in support of proactive cybersecurity that helps to protect and secure medical devices, wherever they are, in expectation of future attacks, complications, or actions.

4.1.1 Hospitals

The intent of the HIPAA Privacy Rule was to safeguard Protected Health Information (PHI) and Personally Identifiable Information (PII). This data is handled by almost every department in a hospital, most likely in multiple systems or applications. Today, most healthcare providers (e.g., physicians, physician assistants, nurses, pharmacists, technicians, dietitians, physical therapists) use EHRs, e-prescribing software, connected medical devices, remote patient monitoring, and laboratory information systems. Revenue cycle employees access this information to work with patients' insurance and medical billing. The list goes on. Unlike many other sectors where this type of data is typically confined within specific departments or units within departments, healthcare data touches almost all operating units within a hospital. In other organizations, access restriction policies can allow for cybersecurity measures to be role-based and more tightly controlled. In a hospital setting, however, even though the data is sensitive and valuable, many departments handle it at least in some manner.

Cybersecurity in hospitals is not only about protecting the data. Healthcare is consistently rated in the top five attacked sectors [3]. Merely protecting data seems a lot less important if all your systems are locked up by ransomware and the people who need both systems and data to provide care cannot access either. Cybersecurity is now one of the top measures required to protect PII and PHI by securing individuals' identities, devices, networks, and data from attacks in these hyperconnected environments.

4.1.2 Clinics, Labs, Testing Centers, Processing Centers, Your Corner Drug Store, and Best Buy

Clinics have not escaped the changes ushered in by COVID-19 and digital health. While the basic environment may be smaller and more contained than a hospital, with better physical access control and monitoring, still, telehealth, phone visits, remote monitoring, and wellness/fitness apps are also being driven to and from this space. As more care is pushed out of the hospital setting and technology advances, more clinics are adding their own small labs and basic imaging capabilities on site. Many retail and office spaces that went vacant during COVID-19 (including some of our own former offices) are turning into labs, neighborhood clinics, urgent care centers, and other opportunities for care close to home. This trend means more devices, more connectivity across public networks, and more opportunities for mishandling, mistakes, and malevolence.

4.1.3 Responses and Next Steps

The healthcare industry has been evolving over the past decade under the label "digital health." COVID-19 significantly accelerated those changes, transforming how care is delivered and consumed, highlighting the need for data integration and new processes across disparate systems, care locations, and stakeholders (patients, care and service providers, and healthcare organizations). The continuum of care is broader and wider than it has ever been. The continuum of data is even more so. This means that healthcare's obligation to protect information and secure the most hyperconnected sector in the world is more complex and more demanding and involves more categories of stakeholders and participants than ever before.

We do see some help for the medical device arena on the horizon with the passage of the Omnibus Bill of 2022 and section 524B of the Federal Food, Drug, and Cosmetic (FD&C) Act. This has provided the FDA with clear authority over MDMs. Improved cybersecurity regulation for medical devices will not solve all the problems and issues that healthcare faces. Medical devices introduce additional complexity into the cybersecurity space in healthcare: Who is responsible for what? Where does authority lie, not only in the provider organization, but across the entire supply chain? While the new authority that the FDA has been given helps with the future, the massive installed base of legacy devices is the more imminent challenge of today [4]. Living with vulnerable outdated technology can be exceedingly dangerous, but replacing tens of millions of functioning devices that were budgeted to have much longer operational lives is also prohibitive on many levels and will require time and a consolidated effort.

Every organization and every facility will have to deal with this new world in ways that work for their culture, with their budget, and across all roles and functions. Cybersecurity, like care, cannot be effective or efficient in silos. It must be collaborative and integrated across users, functions, and departments and even beyond the confines of that organization with all stakeholders across the continuum of data, networks, and the provision of care.

Fortunately, there are also some basic practices that are applicable across all environments, devices, networks, and practices. Here are three security—and thus patient safety—measures that should be taken in and across all healthcare settings to protect patients, devices, digital systems, networks, and data from attacks.

1. *Personnel training:* Lack of IT security understanding and skills represents a major threat to healthcare. Multiple studies over the last 4 years indicate that many employees (40% to 55%) do not have cybersecurity expertise or knowledge of basic security practices and requirements [5]. Professional and regular, ongoing training on cybersecurity is essential. Stakeholders (patients, caregivers, employees) should:
    - Understand how to use equipment in a way that does not expose it to cyber threats and could result in compromise.
    - Be able to recognize specific threats (pop-ups, suspicious/unknown emails, and links).
    - Use strong passwords and multifactor authentication (MFA).
    - Connect to networks using current security best practices.
2. *Data usage and access control:* Clinics should monitor and control malicious file activity by blocking unauthorized actions, restricting sharing of files/emails, and limiting copying outside the network.
    - Capture data and events to identify unauthorized actions (logging).
    - Encrypt data at rest and in transit.
    - Utilize MFA for cloud and other environments based on risk.
3. *Monitoring of mobile and connected devices:* Mobile phones, cloud-based applications, and Internet of Things (IoT) and Internet of Medical Things (IoMT) devices are now standard in healthcare. These all create new opportunities for mishandling, mistakes, and malevolence: theft of data, passwords, and devices themselves. Connected devices can be compromised and used for covert observation and collection of user activities, data, and data flows. In extreme cases, devices can be completely reconfigured to produce incorrect results, outcomes, or assign devices to incorrect employees or patients.

The medical device ecosystem is complex and protecting such a diverse set of equipment against today's sophisticated and targeted threats is challenging at best. Adding to that the ongoing changes in care delivery and the economic challenges as a result of the COVID-19 pandemic, healthcare security professionals have their work cut out for them. Manufacturers of medical devices need to understand these trends so that they can develop and provide devices that offer not only the right capabilities, but also the right security features, for this new and complex care delivery environment.

4.1.4 Cybersecurity in the HDO Acquisition Process

Device manufacturers need to be aware that healthcare buyers, whether large hospitals, small clinics, or group purchasing organizations (GPOs), have started to make cybersecurity part of their vendor selection and purchasing processes. Next is a short, non-exhaustive list of examples of common expectations with which MDMs will increasingly be confronted (for more details, see Chapters 5, 7, and 10). Manufacturers will need to assert compliance with these requirements and provide supporting proof in the form of security documentation, such as MDS^2 or SBOM. Building on templated contract language such as that provided by the Health Sector Coordinating Council (HSCC) [6], device manufacturers and service providers are presented with RFPs and purchasing contracts that include:

- Specific premarket and postmarket security requirements for medical device vendors and service providers. Commonly, these include requirements from a broad range of hospital stakeholders such as information security and IT.
- Default passwords should be avoided or forced to be changed at go-live. Device features should support regular password changes.
- Equipment should be kept current; device end-of-life (EOL) and end-of-support (EOS) (i.e., the device entering legacy status due to an inability to support current security needs or provide expected functionality) should be communicated with sufficient lead time.
- Provide security features that exempt an HDO from breach reporting in the case of device loss or theft, such as encryption and remote lock.
- Provide features and processes so that patches can be applied regularly and on a timely basis. This applies to both routine patches of commercial components (e.g., the operating system) and device/system specific patches.
- An MDM's process for vulnerability identification and remediation should comply with regulatory guidance and offer appropriate support across the diverse range of networked medical devices. Gaps in the process should be avoided, as they may lead to devices not being assessed or patched.
- Devices should support and be capable of operating on segmented or micro-segmented networks. This will reduce threat exposure and cybersecurity risk and will assist with monitoring and/or disabling or disconnecting nonessential or infected devices to support event containment.

- Devices should provide role-based access controls for controlling access by clinical staff, the vendor, third parties, or other systems. This access types should include local and remote access and support, as appropriate for the device use case and access type, MFA, and least-privilege controls.
- Devices should be able to operate in an environment protected by risk and vulnerability management systems as well as support network-based intrusion detection systems; patient safety and data protection depend on identifying and understanding threats and activities on these devices and networks.
- Medical devices should support inventory and monitoring tools to gather inventory data, including data pertaining to security; such data should be accurate and kept current and should match device security documentation such as SBOM and MDS^2.
- Medical device manufacturers should work with their customers as well as third parties (such as internet service providers (ISPs)) in maintaining medical devices' security posture to stay ahead of vulnerability management and patching. This group should, collectively, include medical devices in security operations practices including planning, training, and exercises around the security incident response (see Chapter 9).

4.2 Home Care Environments

Historically, the home was where most healthcare was delivered. Healthcare professionals made regular house calls; some of us may remember this from childhood. However, the practice became less prevalent around the mid-twentieth century with advancements in personal transportation, technology, healthcare specialization, complexity of care delivery, logistics associated with transportation, cost increases, and changes in patient expectations and preferences. Ironically, some of the same reasons that house calls became passé are some of the same reasons that home healthcare is now becoming popular again. For example, the advancements and reduced costs of health technology, patient preference and comfort, convenience and accessibility, and post-acute patient care/rehabilitation have emerged as principal reasons that home healthcare is once again rising in popularity. It is becoming more common to find health technology in the home, such as pulse oximetry (SpO_2), blood-pressure (BP), electrocardiogram (EKG/ECG), temperature, insulin testing, telepresence, and even ultrasound technology, home-based dialysis, or infusion therapy.

With the advent of modern home-based health technology, there also emerged risks specific to the potential compromise of patient data or device functionality. Those risks become even greater when the home care infrastructure includes a bring-your-own-device (BYOD) component, such as an app on the patient's smartphone that acts as an intermediary between a Bluetooth-connected device and the clinical back-end service.

These risks are easily realized as we think of the infrastructure that is required to make it all work. For example, there are many details in multiple domains that must be considered when establishing a reduced risk and secure home healthcare environment. Before we implement a home healthcare system, or before engineers set out to design a new home care device, it is important that we examine the layers of technology that must exist between the patient and a healthcare professional or caregiver. Some of those layers might include:

- Routers (wireless/wired in the home and wired to a modem);
- Switches (managed and unmanaged connecting home devices);
- Wired/wireless access points (wireless local area network (WLAN));
- Hubs/gateways (e.g., controllers for IoT home technology);
- Network repeaters/extenders;
- Public cellular (4G/5G) network;
- Modems, fiber/copper wired communications;
- Antennas (e.g., transceivers that facilitate wireless communications);
- Personal (BYOD) connected devices such as phones, tablets, or laptops;
- Remote support (e.g., services able to remote connect and support user needs);
- Power (unregulated and incorrectly managed power).

It goes without saying that home networks are not risk-free. We must assume that they operate at a lower level of security than professional networks and that they are not well-managed or maintained. Mitch Parker and Peter McLaughlin outlined the divergence of hospital grade versus consumer grade technology and cybersecurity considerations in Chapter 11 of *Advanced Health Technology* [7]. Admittedly, the most secure system would be a system that cannot be used. As we seek to build a system that is as secure as possible but still usable, it is equally important to consider what we are trying to defend against and from what threats we are protecting ourselves. For example, we might first think about "data" as being the most important element that must be protected. However, that data might be in the form of traditional text data, video files, static images (or photos), and audio files, and the threats that we should have in mind could be in the form of direct attacks (both internal and external to the home), brute force attacks (e.g., someone sitting outside your home and hacking your system), accidental disruptions (e.g., loss of power or connectivity, broken technology, system misuse), and disruptions to a home system, such as through a denial of service.

4.2.1 Nonobvious Risks in Home Healthcare

Among the less obvious risks in the home care environment are:

- Non-healthcare tech that might represent a back door or an unmanaged security risk (printer, television, gaming, personal devices);
- Vendor/third-party supplied solutions [7];
- Risks resulting from data colocation (e.g., health data on the same device as geolocation or social media data);
- Unmanaged non-healthcare devices (sensors and passive systems);
- Data corruption through data transformation;
- Injection errors, corruption, man-in-the-middle attacks.

According to Maneesh Goyal, chief operating officer for Mayo Clinic Platform, as Mayo Clinic looks to leverage virtual care models for increasing numbers and greater scope of acuity relative to patients, recommendations for priority areas for engineers to focus on to enable the necessary cybersecurity levels are as follows:

1. Over the air updates;
2. Logging and audit capabilities;
3. Safeguarding of code and data through logical and/or architectural separation (virtualization, segmentation, partitioning);
4. Encrypted tunnel into the device.

4.2.2 Device Security and Home Healthcare Infrastructure

Clearly, security in the home setting has its own set of challenges. To securely deliver care and related services, the device architecture and design need to account for an environment with a lower security posture (the home network) that transmits/receives data via channels that are under neither the manufacturer's, the health care provider's, nor the medical service provider's control (both home and public networks).

Typical infrastructure for home-based devices includes a central hub (gateway) that acts as a data aggregator across all devices, allows management of devices and data flow, and can be used to collect additional patient information (e.g., to confirm identity). A hub can be a custom system, which allows security and privacy features to be designed into it and therefore enforces and assures a certain security baseline. However, the hub function may also be provided by an app on a smart device, which means that security and privacy features would need to be implemented via the app or even by the device itself.

Typical security needs and examples of how they could be realized include:

- Privacy and confidentiality (e.g., encryption);
- Security and protection (e.g., firewall);
- Data flow management (e.g., via configuration interface);
- Data association to device and/or patient (e.g., via configuration interface and message signing);
- Provisioning, maintenance, updates, and patches (e.g., as push function or via user interface);
- Configuration management (e.g., via user interface or remote access);
- Access and authentication (e.g., role-based access control, biometrics).

None of these requirements are necessarily unique to home-based devices, but their location, network connection, and operator capability make their implementation challenging. Ideally, the device as well as the supporting hub should be designed to be sufficiently secure while ensuring privacy, usability (for patients and caregivers), availability, remote control, configuration management, and support for the desired connectivity.

Regulatory requirements affect the degree to which such information needs to be protected, which, in turn, determines the security features and controls required. This may vary based on region (e.g., General Data Protection Regulation (GDPR) for Europe, HIPAA for the United States, or even specific state laws, such as California's Consumer Privacy Act (CCPA)), as well as considerations based on the type of data (e.g., "fitness and wellness" data, general health data, or data related to specific health scenarios). In general, as home care technology is evolving, so is the regulatory landscape [8].

Evolving new technologies will continue to change the picture of home care. IoT-based technologies will enable new devices and solve problems that previously could not be addressed. The advent of fifth-generation (5G) public networks will make it easier to implement some of the connectivity requirements without having to traverse home network gear, but it will also offer new opportunities to adversaries for attacking devices directly.

Consumerization creates another challenge, in that the introduction of consumer devices creates not only new data points into a patient's profile (for example, cardiac data collected by a smartwatch), but also new regulatory and privacy challenges. A patient's exercise history may be relevant to a clinical consult, but it may come from a less controlled device or app. In particular, nonmedical consumer fitness devices have raised significant privacy concerns in the recent past [9].

A more extensive discussion of the spectrum of personal health devices (PHD), their security requirements, architectural models, and standards mapping has been provided in an IEEE white paper [10]. Implementation models for security in the home care environment are evolving as well and will continue to provide useful guidance [11–14].

4.3 Government and Military Agencies

The intent of this section is to address how government providers, including military health services and other federal agencies that provide healthcare services to defined sectors of the population, address cybersecurity, and adhere to specific government requirements. Some of the most unique environments that established similar and very strict cybersecurity requirements for information technology and medical devices include, but are not limited to:

- Department of Defense (DoD)—Defense Health Agency (DHA);
- Department of Veterans Affairs (DVA)—Veterans Health Administration (VHA or VA);
- Department of Health and Human Services (HHS)—Indian Health Services (IHS);
- Department of Health and Human Services (HHS)—Public Health Service (PHS).

The DHA and VHA are two of the largest healthcare providers in the United States. Each of these organizations is responsible for the healthcare needs of specific populations, and there may even be overlap and a need to transition between systems for their patient populations. Yet although the VHA treats veterans, and all veterans were at one time patients of the DHA (DoD), the VHA is not part of the DoD and is instead a separate Cabinet-level agency. However, a relatively common scenario may be an active-duty military member receiving care from a VHA facility local to his hometown where there is not DHA care facility in the area. Similarly, many Native Americans who get care from the IHS are active-duty military, National Guard members, or veterans, and therefore may be getting care from the DoD, VHA, or both. An active-duty military member may receive care from a VHA facility local to his hometown if there is not a DHA care facility in the area. Finally, DHA and VHA often provide care to senior political and other government figures, both within and outside the continental United States. The DHA has medical centers around the world on major military bases, in addition to clinics aboard ships and on shore, while the VHA has medical centers in every U.S. state, plus the Philippines and Puerto Rico. So there is a recognized need for a high degree of communication and cooperation between these healthcare systems, which need to be supported and secured by the medical devices and associated systems and infrastructure involved.

While many cybersecurity factors required by public and commercial healthcare providers and regulators apply to these government agencies and their environments, there are also unique requirements that might indicate the need for a higher level of security. An example might be a service member who is wounded in combat. The portion of the record that covers the location, the mission being conducted, and other operational capabilities that may be noted in the record must be more securely controlled. Think, in this sense, the service member was on a sensitive assignment (assisting with rescue of hostage from a hostile nation), as part of a special unit (such as a Seal Team) and was injured by a foreign combatant. Along with the location, any potentially compromising information will not be entered into, or be redacted from, the medical record. Some of the services rendered, medical devices used, and test results obtained all may have a higher level of sensitivity for operational security purposes. This would also be true for political figures, including embassy staff abroad.

As a result, the cybersecurity requirements for these agencies are generally more stringent than those of most "civilian" HDOs, including cybersecurity requirements for medical devices. We will also note that the IHS, as a part of the HHS, has its own set of requirements for medical device cybersecurity. Additionally, from a logistics standpoint, some of these agencies also have requirements related to maintaining equipment, including medical devices and other equipment that is in held in reserve in case of a future conflict, disaster, or national emergency. These devices may be purchased to go into storage, but need to be ready for rapid deployment in a time of need, including being cyber-ready for conditions that are not currently known. This may require the agency to contract with the MDM to keep these devices maintained and patched with current software even while in storage.

Both the DHA and VHA represent significant purchasing volume. The DHA has over 100 hospitals, while the VHA has over 170 medical centers, plus more than 1,000 medical clinics between them. Both the DoD and DVA began recognizing the significance of having requirements for cybersecurity of medical devices as early as 2003. Both have team members who were involved in presentations at national security conferences, advocating for cybersecurity of medical devices and, for example, the development of the Manufacturers Disclosure Statement for Medical Device Security (MDS^2), and Medical Device Security Network Isolation policies and practices.

Additionally, both organizations have made major strides in research and development in cybersecurity, implementing programs specifically addressing medical devices [15]. For example [16], "at the U.S. Army Medical Materiel Agency, a subordinate organization of the U.S. Army Medical Research and Materiel Command, a team of medical technology experts comprise a cybersecurity cell created in early 2017. This team, part of the Integrated Clinical Systems (ICS) Program Management Office, focuses on ensuring medical devices used by the military comply with strict Department of Defense cybersecurity standards." Meanwhile, in January 2020, the VHA announced [17]: "VA recently signed partnerships with Massachusetts General Hospital and Shepherd University. This collaboration's research will address cybersecurity and compatibility measures needed in devices used for VA patient care. It will also refine existing and emerging cybersecurity standards and practices for network connectable medical devices, medical data systems and other related technology." Over the past few years, the VA has implemented a specialized team within the Office of Information Security known as the Specialized Device Cybersecurity Department (SDCD). According to their website [18], "the Specialized Device Cybersecurity Department (SDCD) is charged with providing enterprise-level common security services to customers and protecting VA Information Technology (IT) assets and customers from the threat of compromise while ensuring the security, confidentiality, integrity, and availability of information, information systems, infrastructures, applications, Medical Devices/Systems (MD/S), and Special Purpose Systems."

Both organizations have developed procurement and implementation requirements, which are continually being updated. Not all of these requirement documents are publicly available, necessitating that MDMs wishing to supply devices to these agencies establish a prior working relationship with agency purchasers in order to gain access. Notable examples include:

- Department of Veterans Affairs, VA Directive 6550: Pre-Procurement Assessment and Implementation of Medical Devices/Systems, June 3, 2019.
- Department of Veterans Affairs, Medical Device Protection Program (sensitive document, not available outside the VA).
- Department of Veterans Affairs: Design Patterns Privacy and Security: Medical Device Security, Version 1.0, January 2017.
- Defense Information Systems Agency (DISA): Security Technical Implementation Guides (STIGs) [19].
- Department of Defense: DOD 5220.22-M, National Industrial Security Program Operating Manual, Change 1, July 31, 1997.

- Department of Defense: Medical Device Security Technical Implementation Guide, Version 1, Release July 1, 2010.
- Department of Defense: DoD Directive 7730.65, Department of Defense Readiness Reporting System (DRRS) [20].
- Cybersecurity in DoD Acquisition Regulations [21].

Federal agencies are also required to adhere to the guidance from the National Institutes of Standards and Technology (NIST), a division of the Department of Commerce. Regarding NIST's role in cybersecurity, the NIST website states that [22]: "NIST develops cybersecurity standards, guidelines, best practices, and other resources to meet the needs of U.S. industry, federal agencies, and the broader public. Our activities range from producing specific information that organizations can put into practice immediately to longer-term research that anticipates advances in technologies and future challenges." NIST has published guidance documents for federal agencies, which may also be used by others, and the following is a small sample of those documents from NIST which DoD, VA, HHS, and other agencies follow:

- NIST Special Publication 800-200, *Minimum Security Requirements for Federal Information and Information Systems*, March 1, 2006.
- NIST Special Publication 800-199, *Standards for Security Categorization of Federal Information and Information Systems*, February 1, 2004.
- NIST Special Publication 800-18, *Guide for Developing Security Plan for Federal Information Systems*, Revision 1, February 24, 2006.
- NIST Special Publication 800-37, *Guide for Applying the Risk Management Framework to Federal Information Systems*, Revision 2, December 20, 2018.
- NIST Special Publication 800-66, *Implementing the Health Insurance Portability and Accountability Act (HIPAA) Security Rule: A Cybersecurity Resource Guide*, Revision 2, July 21, 2022.

Federal agencies are becoming increasingly diligent when it comes to applying cybersecurity requirements and principles for medical devices. An MDM will need to work closely with these agencies throughout their procurement process, but also during the overall lifecycle of the devices that they have in these environments. While the device currently in place on government agency networks may be authorized, as time passes and operating systems and software applications approach end-of-support, the agencies may give MDMs notice of a need to replace or upgrade these devices. Additionally, vendors must provide documentation to support compliance to the agencies, such as updated MDS[2] or other documentation. In general, both agencies require MDMs to provide updated MDS[2] documents for any devices they intend to sell to the agencies.

VA has developed an overlay for the NIST Risk Management Framework (RMF), including requirements that apply to medical devices. Controls and other requirements that the VA has determined for medical devices that will be placed on their networks have been identified, and information is provided on control implementation. When contacting VA procurement for consideration of medical devices to be sold to VA's medical centers and clinics, MDMs should ensure alignment with this document.

As government entities, the various branches of military health services are key participants in executing on the U.S. governments cybersecurity strategies, including using government purchasing powers to move vendors towards more secure technologies. Elements of this strategy have been articulated in the White House's National Cybersecurity Strategy (March 2023) [23] and National Cybersecurity Strategy Implementation Plan (July 2023) [24]. Specific cybersecurity action, including for medical devices and elevated FDA authority, are included in the Consolidated Appropriations Act (H.R.2617, also known as the 2023 Omnibus Act) [25] as well as proposals for vendor cybersecurity self-attestation [26].

4.4 Summary

In this chapter, we outlined how changes in care delivery and different care delivery entities impact medical device operation and procurement with respect to cybersecurity. The wide variance of possible use environments, as well as the potential for dynamism within and between certain environments, impacts how manufacturers design, implement, and support device security.

References

[1] Taylor, M., A. Hollowell, and G. Bruce, "Hospitals Will Be 'Rare Exception': What Healthcare Will Look Like in 100 Years," *Becker's Hospital Review*, June 8, 2023, https://www.beckershospitalreview.com/hospital-physician-relationships/hospitals-be-will-rare-exception-what-healthcare-will-look-like-in-100-years.html.

[2] Best Buy, "Best Buy to Acquire Current Health to Help Make Home the Center of Health," October 12, 2021, https://corporate.bestbuy.com/best-buy-to-acquire-current-health-to-help-make-home-the-center-of-health/.

[3] Verizon Business Resources, "2023 Data Breach Investigations Report (DBIR)," 2023, https://www.verizon.com/business/resources/reports/2023-data-breach-investigations-report-dbir.pdf.

[4] HSCC, "Health Industry Publishes 'Health Industry Cybersecurity-Managing Legacy Technology Security,'" March 2, 2023, https://healthsectorcouncil.org/legacy-tech-security/.

[5] Mageit, S., "Skills Gap in Healthcare IT Industry Causes Security Threats, According to New Report," *Healthcare IT News*, September 16, 2021, https://www.healthcareitnews.com/news/emea/skills-gap-healthcare-it-industry-cause-security-threats-according-new-report.

[6] HSCC Cybersecurity Working Group, "Model Contract-Language for Medtech Cybersecurity (MC2)," March 2022, https://healthsectorcouncil.org/model-contract-language-for-medtech-cybersecurity-mc2/.

[7] Parker, M., and P. McLaughlin, "Hospital at Home: Managing the Risks of Delivering Acute Care in the Patient's Home," *Advanced Health Technology*, 2023, pp. 253–258.

[8] FDA, "CDRH Seeks Public Comment: Increasing Patient Access to At-Home Use Medical Technologies," June 2023, https://www.fda.gov/about-fda/cdrh-strategic-priorities-and-updates/cdrh-seeks-public-comment-increasing-patient-access-home-use-medical-technologies.

[9] Hilts A., C. Parsons, and J. Knockel, "Every Step You Fake: A Comparative Analysis of Fitness Tracker Privacy and Security," *Open Effect Reports*, 2016, https://openeffect.ca/reports/Every_Step_You_Fake.pdf.

[10] Fischer, C., and N. Hamming, "IEEE PHD Cybersecurity Standards Roadmap," *IEEE*, April 30, 2019, https://ieeexplore.ieee.org/document/8703258.

[11] NCCoE, "Securing Telehealth Remote Patient Monitoring Ecosystem," February 2022, https://www.nccoe.nist.gov/healthcare/securing-telehealth-remote-patient-monitoring-ecosystem.

[12] NCCoE, "Mitigating Cybersecurity Risk in Telehealth Smart Home Integration," August 2022, https://www.nccoe.nist.gov/healthcare/mitigating-cybersecurity-risk-telehealth-smart-home-integration.

[13] HSCC, "Health Industry Cybersecurity – Securing Telehealth and Telemedicine (HIC-STAT)," April 2021, https://healthsectorcouncil.org/securingtelehealth/.

[14] American Medical Association, "Telehealth Implementation Playbook," 2022, https://www.ama-assn.org/system/files/ama-telehealth-playbook.pdf.

[15] UL and the U.S. Department of Veterans Affairs, "Strengthening Medical Device Cybersecurity Across the Healthcare Ecosystem," 2019.

[16] Crown, E., "Army Medical Device Cyber Team Balances Benefits and Risks of Technology," May 23, 2023, https://mrdc.health.mil/index.cfm/media/articles/2018/device_cyber_team_balances_benefits_and_risks.

[17] https://vaww.insider.va.gov/va-tackles-medical-device-vulnerabilities-and-cyberthreats/.

[18] https://dvagov.sharepoint.com/sites/OITOIS/KnowledgeService/Pages/SSSD.aspx. (This may be limited to an internal VA network.)

[19] https://public.cyber.mil/stigs/.

[20] Office of the Under Secretary of Defense for Personnel and Readiness, "DoD Directive 7730.65, Department of Defense Readiness Reporting System," May 31, 2023,

https://www.esd.whs.mil/Portals/54/Documents/DD/issuances/dodd/773065p.pdf?ver=2018-05-31-084047-687.

[21] Department of Defense Procurement Toolbox, "Cybersecurity in DoD Acquisition Regulations," https://dodprocurementtoolbox.com/site-pages/cybersecurity-dod-acquisition-regulations.

[22] NIST, "Cybersecurity," https://www.nist.gov/cybersecurity.

[23] The White House, "National Cybersecurity Strategy," March 1, 2023, https://www.whitehouse.gov/wp-content/uploads/2023/03/National-Cybersecurity-Strategy-2023.pdf.

[24] The White House, "National Cybersecurity Strategy Implementation Plan," July 2023, https://www.whitehouse.gov/wp-content/uploads/2023/07/National-Cybersecurity-Strategy-Implementation-Plan-WH.gov_.pdf.

[25] Rep. Connolly, Gerald E., "H.R.2617 - Consolidated Appropriations Act, 2023," 117th Congress (2021-2022), December 29, 2022, https://www.congress.gov/bill/117th-congress/house-bill/2617.

[26] Homeland Security, "Agency Information Collection Activities: Request for Comment on Secure Software Development Attestation Common Form," *Notices*, April 27, 2023, https://www.federalregister.gov/documents/2023/04/27/2023-08823/agency-information-collection-activities-request-for-comment-on-secure-software-development.

5

Supply Chain Management

In today's interconnected world, organizations rely on third-party suppliers and vendors to meet their business needs. However, this increased dependence on external partners also exposes businesses to potential cybersecurity risks. Cybersecurity Supply Chain Risk Management (C-SCRM) has emerged as a critical strategy to mitigate these risks and safeguard the integrity, confidentiality, and availability of sensitive data (used or created by the medical device) and to preserve the device's essential performance, safety, and effectiveness.

Cybersecurity supply chain risks have grown significantly due to the rise in cyber threats and the complex nature of interconnected supply chains. Cybercriminals target third-party vendors as an entry point to launch attacks on their client organization's network, systems, and data. Implementing a C-SCRM system can address the following business imperatives:

- C-SCRM helps to identify and evaluate potential cybersecurity risks associated with third-party vendors and helps to continuously manage said risk as the relationship evolves. By conducting comprehensive risk assessments, businesses can make informed decisions about engaging with suppliers that align with their cybersecurity standards and protocols across their products' lifecycle, thus reducing the likelihood of cyber incidents.
- Cyberattacks involving third-party vendors can severely damage a company's reputation and erode customer trust. By adopting a C-SCRM approach, businesses can demonstrate their commitment to cybersecurity, reassuring customers and stakeholders about the safety of their data.

- Supply chain disruptions due to cyberattacks can lead to significant financial losses and operational downtime. C-SCRM allows organizations to proactively identify vulnerabilities in their supply chain, minimizing the impact of potential cyberattacks and ensuring continuous business operations.

A robust C-SCRM system incorporates several essential components that work together to strengthen supply chain cybersecurity:

- *Vendor assessment and selection:* Thoroughly vetting potential vendors' cybersecurity practices is vital. This involves evaluating their security policies, incident response capabilities, and track record in dealing with cyber threats.
- *Risk assessment and monitoring:* Continuous risk assessments are essential to identify potential vulnerabilities and monitor third-party vendors' security posture. Regular audits and assessments help organizations to stay ahead of emerging threats.
- *Contractual obligations:* C-SCRM should be embedded into vendor contracts to ensure compliance with cybersecurity standards and facilitate incident response and responsibility allocation in case of security breaches.
- *Communication:* Vendors and clients should be contractually responsible for timely communication of newly discovered cybersecurity risks through predefined channels and contact points and within agreed-upon time frames.
- *Monitoring and management of cybersecurity signals:* Device manufacturer and suppliers should work collaboratively to develop plans to monitor, identify, and manage cybersecurity signals such as newly discovered vulnerabilities, evolving threats, new risks, and cybersecurity incidents to ensure that any new cybersecurity signals will be addressed promptly and effectively.
- *Continuous improvement:* C-SCRM is not a one-time effort; it requires continuous improvement and adaptation to evolving cybersecurity threats and changes in the supply chain.

Implementing a C-SCRM system helps organizations mitigate cybersecurity risks, ensure regulatory compliance, safeguard their reputation, and maintain the trust of customers and stakeholders. By making strategic investments in

C-SCRM, organizations can proactively protect their supply chains, data, and operations from continuously evolving threats.

Manufacturers tend to think of themselves as being at the end of a long supply chain, but, in truth, manufacturers are in the middle of the chain, with the healthcare delivery organization being at the end. MDMs' position near the middle means that, out of all the entities on the supply chain, MDMs are best placed to understand, monitor, and support the security of their device throughout its total product lifecycle. However, the potential of this position can only be actualized by implementing a robust C-SCRM program, with appropriate controls both upstream and downstream from the MDM.

5.1 Upstream Supply Chain Management

The U.S. Code of Federal Regulations part 820 section 50 (CFR 820.50) states that [1]: "Each manufacturer shall establish and maintain procedures to ensure that all purchased or otherwise received product and services conform to specified requirements."

For the last several years, securing the software supply chain (Figure 5.1) has been an area of focus for both the FDA and international regulatory bodies. Activities expected under this broad conceptual umbrella include:

- Formal identification, inspection, and documentation of "Critical to Quality" attributes of purchased or received product and services;
- Establishing and enforcing processes to control and manage vendors, including collecting consistent and reliable data on all suppliers.

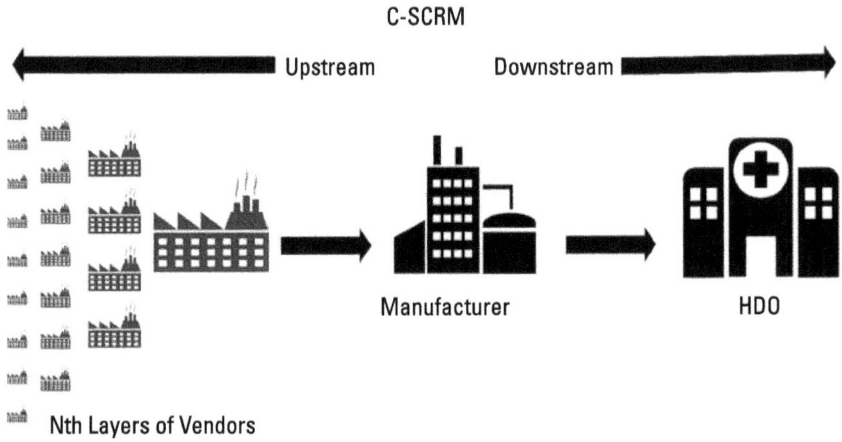

Figure 5.1 Conceptualizing the software supply chain.

More recently, security has been added to the list of supplier attributes that manufacturers are expected to monitor. This has become an attribute of heightened concern and scrutiny because when it comes to security, the supply chain is just that, a chain: only as strong as its weakest link. Attackers know this and are more likely to direct attacks against smaller, less-resourced companies in the supply chain, hoping to leverage an easier exploit opportunity against the larger, better-resourced, more attractive target that the smaller company supplies.

One of the authors witnessed a prime example of this tactic when one of our consulting clients was acquired by a Fortune 500 company. On the day that the acquisition was announced, our client experienced a thousandfold increase in attempts to breach their security. The announcement had alerted malicious actors to the fact that our client was part of the supply chain of a prime target.

Manufacturers need to take steps to protect their devices from the following supply chain attacks:

- Computing platforms pre-infected with malware;
- Malware insertions into subassembly firmware, device software, or electronic components;
- Vulnerabilities introduced into software;
- Hardware-based vulnerabilities and exploits introduced into hardware components;
- Counterfeit components and subassemblies.

Upstream supply chain management is a difficult and expensive endeavor and is divided into two main areas: electronics and third-party software components (TPSC).

5.1.1 Counterfeit Electronic Components

Several years ago, components started appearing on the market that were branded as the correct component in the correct type of packaging. However, the internal die was either completely different, or the component had been relabeled and up-rated (e.g., labeling a standard component as MIL-Spec) or, in some cases, just a film of copper.

These components were usually repurposed components scavenged from e-waste recycling dumps. These scavenged components are stripped from a PCB assembly (PCBA) by noncontrolled heating methods, washed in whatever water is available, and handled without any electrostatic discharge (ESD) protection. Components are then sanded or subjected to an acid bath, re-coated, and re-marked with the counterfeit component's branding. This process is typically referred to as "blacktopping."

Visual inspection of incoming blacktopped components proved minimally effective. Inadequate quality of the rebranding or misalignment of the leads (for leaded components) sometimes resulted in detection during receiving. However, many blacktopped components made it through initial inspection and onto the final assembly floor, where they were generally detected by testing during manufacturing.

Because blacktopped components generally cannot perform to specification (or at all), they have rarely made it past the manufacturing floor and therefore rarely posed problems for end users. However, they have been responsible for expensive and sometimes embarrassing delays in manufacturing.

5.1.1.1 Newer Counterfeits

The second form of counterfeit component appeared in the market more recently, one that not only appears to be correct but performs the correct component's intended operation, at least to some degree. This proved to be a much harder counterfeit to detect. For example, it is believed that many FTDI FT232RL USB-to-serial components were counterfeits containing a ROMed microcontroller that was emulating the original FT232 functionality [2, 3]. This emulation was not completely accurate to the original FT232 component, which could cause impacts on many of the devices in which it was placed.

The FTDI component is just one example of many [4], including the ATMega328 [5], and MicroSD cards. Probably most surprising is the counterfeiting of extremely low-cost (but high-volume) components such as diodes, capacitors, and resistors. Usually, components are counterfeited when the cost of the original, genuine components has increased. This can be due to availability shortages, natural and other disasters impacting production or transportation, or fluctuations in the costs of raw materials. The bottom line is that no component can be assumed safe from counterfeiting.

In addition to the cost-skimming, newer counterfeits may also be created as a novel attack vector for supply chain attacks. These counterfeits would closely conform to the original specifications, or seek to convincingly emulate it as in the previous example, altered to convey an exploitable vulnerability (e.g., a "back door") or vulnerabilities into the supply chain.

5.1.1.2 Testing for Counterfeits

Besides visual inspection of the components, there are other means to confirm the validity of the component. These include both destructive and nondestructive tests.

An example of destructive testing (applied to a representative sample from each component lot) is decapping the device and visually comparing it with the target using a microscope to ensure the die surface is consistent with the specification.

X-ray stations allow for the verification of internal lead frames, bonding wires, and die construction. This can be performed by third-party testing houses or on-premise X-ray machines, such as the Glenbrook Technologies JewelBox product line. Some types of components may be destroyed by X-ray testing, while other types may not.

A nondestructive test method is to analyze multiple unique electrical characteristics of a component. This is the approach taken by Battelle's Barricade test station. This method is nondestructive and low-cost per component and can be applied to all components (not just a lot sample) and can be configured for high throughput rates.

5.1.1.3 Inspection Triggers

The following are some red flags that should trigger inspections or, depending on the when in the process the red flag is encountered, prompt further action to verify that the component is genuine and trustworthy:

- The item comes from a source other than the original equipment manufacturer (OEM) or the authorized source.
- The item comes from an unknown source.
- The item comes from a source in a suspect geopolitical location (e.g., component stores in China).
- The item is priced too low or is significantly different from historical pricing.
- Scarce components are suddenly available in large quantities.
- The chain of ownership is unverifiable.
- There is no certificate of conformance.
- There are aberrations of size, shape, color, finish, or materials.
- There is evidence of rework, repair, refinishing, or resurfacing of the component.
- The component leads are deformed or exhibit alignment issues.
- There is an increase in failure rate during testing or of fielded devices.
- There are component labeling or branding issues:
    - Does not match previous items of the same type;
    - Incomplete orders;
    - Incorrect size of packaging;
    - Lot number/date code issues;

- Shoddy package/construction issues;
- Unusual packaging medium (e.g., reel versus tray).

For advanced needs for electronic component inspection, including computer chips, several research institutions have developed advanced reverse engineering capabilities, for example, the REFINE lab at the University of Connecticut, as well as many other commercial service providers [6, 7].

The management of your Approved Supplier List (ASL) needs to include results from counterfeit detection events. In general, avoid gray market suppliers of electronic components and contractually hold manufacturers responsible for sourcing components from reliable suppliers and not migrating to new sources without prior approval.

5.1.1.4 Subassemblies

Subassemblies pose a much harder threat to mitigate than individual components, because they expose MDMs to a tree of potential counterfeit sources, most or all of which may be hidden from your knowledge.

Compliance with key security requirements has to be contractually specified between an MDM and the vendor of the subassembly. The vendor likewise needs to contractually enforce similar security requirements with his or her vendors, and so on down the supply chain, until the component level is reached. This chain of contractually clarified trust must also be specified in the contract between the MDM and the subassembly vendor.

The mechanism used to convey this trust is a Certificate of Conformance (CoC) for each shipment. The CoC is a formal document signed by the vendor and given to the MDM. It affirms that all purchase order requirements have been met and these requirements should include not only specific information about the parts, such as quantity, lot, and date codes, but also sourcing information about the vendor's vendor, including OEMs and all distributors contributing to the buildout of the subassembly.

5.1.1.5 Disposal of Counterfeit Components or Subassemblies

Counterfeiting invariably involves fraud, often from an upstream supplier several tiers removed from the MDM itself. Therefore, coordination with appropriate officials must take place prior to disposing of counterfeit parts.

If the suspect items are contained within its facility, the MDM should:

- Maintain counterfeit parts or materials in quarantine, clearly identified as counterfeit materials.

- Never return counterfeit materials to the vendor under any circumstances, as they could be reintroduced into the supply chain to be sold again to another victim.
- Contact legal authorities to initiate an investigation into the counterfeiting activity. If such an investigation is performed, then the counterfeit materials may be required as evidence.

5.1.2 TPSC

The other primary area of upstream supply chain management concerns TPSC. Alternatively known as "SOUP" (an acronym variously defined as "Software of Unknown Provenance/Pedigree/Parentage"), TPSC includes open-source or closed-source libraries, frameworks, hardware drivers, operating systems, and run-time engines (e.g., .NET Framework) used in the medical device system. These are referred to as TPSC in this book and in supply chain cybersecurity generally because the set of software components that needs to be controlled from a security standpoint includes all third-party components from any source, whether known or unknown.

TPSCs are software components integrated into the actual finished product. These are not to be confused with:

- The toolchain used to compile the code;
- The operating system on the workstation used to build the code;
- The programming editor/integrated development environment (IDE);
- The Continuous Integration tools;
- The bug tracking tool;
- The version control tool.

Each of these merits its own security considerations and should be controlled under the purview of the MDM's infrastructure and IT security framework. However, none of them fall under the category of TPSC.

The quantity of TPSCs used in a product depends on the nature of the embedded medical device. For example, a small coin cell, battery-operated device containing an 8-bit, 16-bit, or 32-bit microcontroller will usually incorporate many fewer TPSCs (typically fewer than 10) compared to an embedded PC or a smartphone app, where hundreds of TPSCs are typical.

TPSC-based attacks have been occurring with increased frequency since 2018. Twelve popular open-source libraries were compromised through a variety of exploits [8], including in several cases convincing a previous project moderator to turn over control of the TPSC to a new moderator, who then promptly injected exploit code into the repository [9]. However, attackers are targeting not only open-source libraries, but closed-sourced programs as well [10, 11].

(Interestingly, those referenced cases are linked. The CCleaner malware is thought to have given the attacker access into ASUSTeK Computer Inc., which then allowed them to steal the signing credentials for the ASUS firmware update server, making this an example of a chained attack on the supply chain.)

The rapidly growing threat of supply chain attacks has even prompted the U.S. Department of Homeland Security to get involved, establishing a Supply Chain Risk Management Task Force in November 2018 [12].

Practically speaking, MDMs can do very little to avoid as-yet undiscovered or undisclosed TPSC-related problems. Instead, they need to prepare for the inevitable by:

- Layering system mitigations in a defense-in-depth approach;
- Leveraging SBOMs to monitor for disclosed vulnerabilities resulting from a compromised TPSC;
- Designing their device and its associated FOTA update system for rapid, controlled response to newly disclosed TPSC vulnerabilities through software patches and updates (further discussed in Chapters 8 and 9).

At the time of regulatory submission, a required element of the submission package is a review of the TPSCs utilized in the system and a justification for the inclusion of any TPSC with a known vulnerability. This justification should take the form of a cybersecurity risk analysis. Outcomes should be documented in the device design history file (DHF) as well as included in the submission package. Please refer to Chapter 7 for more information about TPSC disclosure and justification. If applicable, the TPSC review should also include any subsequent versions of the TPSC that have been released since the version being used in the medical device system, including information about whether the more recent version(s) include security updates.

Information about all utilized TPSCs will also be codified in a machine-readable format as an SBOM.

Supply chain management cannot be achieved with just a technical solution; most controls are "people-powered." Similarly, this is not just an engineering department process; rather it is a process that involves many groups within the company to perform various aspects of securing the supply chain.

Supply chain management is difficult and costly, but not as costly as failing to secure the supply chain. It should not be taken lightly or assumed that some other group in your organization is performing it. The processes need to be documented, performed, and tested before any confidence in the supply chain or your products is warranted.

5.2 Security Criteria for ASLs

In controlled research and development and controlled production, all suppliers are vetted against established criteria for inclusion on an ASL. Every item on a proposed bill of materials (BOM) must be sourced from a supplier in good standing on the ASL, swapped out for one that is, or designed out of the system.

Typical ASL vetting criteria includes such items as:

- Quality and maturity of the component production line;
- Responsiveness and effectiveness of customer service;
- Contract and confidentiality terms that the supplier considers acceptable or unacceptable;
- Disclosure of details about the supplier's facilities, operations, business culture, quality system, and outsourcing;
- Control of subsuppliers and other business partners;
- Disaster recovery and incident response plans and readiness;
- Responses from supplier-provided references.

Per current interpretations of CFR 820.50, MDMs must expand their existing ASL practices to include cybersecurity if they have not already done so. This expansion will impact their existing ASL auditing procedures, trigger reevaluation (and possibly reclassification) of active suppliers already in good standing on the list, and could ultimately have impacts on the BOMs of projects currently in development. Allowances should be made for the potential ripple effect, keeping in mind that these changes are necessary to achieve proper supply chain security management.

Cybersecurity considerations for ASLs include the following:

- Expansion of the list to encompass not just suppliers on the hardware BOM, but the software BOM (SBOM) as well. Traditional ASL considerations will need to be evaluated and perhaps modified or excluded, based on their applicability to TPSC suppliers.
- Each supplier, whether of hardware and software, is considered untrusted when added to the ASL. Only sustained compliance and auditing can raise the trust level of a supplier.

- Auditing should be performed on the entire supply chain of dependencies for each BOM entry, not only on the supplier immediately above the MDM.
- When audits are performed, they should include a focus on security (both physical security and cybersecurity).
- Each of the suppliers in the chain must have standard processes for security and be able to provide confirmation that these standards are being practiced.
- Each of the suppliers should perform regular "Red Team" tests of their organization. Red Teaming should stress:
    - *Physical:* including access control, temporary access, warehouses, offices, waste, server rooms, wiring closets, elevators.
    - *Cyber:* including routers, Wi-Fi, networks, IoT devices, printers, and public and private access levels.
    - *People:* staff, contractors, business partners, vendors, social engineering, coercion, bribery, data authorization levels.
- Open source:
    - Open-source utilization at each vendor should be evaluated for license usage and adherence to each specific license type's constraints and requirements.
    - Open-source projects should not be inherently trusted by any supplier in the chain, including the MDM.
    - Attestation of the origin of the open-source project being utilized should be confirmed (i.e., verify whether any supplier is using a forked variant of the original open-source package).
- Changes in ownership, authorship, or moderation of a component being provided via the supply chain should reset that supplier to a non-trusted status until trust can be reestablished.
- Supplier rapid response to detected and/or disclosed vulnerabilities increases the preference for continued or additional engagements with the vendor.
- Vendor relationship to the MDM in notifying the MDM of all cybersecurity events, including breaches and vulnerabilities.
- Audits should physically confirm the disposition of any scrap or media containing the intellectual property of the vendor or the MDM.

(For a complete list of vendor questions, please refer to CISA's Vendor Supply Chain Risk Management (SCRM) Template [13].)

It should be noted that, although MDMs should prioritize the ASL according to inflexible "must meet" criteria, there may be cases where the MDM must work with a supplier of a unique component that has no viable substitute even though the supplier is unable or unwilling to meet approval according to the MDM's established ASL procedure. In such cases, the ASL procedure can allow a secondary, less-preferred approval pathway that waives the standard requirement(s) in favor of additional compensating controls and safety measures. These might include additional audits, inspection, contractual mitigation, and source code escrow.

5.3 Software Bill of Materials

An SBOM is a detailed inventory of all the software components and dependencies used in a particular software application or system, whether open-source or commercial. It provides a clear list of all the software elements, including libraries, frameworks, operating systems, communication stacks, and other third-party software components that become part of the finished software. It does not contain any software used to develop or deploy the software, such as compilers, software configuration management, or the development system's operating system.

SBOMs are realized in different forms, both human-readable and machine-readable. Human-readable SBOMs are essentially a list of "ingredients" used in the system, and are typically rendered only one level deep. If ACME Ventilator used the Java logging library "Log4j," it would have the ventilator software release as the parent, and Log4j as the child, with no further dependencies represented below Log4j, even though in reality there are 294 dependencies required for Log4j. Machine-readable SBOMs go many levels deeper and are developed and released in industry data formats, such as XML or JSON. They are defined by one of two industry syntax standards, CycloneDX or SPDX. Regardless of which standard is used, as much of the complete list of ingredients and dependencies as possible should be included. Both formats also support attributes where it is possible to indicate that there are additional dependencies but they are not known to the author of the SBOM ("known unknowns").

The SBOM is an essential component of supply chain security, particularly in the context of software supply chains. Here is how an SBOM helps secure the supply chain:

1. *Identifying vulnerabilities:* By having an SBOM, organizations can quickly identify and assess any known vulnerabilities associated with the utilized software components. They can regularly check security databases for recently disclosed potential vulnerabilities in these

components and, if found, initiate an additional investigation to determine if the vulnerability in the component affects their product (due to differences in configurations and environment, not all vulnerabilities in TPSCs create vulnerabilities in the finished product). The results of this investigation then feeds into a risk assessment process; depending on the risk, this could result in corrective actions and/or FOTA updates to devices in the field.

2. *Customer (or downstream) risk management:* By utilizing an SBOM for all of the medical devices in a hospital, HDOs can evaluate the risk associated with each medical device that stems from its respective software components and prioritize their security efforts accordingly. Critical components with known vulnerabilities can be addressed with higher priority, reducing the chances of exploitation by cyber attackers. The SBOM can also be utilized by hospitals during the acquisition process to prevent the hospital from acquiring known vulnerable devices.
3. *Regulatory compliance:* Most medical device cybersecurity regulations and industry standards around the world now require manufacturers to maintain (create, distribute, and keep current) SBOMs as part of their software development and deployment practices. By adhering to these requirements, businesses can ensure compliance and avoid potential legal and financial consequences. Notably, the FDA requires both human-readable and machine-readable SBOMs as part of the premarket submission process.

SBOM use and procedures continue to be in transition and have not become standard practice. Many organizations are creating and monitoring each device's SBOM across product lines, resulting in large numbers of SBOMs being scanned and analyzed on a regular basis. Generally, we are seeing the biggest adoption of SBOM by manufacturers seeking to manage their supply chain, but there is not yet widespread demand from health delivery organizations for SBOMs to use in their own monitoring (which has been one of the principal use cases for SBOMs as stipulated by the FDA, for example).

However, this appears to be changing with the Department of Health and Human Services (HHS) HICP releasing its *Technical Volume 2: Cybersecurity Practices for Medium and Large Healthcare Organizations, 2023 Edition,* detailing best cybersecurity practices for the hospitals to follow, which includes medical device SBOM monitoring [14]. While these best practices are currently voluntary (meaning that adoption may be slow), HHS has also published a four-step plan that starts with voluntary cybersecurity performance goals but progresses toward [15] "enforcement and accountability." The HHS plan seems to be very levelheaded with regard to voluntary adoption, stating that [16]: "Funding and voluntary goals alone will not drive the cyber-related behavioral change needed across the healthcare sector." Once these best practices are mandated by HHS, we expect to see widespread adoption and use of SBOMs by HDOs.

4. *End of life/support management (EOL/EOS):* By providing full visibility of all software components in a product, SBOM is also an effective tool to manage component EOL/EOS and to mitigate the security risks that come with unsupported software components. Also, because SBOM offers a viable means to proactively communicate the product's EOL/EOS horizon to customers, as it is now being expected by regulators.

Recognizing these benefits, other industries are actively adopting and monitoring SBOMs, including automotive and aerospace. It has, for example, been reported that Lockheed Martin will no longer accept products from its vendors unless SBOMs are provided, a titanic shift that will result in millions of SBOMs being created and distributed and prompt other entities to follow suit.

An SBOM is a critical tool that enhances supply chain security by providing transparency, identifying vulnerabilities, enabling risk management, and promoting secure software development practices. Its use allows organizations to strengthen their overall cybersecurity posture and build more resilient supply chains in the face of ever-evolving cyber threats.

5.4 Build Integrity and Attestation

How can we be sure that the software that we write and build, the development platforms that we use, and all of the external dependencies that we leverage are intact, and have not been corrupted to include malware? Moreover, while we all prefer to think of security threats as originating from outside an organization, the reality is that roughly 50% of all breaches originate from a member of that organization, either from malice, opportunism, or simple negligence (an "insider attack") [17]. How can MDMs mitigate that insider risk?

Fortunately, Google has for several years been using an internally created system to address such concerns. Originally called "BAB," it has since been released to the public as "SLSA" (pronounced "salsa") and is now maintained by the Open Source Security Foundation [18]. The SLSA framework is a set of guidelines for supply chain security organized in tracks and levels. Currently there is only one track, the "build track," but that will increase in future versions of SLSA to specifically cover other parts of the supply chain. The levels are a form of maturity that indicate the use of increasingly more stringent security practices. Levels range from 0, no security, to 4, currently left undefined for future enhancements. These levels are summative, meaning that Level 3 assumes that Level 1 and Level 2 are fully implemented.

Level 1 is focused on provenance (showing how the package was built) and is used to prevent mistakes. Adherence to Level 1 should not be considered a security mitigation. Level 2 creates a signed provenance, which is performed on a hosted build platform such as CI/CD. Builds are not produced on a development workstation. This allows for downstream verification of the authenticity of the provenance and prevents tampering after the build. Level 3 prevents tampering from occurring during the build process and preventing separate build runs from affecting one another. This is performed by providing hardened build platforms that are ephemeral in nature.

Some of the future features we can assume to be present in Level 4 will be "hermetic" and "reproducible" builds. In this context, "hermetic" means that every step, source code item, and dependency are uniquely identified to be an immutable artifact, with no unverified references. We can likewise assume that these artifacts will have been subjected to other security testing (SAST, SCA) before they can be considered to be a verified reference. We will also probably see "reproducible" builds, where these tightly controlled inputs to the build will produce the exact same binary outputs. This is a superstate, building upon "isolated" (Level 3) and hermetic (Level 4) builds. Automatic verification on several separate build platforms to confirm the resultant binaries are identical may be used as well.

Many of these SLSA features can be enabled in current standard build platforms, and as we move forward, we should expect to see SLSA Level 3 and 4 build systems become commonplace in development environments.

5.5 Downstream Supply Chain Management

Relative to MDMs, the downstream supply chain typically refers to consumers of a manufacturer's products, either HDOs or patients and caretakers, as with a home healthcare service. However, where cybersecurity is concerned, it is also inclusive of production, delivery, communication, documentation, and channel partner considerations. These considerations change to some degree when reflecting differences between the geographic location of manufacturing versus the market in which the product is sold. For example, a manufacturer may sell through its own sales organization in its domestic market, but through distributors in other regions. Similarly, devices may be manufactured in-house or by a third-party contract manufacturer who may even ship directly to the MDM's customers. In either setting, device integrity could be compromised or intellectual property could be exposed, but the MDM must take a different approach to ensuring the security of contract manufactured goods than would normally be available in-house.

In another common scenario, fulfillment houses can be involved with configuring and commissioning PCs, mobile devices, or other off-the-shelf (OTS) widgets intended for use as part of a medical device system, creating opportunities for malware to be installed—intentionally or unintentionally. Even OEM suppliers should not be trusted without verification [19] and should instead be ordered and supplied as a "bare metal" configuration. Failing that, OTS components should be completely wiped and reimaged before reaching the factory line.

Finally, the larger distribution system and warehousing can also be areas where loss of integrity, IP exposure, and malware injection can take place [20, 21]. For an attacker, these are highly desirable targets, as they can infect many HDOs by only infecting one manufacturing or fulfillment location. This attack technique would be commonly referred to as a supply chain attack [22].

For the MDM, strong relationships and audits of the suppliers, processes, and components that may make up a complex downstream supply chain are necessary to maintain any control of risk and reduce potential exposure to security compromise. Likewise, as devices are returned from the field for service, if any intentional corruption is discovered, infected devices should be logged for further investigation and root-cause analysis, including assessment of the downstream supply chain to identify potential sources of security compromise.

Also, device operators and/or users (HDOs, at-home healthcare providers, patients) should have access to functionality that enables them to confirm the device's integrity and security status. As appropriate for the user role, this can include items such as log entries related to security issues, malware scanners, tamper-evident packaging, cryptographic signatures, hashes, and even an online presence that supplies non-repudiation evidence to the end user. Given the possible wide range of competence and comfort with this type of functionality among operators and users, the MDM should ensure that clear guidance is included as part of the setup process and is easy to understand and perform. This approach enables the end user to have trust in the received device and facilitates the MDM's ability to detect any systemic compromise early on. However, these types of instructions, notifications, and required tasks will widely depend on the role of the end user, their technical capability, and their ability to respond to the message in a meaningful and constructive manner.

Much of the postmarket relationship is now being standardized in contracts (further discussed in Chapter 7) between the MDMs and the HDOs. This means that the relationship between MDMs and HDOs will be tighter than ever, such that ensuring the integrity of the delivered medical device will be in the best interests of the MDM. Additionally, proper security documentation and open communication will further aid the building of a trusted relationship and allow for constructive security management by the HDO as part of their device lifecycle (see Chapter 10).

Lastly, we need to recognize that today's devices are often complemented by cloud-based services, managed either by the MDM or a third-party provider, with the goal to improve clinical utility, enhance serviceability, or provide business analytics. Although the benefit of these complementary services is undisputable, they further increase the potential attack surface of the device and thus add another wrinkle to the downstream supply chain. Steps must be taken to ensure appropriate security measures are in place on complementary services, either through the MDM's internal processes or, in the case of a third-party provider, contractually ensured and controlled through the ASL as with upstream suppliers. Also, such complementary cloud-based services often expose the MDM to sensitive data, such as data that might otherwise only be available to the HDO, and would require the MDM to comply with additional laws and regulations and may require additional contractual agreements, such as a HIPAA Business Associate Agreement (BAA).

5.6 Endemic and Pervasive Vulnerabilities

All cybersecurity professionals remember December 2021 and the rather untimely appearance of the Log4Shell vulnerability on the international software stage. Many of us spent a good portion of our holiday break trying to figure out whether our IT infrastructure or systems ran the affected version of the widely used Log4j Java library. Cyber adversaries were quick to utilize the opportunity and launch attacks, at rates of up to 10 million per hour, including against critical infrastructure industries like healthcare [23]. It was estimated that 10% of all worldwide digital assets were susceptible to this vulnerability [24], and that globally, within 5 days of discovery, up to 50% of corporate networks had been targeted [25]. It was estimated that it will take years to fully eradicate Log4Shell.

The term "endemic" is borrowed from epidemiology, where it describes an infection that is present or maintained at a baseline in a specific population or geography. Log4Shell was the first vulnerability that officially earned this designation from Homeland Security's Cyber Safety Review Board (CSRB) as one of the most serious vulnerabilities discovered [26].

The official designation of "endemic" is reserved for a few widely distributed and difficult-to-eradicate vulnerabilities. However, looking back at the past few years, we have seen a number of vulnerabilities that have not received the official designation of "endemic" but were nevertheless pervasive in nature and of significant consequence. The main concern that would make a vulnerability endemic or pervasive is its distribution (i.e., how many systems it is present in), which, in turn, is driven by its age. For example, the Log4Shell vulnerability was discovered in 2021, but had been present in various versions of Log4j since 2013, which means many thousands of software products have been built using vulnerable libraries.

We have many other examples of pervasive vulnerabilities and their impact. EternalBlue (CVE-2017-0143) exploited by the WannaCry and NotPetya malware attacks and shut down shipping companies and pharmaceutical manufacturers, costing the global economy an estimated $14 billion. Then there was Heartbleed (CVE-2014-0160), ShellShock (CVE-2014-6271), Spectre (CVE-2017-5753 and CVE-2017-5715), SweynTooth (a total of 12 CVEs), Urgent/11 (11 CVEs), BadAlloc (multiple CVEs), and many more headline-worthy vulnerabilities.

Although many of these high-impact vulnerabilities are related to open-source software, we also have many examples of high-impact vulnerabilities affecting commercial software. Recently, open-source has received a lot of attention by security researchers, a trend dubbed by some as "open season on open source." There is a fundamental difference between commercial and open-source vulnerabilities that is important to understand, which is that when notification of a new and critical vulnerability is issued, it is much easier for device operators to determine which devices may run an affected version of commercial software, as compared to determining which devices are using some obscure communications library or logging tool that the operator may not even know existed prior to getting the notification, let alone whether it was used to develop their device. (This challenge is what SBOMs help to solve and is part of why they are so important.) Similar considerations apply to MDM engineers when confronted with the question of whether or not a newly disclosed vulnerability is present and exploitable in their product, or for large MDMs, is or is not present and exploitable in dozens or hundreds of their products or product versions.

In the end, an organization's response and urgency, or level of panic, is determined by a number of factors:

- Analysis:
  - Is the vulnerability present and exploitable?
  - Is the vulnerability widely known? Are there known exploits? Have customers reported incidents?
  - How easy is it to exploit (analogous to the "exploitability" metric of a design vulnerability assessment)?
  - How much damage can be done by exploiting this vulnerability (analogous to "impact" in risk assessment)?
  - What communication is required immediately (customers, regulators, public)?

  - Is there a need for external engagement (ISAC/ISAO, industry, external expertise, law enforcement, regulators, government)?
- Response:
  - Is there a temporary (internal or external to the device) mitigation that can reduce the risk in the short term (exploitability and/or impact)?
  - How difficult is it to develop the temporary mitigation? (This can range from simple software patch all the way to practically impossible (e.g., a hardware vulnerability that necessitating component replacement).)
  - How difficult is it to deploy the mitigation (a combination of technical factors (e.g., FOTA capability), plus organizational logistics (e.g., do I need to work with distributors or other third parties))?
  - What is the resulting legal exposure and risk and what legal or regulatory action is required?

Specific processes and practices that become applicable here are discussed in Chapters 8 and 9. In order to be able to respond to vulnerabilities in general, and endemic/pervasive vulnerabilities in specific—efficiently, timely, and reliably—good lifecycle practices (Chapter 6), supply chain management, and software transparency (e.g., SBOM visibility) are required.

CISA's Cyber Safety Review Board provided a helpful resource and a good list of recommendations of how to deal with Log4Shell [27], which can be generalized as recommendations for how governments as well as producers and maintainers (operators) of software and software-based products can reduce the risk of endemic/pervasive vulnerabilities and how to form effective response:

- Government and community:
  - Government should produce, and producers and operators should consume, authoritative cyber risk information.
  - Federal and state regulators should drive implementation of CISA guidance through their own regulatory authorities.
  - Explore a baseline requirement for software transparency for federal government vendors.
  - Examine the efficacy of a Cyber Safety Reporting System (CSRS).
  - Explore the feasibility of establishing a Software Security Risk Assessment Center of Excellence (SSRACE).
  - Establish a government-coordinated working group to improve identification of software with known vulnerabilities.

- Increase investments in open-source software security.
- Study the incentive structures required to build secure software.
- Open-source software developers should participate in community-based security initiatives.

- Operators and maintainers:
    - Be prepared to address endemic/pervasive vulnerabilities for years to come.
    - During that time, continue to report (and escalate as needed) observations of exploitation.
    - Invest in capabilities to identify vulnerable systems.
    - Develop the capacity to maintain an accurate digital asset and application inventory.
    - Have a documented vulnerability management and response program.
    - Have a documented vulnerability disclosure and handling process.
    - Software developers and maintainers should implement secure software practices.
    - Invest in training software developers in secure software development.
    - Improve SBOM tooling and adaptability.

In principle, dealing with endemic and pervasive vulnerabilities should fall within the scope of any premarket and postmarket vulnerability management program. However, the potential for high-impact, long-term exposure makes this a unique class of vulnerabilities that warrant dedicated attention and additional focus beyond a typical vulnerability program, from both the design and development organization as well as the postmarket support organization. For the MDM, this requires tight control of security processes, including supply chain management, good internal and external communication, as well as considerations of where and how their own IT or OT infrastructure may be affected in addition to their responsibility towards their customers.

With increasing reliance on a complex software supply chain (including commercial, open source, contracted, and embedded), managing its cybersecurity risks becomes a critically important but also increasingly complex and challenging task. Software transparency, through SBOM and related processes, provides visibility into many aspects of the supply chain, for example:

- Cyber vulnerabilities;
- License consumption;

- Component end-of-life and end-of-support (see Chapter 8);
- Export control.

Consequently, mature software supply chain processes help reduce both cyber and business risks and provide organizations with the ability to control the impact of adverse events, tailor their responses, optimize communication, and minimize impacts on business and customers.

References

[1] The U.S. Code of Federal Regulations, Title 21, Volume 8, Part 820, Section 50, April 1, 2019, https://www.accessdata.fda.gov/scripts/cdrh/cfdocs/cfcfr/CFRSearch.cfm?fr=820.50.

[2] Chirgwin, R., "FTDI Boss Hits Out at 'Chinese Criminal Gang' Pumping Knock-Off Chips," *The Register*, February 10, 2016, https://www.theregister.co.uk/2016/02/10/ftdi_says_knockoff_chips_part_of_criminal_operation/.

[3] "FTDI FT232RL: Real vs Fake," *Zeptobars*, February 17, 2014, https://zeptobars.com/en/read/FTDI-FT232RL-real-vs-fake-supereal.

[4] Villasenor, J., and M. Tehranipoor, "The Hidden Dangers of Chop-Shop Electronics," *IEEE Spectrum*, September 20, 2013, https://spectrum.ieee.org/semiconductors/processors/the-hidden-dangers-of-chopshop-electronics.

[5] Grady, E., "Revisiting the Counterfeit ATMega328s," *SparkFun*, May 17, 2010, https://www.sparkfun.com/news/364.

[6] Smith Counterfeit Detection Lab, https://www.smithweb.com/counterfeit-detection-lab/.

[7] Micross Counterfeit Mitigation Services, https://www.micross.com/electrical-test/counterfeit-mitigation/.

[8] Goodin, D., "The Year-Long Rash of Supply Chain Attacks Against Open Source Is Getting Worse," *Ars Technica*, August 21, 2019, https://arstechnica.com/information-technology/2019/08/the-year-long-rash-of-supply-chain-attacks-against-open-source-is-getting-worse/.

[9] Perekalin, A., "A Bad Link in the Cryptochain," *Kaspersky Daily*, November 29, 2018, https://www.kaspersky.com/blog/copay-supply-chain-attack/24786/.

[10] Goodin, D., "CCleaner Malware Outbreak Is Much Worse Than It First Appeared," *Ars Technica*, September 21, 2017, https://arstechnica.com/information-technology/2017/09/ccleaner-malware-outbreak-is-much-worse-than-it-first-appeared/.

[11] Zetter, K., "Hackers Hijacked ASUS Software Updates to Install Backdoors on Thousands of Computers," *Vice*, March 25, 2019, https://www.vice.com/en_us/article/pan9wn/hackers-hijacked-asus-software-updates-to-install-backdoors-on-thousands-of-computers.

[12] Department of Homeland Security, "DHS Announces ICT Supply Chain Risk Management Task Force Members," November 15, 2018, https://www.dhs.gov/news/2018/11/15/dhs-announces-ict-supply-chain-risk-management-task-force-members.

[13] CISA, "Vendor Supply Chain Risk Management (SCRM) Template," April 2021, https://www.cisa.gov/sites/default/files/publications/ICTSCRMTF_Vendor-SCRM-Template_508.pdf.

[14] HHS Healthcare & Public Health Sector Coordinating Council, *Technical Volume 2: Cybersecurity Practices for Medium and Large Healthcare Organizations, 2023 Edition*, 2023, https://405d.hhs.gov/Documents/tech-vol2-508.pdf#p125.

[15] DHHS, "HPH Cybersecurity Performance Goals," January 2024, https://hphcyber.hhs.gov/performance-goals.html.

[16] HHS Cyber Team, "Healthcare Sector Cybersecurity," 2023, https://aspr.hhs.gov/cyber/Documents/Health-Care-Sector-Cybersecurity-Dec2023-508.pdf.

[17] van Zadelhoff, M., "The Biggest Cybersecurity Threats Are Inside Your Company," *Harvard Business Review*, September 19, 2016, https://hbr.org/2016/09/the-biggest-cybersecurity-threats-are-inside-your-company.

[18] The Linux Foundation, "SLSA Specification Version 1.0," 2023, https://slsa.dev/spec/v1.0/.

[19] "Superfish Adware in Lenovo Consumer Laptops Violates SSL, Affects Companies Via BYOD," *Trend Micro*, February 20, 2015, https://www.trendmicro.com/vinfo/us/security/news/cybercrime-and-digital-threats/superfish-adware-in-lenovo-consumer-laptops-violates-ssl.

[20] Simpson, C., "The NSA Intercepts Laptops Purchased Online to Install Malware," *The Atlantic*, December 29, 2013, https://www.theatlantic.com/technology/archive/2013/12/nsa-intercepts-laptops-purchased-online-install-malware/356548/.

[21] Masnick, M., "NSA Interception in Action? Tor Developer's Computer Gets Mysteriously Re-Routed to Virginia," *Tech Dirt*, January 24, 2014, https://www.techdirt.com/articles/20140124/10564825981/nsa-interception-action-tor-developers-computer-gets-mysteriously-re-routed-to-virginia.shtml.

[22] NCSC, "Supply Chain Security Guidance," November 16, 2018, https://www.ncsc.gov.uk/collection/supply-chain-security/supply-chain-attack-examples.

[23] Wilburn, D., and C. Schmidt, "Log4Shell and Endemic Vulnerabilities in Open Source Libraries," MITRE, March 2022, https://www.mitre.org/sites/default/files/2022-03/pr-22-0797-log4shell-and-endemic-vulnerabilities-in-open-source-libraries.pdf.

[24] Yoran, A., "One in 10 Assets Assessed Are Vulnerable to Log4Shell," *Tenable*, December 22, 2021, https://www.tenable.com/blog/one-in-10-assets-assessed-are-vulnerable-to-log4shell.

[25] Palmer, D., "Log4j Flaw: Nearly Half of Corporate Networks Have Been Targeted by Attackers Trying to Use This Vulnerability," *ZDNet*, December 14, 2021, https://www.zdnet.com/article/log4j-flaw-nearly-half-of-corporate-networks-have-been-targeted-by-attackers-trying-to-use-this-vulnerability/.

[26] DHS, "Cyber Safety Review Board Releases Unprecedented Report of Its Review into Log4j Vulnerabilities and Response," July 14, 2022, https://www.dhs.gov/news/2022/07/14/cyber-safety-review-board-releases-report-its-review-log4j-vulnerabilities-and.

[27] CISA Cyber Safety Review Board, "Review of the December 2021 Log4j Event," July 11, 2022, https://www.cisa.gov/sites/default/files/publications/CSRB-Report-on-Log4-July-11-2022_508.pdf.

6

Secure Development and Production for Medical Device Manufacturers

6.1 Introduction

What is medical device cybersecurity? What are we trying to protect, and why?

If you only read regulations and security standards, such as those provided by the FDA, AAMI, UL, Australia's TGA, IMDRF, ISO/IEC, and so forth, you would believe that the process of securing a medical device is only about protecting the device's safety and efficacy for the benefit of patients and end users. However, that is not the whole story. Stakeholders across the industry, including regulators, are recognizing the gaps resulting from narrowly defined security objectives and moving towards a more comprehensive approach.

It may be true that many MDMs would not have adopted a Secure Development Lifecycle (SDLC) for medical devices without a mandate from regulatory agencies in major markets. Oversight from these regulators is driving MDMs to perform certain security activities and to create documentation to substantiate that their devices are sufficiently secure, with residual risk sufficiently low, for patients to use. The approach to secure development described in this chapter is the basis for these activities and artifacts (which are submitted during premarket approval and scrutinized by regulators; for more details, see Chapter 7). These activities and artifacts are not difficult to perform and create, and they help MDMs to ensure that they develop secure systems that risk neither end users' safety nor the MDM's business interests.

MDMs are very concerned with safety and efficacy, so that patients are not harmed through malicious action or failure of the essential performance of the medical device. However, there are many other areas and activities that are highly relevant to securing a medical device; medical device security should not be driven solely by regulatory consideration. Collectively, these other attributes of a secure medical device can be addressed as "threats to the manufacturer's business model" or, in contexts where we are clear about two uses of the term "risk," we can use the more familiar phrase "business risk."

Business risks may include:

- Loss of customer confidence and subsequent market share;
- Delay in market entry due to regulatory scrutiny or rejection;
- Cloning/counterfeiting of durable or disposable accessories;
- Reenabling the use of consumed accessories that have reached their end of operational life;
- Extracting intellectual property, such as software and algorithms, from the medical device;
- Extracting security mitigation elements, such as cryptographic keys and shared secrets, to leverage in a subsequent attack against a larger target, for example, injecting ransomware into an HDO, MDM, or government agency's critical infrastructure.
- Government or shareholder legal action due to violation of corporate or financial laws, for example, those enforced by agencies such as SEC or FTC.
- Loss of sales to US federal government entities, such as the Veterans Administration (VA) hospitals or the military.

In addition, there are business risks to the operator of the device (typically the HDO) that, should they manifest themselves, may lead to loss of market share or the affected party seeking compensation or file lawsuits against the MDM. These risks are largely related to the device not meeting end user needs and requirements, including those that support the user's legal obligations to their customer:

- Prevent exposure of PHI/PII and the associated legal fines for such a breach;
- Loss of business due to inability to deliver care;
- Loss of business due to reputational impact;
- Harm of patients or users due to a targeted or nontargeted attack on a device;

- Weaponizing the medical device to attack other entities or to act as a pivot point into an intranet, such as a manufacturer or HDO's intranet, or a patient's home network.

It should be noted that in conjunction with some of these business risks, the patient can also be negatively affected, for example, by the potential poor performance of a cloned or reenabled accessory, or due to a device being utilized as a pivot point to attack other systems or the infrastructure on which those devices depend, thereby placing the entire organization and other patients at risk of harm [1].

In order to gain control across the entire range of risks, medical device manufacturers (MDMs) are advised to adopt a Secure Development Lifecycle for medical devices, as shown in Figure 6.1.

6.2 Secure Lifecycle Diagram Overview

In Figure 6.1, the four rings illustrate all aspects of a medical device's lifecycle for the two primary stakeholders and the relationship between them: MDMs as the device producer and HDOs as its consumer and operator (and, typically, the entity legally responsible for device maintenance) [2]. Each separate ring progresses asynchronously through its activities. These activities are oriented towards specific purposes, and, in some cases, they inform or are themselves informed by activities in the other rings, as indicated by the arcs and arrows interconnecting the rings. Note that one of these rings is primarily performed by the HDO (further discussed in Chapter 10), while the other three rings, the focus of the remainder of this book, are primarily performed by the MDM.

The largest ring in the diagram, the MDM Development Lifecycle, is probably the most familiar due to its alignment with ISO 81001-5-1 activities. It represents establishing the viability of the project with a business case, market forecasts, schedules, and budgets, setting off on the development path, which, if successful, leads to the release of the device into the market and then starts the whole process over with a new product. However, including cybersecurity considerations, activities, and artifacts changes many things about the development lifecycle.

The first and easiest change that we must now adopt is the concept of the ASL, governed by the topmost circle, the MDM Supply Chain, as mainly discussed in Chapter 5. ASLs are not a new concept, but historically they have been hardware-oriented and quality-focused, used for controlling sources of electronic parts and other components that met our needed functionality and tolerances and were based on our organizations' past experiences of interacting with a particular supplier. Now the process also includes TPSCs, libraries,

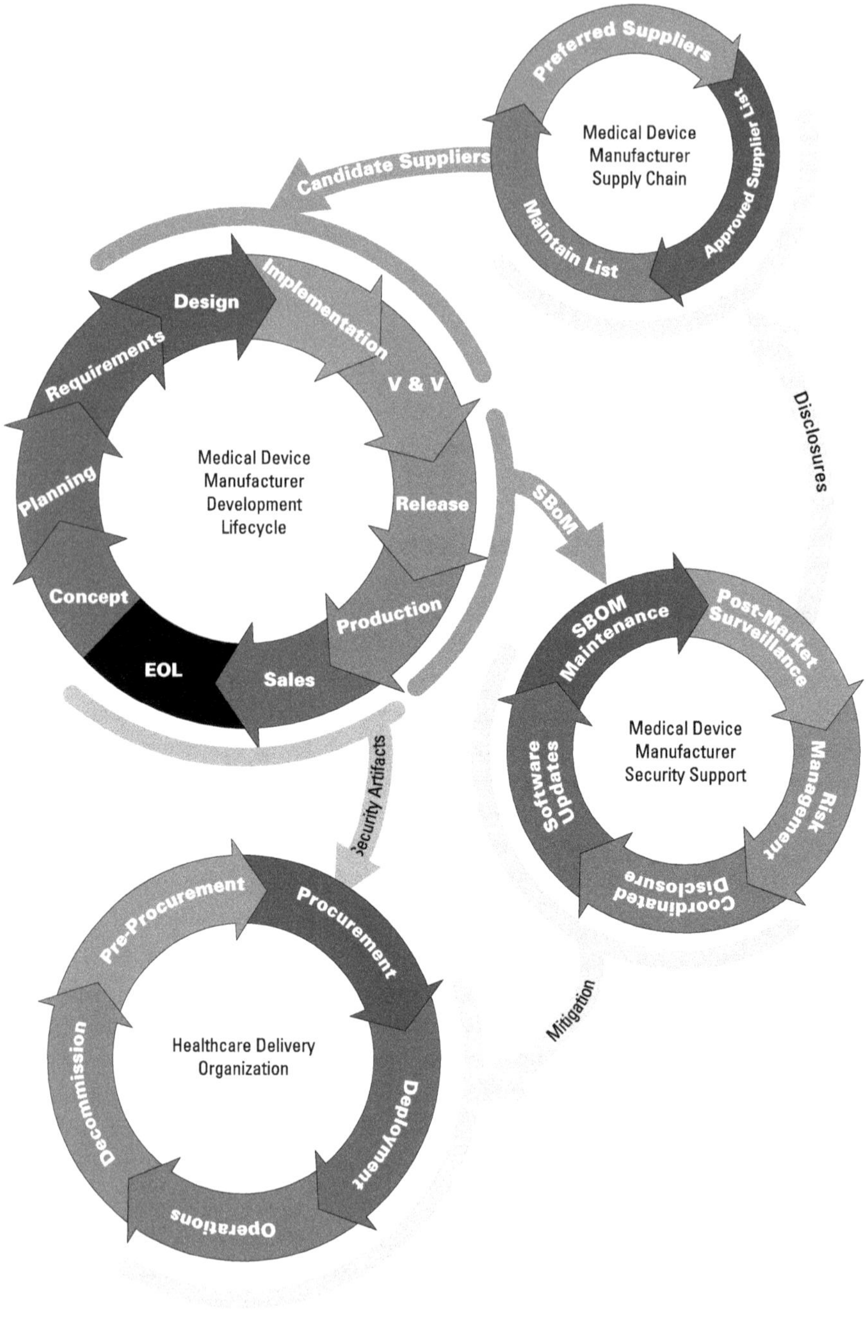

Figure 6.1 Secure development lifecycle of medical devices.

frameworks, and operating systems, and suppliers are graded based upon those same traditional criteria plus a new addition: the supplier's approach to and alacrity in addressing security issues. (Note that TPSCs are similar conceptually to "SOUP" (Software of Unknown Pedigree/Provenance), but the TPSC concept has been broadened to include all external software components, whether of known or unknown origin.)

The MDM Security Sustaining ring addresses ongoing activities pertaining to the monitoring and collection of cybersecurity signals (new vulnerabilities, threats, and risks) through postmarket surveillance of the device. Disclosures from postmarket surveillance activities provide the data that helps to drive the new ASL process (Chapter 5) as well as the secure update process for devices in the field (Chapter 8). In addition to creating timely awareness of security issues, the Security Sustaining Ring also drives the mitigation and patching strategy employed in partnership with HDOs.

Ongoing cybersecurity surveillance is a new concept for traditional MDMs. Having to support the security posture of a product for its entire supported life is foreign and perhaps even slightly scary for some. Besides acquiring the expertise needed for these new activities, there is the question of who is going to pay for the new costs. Postmarket surveillance is rapidly becoming the new normal, an expectation of both customers and regulators. For example, as per the FD&C Act section 524B and implemented in the FDA Premarket Guidance, which stipulates a Postmarket Management Plan to be submitted as part of market approval application, as well as FDA Postmarket Cybersecurity Guidance, and internationally, for example, under EU MDR/IVDR (with cybersecurity guidance provided in MDCG 2019-16). Yet few discussions about the costs and maintenance burden that will have to be borne by and negotiated between the MDM and HDOs are taking place.

Additional design considerations arise from the expectation that MDMs will support HDOs for the entire operational life of the medical device (as part of the HDO lifecycle), including supporting the HDO's decommissioning scenario (which is different from and independent of the MDM's EOL/EOS scenario). This means including functionality that securely implements a "factory reset" and/or "data purge" features in the device, which clears patient data (both PHI and PII), acquired values, settings, access control lists, user credentials, access credentials to networks, and Wi-Fi.

Security artifacts used to be driven by the questions and requirements of the customer, typically requesting SBOMs, MDS^2, security questionnaires, and negotiations over contract language. Some of this has now moved into the premarket submission of your device, where, as part of the approval process for cyber devices, documentary evidence such as SBOMs and MDS^2 are required (further discussed in Chapters 5 and 7) as well as "a plan to monitor, identify, and address, as appropriate, in a reasonable time, postmarket cybersecurity vulnerabilities and exploits."

Throughout this book, we refer to this four-rings diagram both to help the reader to remain oriented and to ground each cybersecurity activity discussed within the overall lifecycle of secure medical devices. The present chapter focuses on the cybersecurity aspects of what is probably the most familiar ring, the MDM Development Lifecycle, from Concept through Transfer to Manufacturing.

6.3 Securing Development Environments and Activities

A frequent mistake made by organizations new to creating secure medical products is to try to somehow apply the same IT infrastructure security standards to secure medical device development procedures as well as their IT environment. Although such standards are readily available for PCs, printers, networks, routers, and Wi-Fi access points, they simply do not represent the best approaches to development activities for creating a secure product.

However, IT security standards can play a helpful role in secure product development. The two most useful secure development framework standards are IEC 81001-5-1 "Health Software–Activities in the Product Life Cycle" and NIST SP 800-218 "Secure Software Development Framework (SSDF)." Both of these secure product development standards reference the need to address infrastructure security in the development environment. They share several commonalities, such as the need to define roles and responsibilities, training, and adoption of procedural and technical controls for protecting the infrastructure that is utilized for product development (including transfer to manufacturing), and maintenance. The SSDF is cross-referenced to other standards, one of which is the OWASP SAMM organizational security metrics (further discussed in Chapter 2). In its "Policy and Compliance" section, SAMM calls out the need for such infrastructure security procedures. SSDF's "Protect the Organization" section provides additional specificity concerning what security measures should be adopted in the development environment, including:

- Implementing multifactor authentication (MFA);
- Implementing network segmentation, especially around build systems;
- Restricting access to the internet, especially from build systems;
- Restricting physical access to build systems;
- Continuously monitoring and logging all alerts in the development environment;
- Implementing a zero-trust architecture in development;

- Configuring development endpoints for encryption of all data at rest and in motion;
- Configuring development endpoints to enforce the principle of least privilege.

Typically, the best (or least disruptive) infrastructure security standard to be implemented in a product development environment (potentially with adjustments, as needed) is the same one that has been adopted by the rest of the organization. This is usually ISO 27001 or NIST 800-53 R5 or later. Apply the infrastructure standards to your development endpoints and IT environment and consider elevating the security level of development environment systems.

6.4 Introduction to Vulnerability Scoring and Its Application During Early Lifecycle Activities

The list of potential vulnerabilities in a given design topology can easily reach into the hundreds, given that it will probably include false-positives as well as some duplication. Designers could attempt to mitigate every listed potential vulnerability; however, that approach would most likely waste time and money, as not all potential vulnerabilities are applicable to the type and use profile of the medical device. Moreover, not all potential vulnerabilities will be exploitable in the real-world device.

For example, a design might use a Bluetooth low energy (BLE) radio to broadcast the battery status of your medical device. Usually, this is information that nobody cares about exposing to anyone. However, if that same BLE interface could modify the therapy delivered to a patient, that is a completely different use case that requires security mitigations to protect the BLE interface, thereby protecting the essential performance of the medical device.

MDMs have limited time and budget to identify and mitigate design vulnerabilities that will matter to the eventual product. Evaluating the list of potential vulnerabilities requires a metric, created by a rubric, that facilitates quick and easy examination of each potential vulnerability's impact on confidentiality, integrity, and availability (CIA) relative to its exploitability (which is a combination of how difficult it is to exploit the vulnerability plus the scope or range to perform the exploitation, such as "physical contact," "nearby," or "internet") and severity (how much harm to patients, the environment, and/or businesses a successful exploit could cause). To achieve this level of prioritization, the following steps should be performed on the list of potential vulnerabilities that:

- Identifies each potential vulnerability;

- Identifies and removes any false positives and duplicates;
- Estimates each potential vulnerability's exploitability;
- Estimates each potential vulnerability's severity;
- Flags each potential vulnerability that could impact the safety and efficacy of the device for end-users (note, however, that any vulnerability not flagged for impacts on safety and efficacy should still be considered a vulnerability that could threaten the business model of the MDM or its customers).

In the case of vulnerabilities that pose a potential safety risk to patients, security engineers do not have the tools, experience, or education to make the judgment call as to whether it does represent an impact on safety and efficacy. Instead, flagging such a vulnerability indicates that this item needs to be introduced into the company's existing (safety) risk management system for review and that a link needs to be established between the safety risk management system and security risk management system. These two separate risk management systems (safety and security) are performed by different people, in different time frames, using different tools. However, they can interact as described above and in addition they also need to cross-check any mitigating controls as implemented by the "other" risk management system. AAMI's TIR57 has an excellent description of this process in Section 3.1.1 as well as Figure 4, "Relationships between the security risk and the safety risk management processes."

A good example of why cybersecurity risks that potentially impact safety and efficacy are referred into existing risk management processes is illustrated by considering the temporary cessation of the medical device's essential performance. Let's say that the engineer or security professional identifies a vulnerability that could allow an attacker to disrupt the functionality of this device for 5 minutes and so flags it as a possible patient risk. This is then transferred to the RMS, where it is reviewed and determined that either:

(A) This is not a risk to a patient (e.g., the device is a personal insulin infusion pump, 5 minutes of downtime would be an acceptable risk to remain in the released product)).

(B) It is a risk to the patient (e.g., the device is a ventilator, 5 minutes of downtime is life-threatening).

Staff engineers and security professionals should not be performing the tasks that the safety risk management process exists to perform.

So how do we know which of these potential vulnerabilities matter to the security of our device? Up to this point, we have been using a threat modeling approach and thinking like an engineer. Now we have to start trying to think like a hacker. As we move to score the list of potential vulnerabilities, being aware of common attacks is necessary for scoring the ease of attack (exploitability) of the vulnerability. A good example of how exploitability can be scored incorrectly, and one that we hear frequently is, "Once the software source is compiled, no one can read it." This is not true; however, it remains a surprisingly widespread belief. If you are unaware of exploit methods and tools, such as decompilers, you cannot accurately score the exploitability of a vulnerability.

One potential rubric that can provide the necessary insights into design vulnerabilities is a modification of the existing CVSS v2 vulnerability scoring rubric. Following the application of CVSS to each of the potential vulnerabilities, we determine that any potential vulnerability with, let's say, a score greater than 4 should be investigated with the intent of implementing a mitigating control or making a change to the design, or both, which either eliminates the vulnerability completely or reduces it to a score lower than our target, in this example to less than 4. It is also at this point where any false-positives are identified and removed from further vulnerability processing.

The discussion on scoring methodologies below is generally scoped to early activities in the development lifecycle, specifically during concept, requirements analysis, and architectural design activities. However, we would be remiss not to first address the broader applications of vulnerability scoring and the issues surrounding its use. Vulnerability scoring needs to be utilized across the total product lifecycle of a medical device. Unfortunately, most of these life phases have radically different needs for the output of a scoring rubric. Similarly, each set of stakeholders has different perspectives and needs for its interpretation and application, as well as limited access to critical information available to other stakeholders. For example, the MDM's systems engineer is familiar with the technical design and inner workings of the device, but has next to no actionable information about the HDO's IT and cybersecurity infrastructures, whereas the HDO's CISO can access very detailed information about the device's operating environment in the hospital, but may have to treat the device itself as a proverbial "black box." This diversity of perspectives and information siloing also implies that we should be wary of uncritically accepting and adopting scoring performed by other entities. Most likely, they do not have the detailed knowledge or perspective needed to support your situation.

There is no single scoring methodology that can serve all possible stakeholders across all phases of the device's lifecycle. This has led to a multitude of critiques and complaints about any given scoring rubric, most of them legitimate, as each stakeholder is evaluating the rubric according to a particular perspective and needs for the resultant score. It has also resulted in a wide selection of possible scoring rubrics (nearly 20 at last count, not all of which are still supported).

So do not fixate on the exact results of scoring, such as ignoring a vulnerability with a CVSS score of 3.9 while mitigating a vulnerability with a CVSS score of 4.0. At best, these rubrics create an approximation of the severity, exploitability, or risk of a given vulnerability. They offer a predictive model that will never be able to precisely reflect reality. Much like assigning baseball players their positions based on past performance statistics, using a good rubric helps to manage probabilities and optimize the odds of success, yet actual outcomes will likely differ.

Finally, the whole process of vulnerability management is based on a score that is derived from only a handful of attributes (and even the formula calculating the total score is subjective). Is that really enough? We should always be critical of the process and the humans that perform it, given that we all have conscious or unconscious biases towards underweighting certain risks ("gaming the outcome") due to ignorance, business pressures, denial, and fear. We need to question the outcomes, especially if they seem to point toward letting problematic vulnerabilities through the screening and into the final product. Did we include critical technical concerns, such as impacts to the device's operational timing? Is this scoring approach really effective? Or are we just trying to avoid doing the hard work of creating mitigations for a discovered vulnerability?

6.4.1 Mitigations

Mitigating controls are always partial solutions, never a complete solution to the vulnerability, and, in some cases, mitigation may not even be possible. One such example is a denial-of-service (DoS) attack against a wireless connection, where so much radio frequency (RF) noise is blasted out with a high enough amplitude that the device's RF connection will not be available until the disruptive broadcast ceases. The only real mitigating control for this is probably already planned for in the design, namely, to notify the operator of the device's inability to make or sustain a connection to the other device. This is not a perfect mitigation but given the reality of the communications medium, this may be the best that an MDM can achieve.

Wherever possible, mitigations are created for each of the high-scoring vulnerabilities. Then the scoring process is repeated on each of these mitigated vulnerabilities, this time considering the impact of the implemented mitigating control. Outputs of this process inform the project's architecture control documents, requirements, design, and risk management activities—in short, the remainder of the development lifecycle, as shown in Figure 6.2.

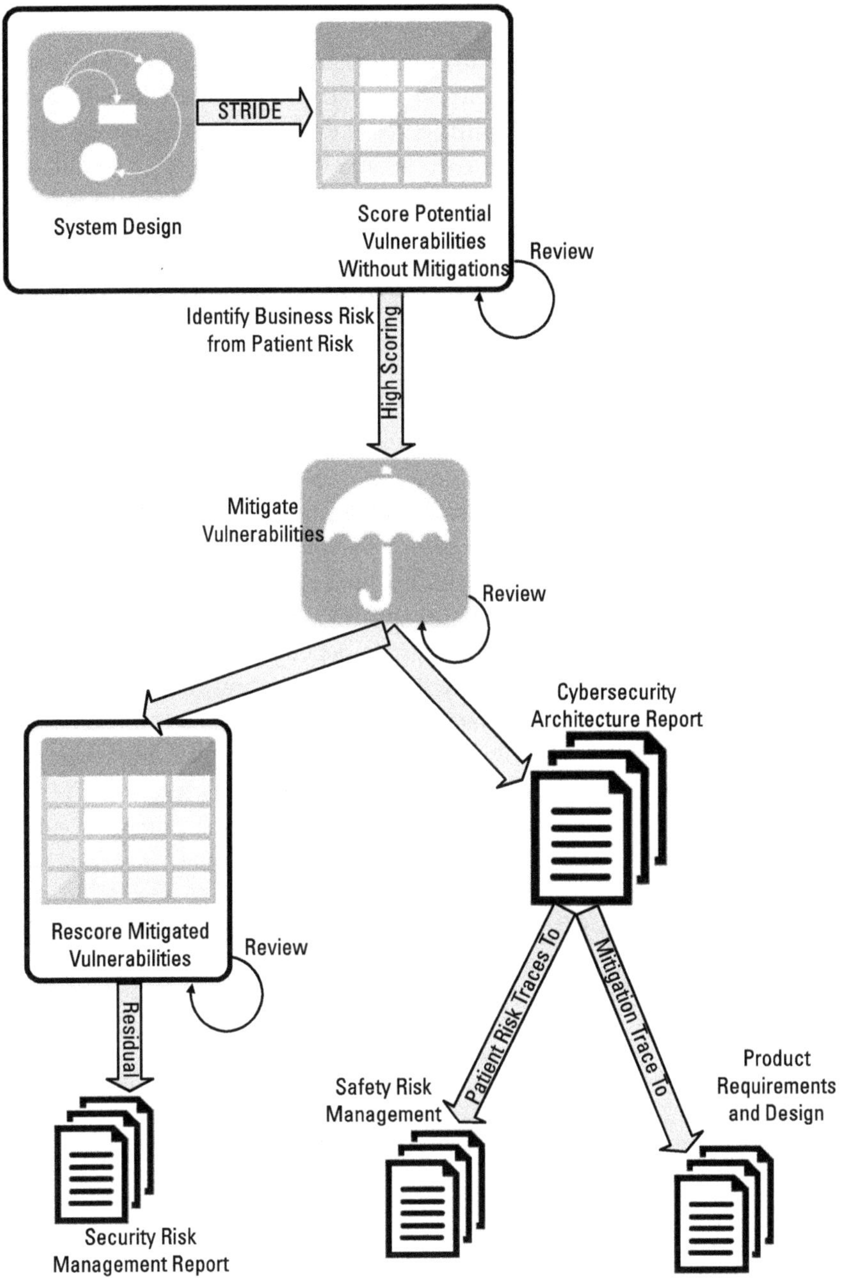

Figure 6.2 Cybersecurity activities during design.

6.4.2 Threats Versus Vulnerabilities

The cybersecurity industry frequently uses the terms "vulnerability" and "threat" as if they were interchangeable, which leads to confusion when discussing tools and techniques. This confusion further leads to fuzzy thinking about how to approach security, especially during the system design phase. This is due to the prototypical cybersecurity scenario being a live, in-the-field IT installation that is exposed to a threat because of an inherent vulnerability. In this scenario, the practical distinction between the two terms amounts to a finely split hair, so the frequent conflation, although incorrect, is understandable (for a more granular representation of cybersecurity terms and their relationship, see Figure 6.3). However, when applied to the lifecycle of a medical device, they are not even close to the same thing.

If the cybersecurity controls implemented in your device were based on the specific threats and attack techniques known to be in use during the time of development, your device inevitably would be behind the curve. Because threats continually evolve, it would be insecure by the time it reaches the field.

Take ransomware as an example. The first ransomware appeared in 1989 and was distributed via floppy disks (at an AIDS conference, hence named the "AIDS Trojan" [3]). Compare those humble beginnings to the ransomware of today, with so many variations on the theme and so many vulnerabilities being exploited to infect the various platforms. What mitigations could have been implemented in 1989 that would still have any validity today? However, a 30-year functional lifespan is not uncommon for medical devices.

During the design activities of a medical device, a requirement such as "The medical device shall not be susceptible to ransomware" (i.e., protecting against a specific threat) initially sounds like a pretty good idea. Who wants their medical device to fall victim to ransomware? Is that requirement actionable? No, it is not. There are many different types of ransomware; however,

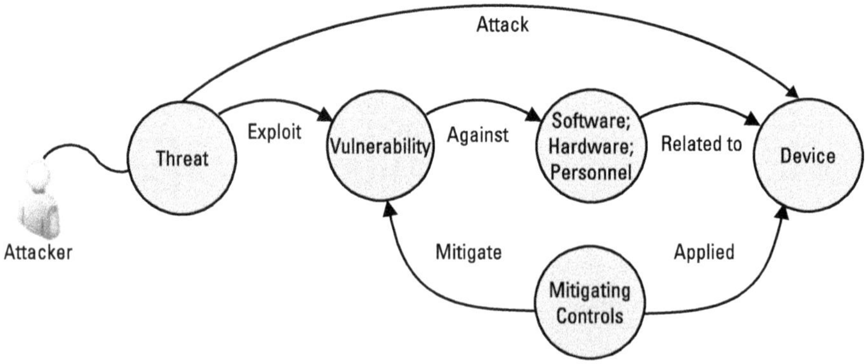

Figure 6.3 Applied terminology.

most literature focuses on what the ransomware does (encryption of files, locking the user out of the system, wiping the hard disk) [4]. Few studies focus on how the ransomware arrived in the device. Common infection vectors include [5]:

- Links in email or social messaging;
- Drive-by infections from a compromised website;
- Pay-per-install infections from botnet-compromised systems;
- Intrusion by exploiting vulnerabilities in the operating system or services, such as malformed SMB packets and remote desktop protocol;
- Wormable intrusions that exploit network vulnerabilities;
- Unsecured local communication ports, such as USB [6].

So how does knowing any of this inform the medical device design? Short of removing all forms of communication from the design, few actions could be taken as a result of such threat-based requirements. Threat-based requirements do not answer any design questions, such as:

- Should hardcoded credentials be stored in an off-die external flash memory?
- Should a secure element be included in my design?
- Does the RDS connection need additional layers of protection? What about data traveling via Wi-Fi?

What activities, then, are needed to inform and guide design work? The common theme is that a threat relies on the existence of vulnerability that can be exploited. However, "preventing burglary" is not the kind of job that cybersecurity requirements do. Their job is to secure the device against the burglar's most viable attack vectors, for example, by identifying better doors, installing better locks, and using them properly to lock all the doors, a task that generally starts with understanding where all the doors are.

A *vulnerability* is a weakness in design or implementation, and a *threat* is the potential to exploit that vulnerability to achieve a certain goal. A weakness in the design cannot be exploited because it does not exist yet; it is just a concept developed during the design activities. Consequently, there exists no actual threat to the design of a device (until that design has been implemented and the device has been produced). However, one can make assumptions about how typical threats could or would exploit a vulnerability, and then analyze and improve the design based on this type of abstraction. This is the approach commonly taken through threat modeling during the design and architecture activities.

6.4.3 Vulnerability-Related Terminology

Here are some further examples of key terms used in this chapter (see the Glossary):

- Design vulnerabilities, such as:
  - No authentication for a user to access the device;
  - No integrity checking of a data stream;
  - No encryption of PHI.
- Implementation vulnerabilities, such as:
  - Did not enforce the length of an input buffer;
  - Utilized intrinsic functions;
  - "scanf()" and "fscanf()" format strings did not provide a field width for the "%s" string.
- TPSC vulnerabilities (i.e., vulnerabilities inherited through the selected third-party software components), such as:
  - Heartbleed;
  - Urgent/11;
  - Swenytooth;
  - Log4Shell.
- Threats (to consider during design and architecture and to test for during implementation, integration, verification, and validation), such as:
  - Ransomware;
  - Worms;
  - DOS;
  - Physical and side-channel attacks.
- Exploits (specific implementations of threats), such as:
  - Cryptolocker (ransomware);
  - NotPetya (ransomware);
  - Zcryptor (ransomware);
  - SYN flood (DOS);
  - Sasser (worm).

These concepts interact as in the following example. At the design time, a design vulnerability is created, such as allowing access to the lowest levels of a medical device via a communications medium. This occurred because access to this communications medium was left open, without requiring authentication. At the time of design, all we can say is that we have a design vulnerability. If that design vulnerability cannot be mitigated through a design change, it should be identified and addressed during the development lifecycle, but, if it is not, once that design vulnerability exists in the field, it could be exploited in numerous ways, including by threats such as ransomware, worms, botnets, and used for various purposes, such as extraction of intellectual property, extraction of PHI, and compromise of design performance, to name a few of the risks.

This is a different approach and way of thinking about the problem, and for people with IT backgrounds accustomed to dealing with deployed systems and a more continual DevSecOps approach, this feels incorrect (and, for their environment, it is), but it is the correct approach for device developers.

During the design activities, vulnerabilities can be identified in a formal, repeatable process that has been codified in various tools. Vulnerabilities can be created or introduced not only during design, but also during implementation and through TPSCs.

When distinguishing between vulnerabilities and threats, a TPSC is a special case because it is both a product already in the field and a component or subcomponent of the device under development. Therefore, it may introduce vulnerabilities for which threats already exist (for example, consider a commercial operating system that may be implemented on the device or one of its accessories). This is one reason that we treat TPSCs differently during development. We look for disclosed vulnerabilities as well as known threats to these components. TPSCs are discussed in greater detail in subsequent sections and chapters.

6.4.4 Vulnerability Scoring

Initial vulnerability scoring assumes that no mitigation exists for the vulnerability. The list of scored vulnerabilities is what guides the priority and efficacy level for each of the required mitigations.

After selecting and/or designing mitigations for each scored vulnerability, the list of vulnerabilities is rescored. In this second round of scoring, mitigations are included to produce a post-mitigation score for each vulnerability. This second round highlights whether all mitigations are sufficient as designed or whether additional mitigations are needed for any residual vulnerabilities.

If no further mitigations are possible, then these high-scoring vulnerabilities are considered residual vulnerabilities that will be present in the marketed version of the medical device.

Note that the vulnerability assessment (VA) process that we are describing does not (and logically cannot) address vulnerabilities that may be introduced during implementation. It is solely focused on applying a rigorous formal method to expose areas in the initial design that necessitate mitigating security controls.

6.4.5 Scoring Rubrics

It is critical to apply the right scoring tool to the appropriate point in the product lifecycle: threat modeling during design (and scoring of the identified design vulnerabilities); vulnerability assessment and reassessment during implementation (for supply chain components and MDM application software), pre-mitigation and post-mitigation; and periodic ongoing postmarket vulnerability assessment (including, if beneficial, revisiting threat models) for products in the field as part of cybersecurity sustaining activities.

There are many documented criticisms of existing vulnerability scoring rubrics, and while these criticisms may be valid, they are often made in the context of finished, installed devices in the field. For such devices, cybersecurity assessments should be made against security risks, which are scored and mitigated differently than vulnerabilities. Our focus for this section is on vulnerability assessment rubrics used to evaluate the security of a proposed design. Until a universally accepted standard for scoring design activities vulnerabilities is created, we recommend modifying CVSS v2 as described in the following sections.

The common vulnerability scoring system (CVSS) is defined by FIRST's CVSS Special Interest Group [7]. FIRST is the Forum of Incident Response and Security Teams; as indicated in their name, this group is composed of responders to security incidents, primarily breaches of postmarket products and deployed infrastructure. Consequently, CVSS is segmented into three major groupings of metrics: Base Metrics, Temporal Metrics, and Environmental Metrics.

This organization system is indicative of FIRST's postmarket viewpoint. In particular, the Temporal group and the Environmental group consist of metrics that have no relevance to scoring a vulnerability during the design activities of a medical product. By contrast, in the Base group, all of the metrics are relevant to vulnerabilities discovered during the design activities of medical device development.

A metric that exists in the Environmental group of CVSS v2 (but not in v3) is "Collateral Damage." This metric is semantically equivalent to a more commonly used term in medical device safety risk assessment, "Severity." Severity represents the potential harm to users, as well as the potential risk to the device manufacturer's business model, which could result from a given vulnerability being exploited. It is the single most important metric when scoring a medical device design.

During the design activities, our purpose is to assess vulnerabilities and assign a score to create a meaningful overall score that includes severity ("collateral damage" from CVSS v2) as well as exploitability (base metric group), and measures their impacts to Confidentiality, Integrity, and Availability (base metric group). Our final score must be one that prioritizes which vulnerabilities to mitigate in the device design. Since 1:1 adherence to FIRST's CVSS specification is only required for published vulnerabilities, such as those cataloged in the NVD database, it is perfectly acceptable to customize CVSS for any internal use to create a more useful and suitable scoring mechanism for evaluating design vulnerabilities which have not yet been actualized. (Obviously, should a manufacturer publish a real or "live" vulnerability via NVD or ICS-CERT, it would need to be rescored per the CVSS specification.) Customized CVSS approaches are already in use by several MDMs for premarket assessment activities and regulators are familiar with this methodology.

6.4.5.1 Modifying CVSS v2

While CVSS v2 comes closer to what we need for scoring design vulnerabilities, there are several aspects to it that need to be extended and modified to support this effort. Several attributes are not valid and thus must be set to values that nullify their effect on the overall score. Table 6.1 shows the nullified attributes and their assigned values.

6.4.5.2 Redefining CVSS v2 Attributes

Redefinition of the remaining (non-null) CVSS attributes is necessary to better address the embedded device domain as compared to the original IT domain. Our recommended modifications are described here:

- Access vector:
  - *Local:* A locally exploitable vulnerability requires the attacker to be in physical contact with the device. Examples of locally exploitable vulnerabilities are USB ports, serial ports, and JTAG debugging ports.
  - *Adjacent network:* A vulnerability exploitable via adjacent network access requires the attacker to have access to either the broadcast or col-

Table 6.1
Nullified Values for CVSS v2

Attribute	Value
Exploitability	high (1)
RemediationLevel	unavailable (1)
ReportConfidence	confirmed (1)
TargetDistribution	high (1)

lision domain of the vulnerable software. Examples of local networks include local IP subnet, Bluetooth, BLE, IEEE 802.11, local Ethernet segment, and low power radio communications, with a maximum range measured in hundreds of feet.
  - *Network:* A vulnerability exploitable via network access means the vulnerable software is accessible via a network stack and the attacker does not require local network access or local access. Such a vulnerability is often termed "remotely exploitable." Wide area wireless networks (such as cellular radio) and broadcast radios would also be included in this category. In certain cases, a link to a smartphone could also be considered a network attack.
- Access complexity:
  - *High:* Extremely difficult to exploit, consisting of hundreds of man-hours and/or expenses running into the hundreds of thousands of dollars. Specialized access conditions exist. For example:
    - The attack depends on the victim performing several suspicious or atypical actions.
    - The vulnerable configuration is seen very rarely in practice.
    - The vulnerability depends on a narrow window in time.
    - Exploiting the vulnerability requires expensive tools.
  - *Medium:* The access conditions are somewhat specialized. Examples include:
    - The attacking party is limited to a group of systems or users at some level of authorization, possibly untrusted.
    - Extensive information must be gathered before a successful attack can be launched.
    - The affected configuration is a nondefault configuration, which is not commonly utilized.
    - The attack requires a small amount of social engineering that might occasionally fool cautious users.
    - The attack requires insider knowledge of the device's hardware or software design.
  - *Low:* Specialized access conditions or extenuating circumstances do not exist. Examples include:
    - The affected product typically requires access to a wide range of systems and users, such as anonymous and untrusted connections with no authorization.
    - The affected configuration is default or ubiquitous.

    - The attack can be performed easily and requires little skill or additional information-gathering.
    - A race condition exists; however, it is a lazy one (i.e., it is an easily winnable race).
    - A distributed system gives attacker all the tools and/or knowledge to exploit the vulnerability.
    - Commonly available tools, such as those easily downloadable from the internet, can exploit the vulnerability.
- Authentication:
    - *None:* Authentication is not required to exploit the vulnerability.
    - *Single instance:* Authentication is easily defeated or uses a weak method for vetting.
    - *Multiple instances:* Authentication employs the industry's best practices for vetting the authenticity of the user or device.
- Confidentiality impact:
    - *None:* There is no impact on the confidentiality of the system.
    - *Partial:* There is considerable informational disclosure. Access to some data (including executable) is possible, but either the attacker does not have control over what is obtained, or the scope is constrained. An example is a vulnerability that divulges only certain fields of data out of the set of what is available.
    - *Complete:* There is total information disclosure, resulting in all data residing on or accessible via the device, including its executable, being revealed. The attacker is able to read all possible data (volatile or nonvolatile memory, data streams, external communications, data at rest, data in motion).
- Integrity impact:
    - *None:* There is no impact on the integrity of the system, including on the operation of the system.
    - *Partial:* Modification of some data (including executable) is possible, but either the attacker does not have control over what can be modified or the scope of what the attacker can affect is limited compared to what is available to this vulnerability. Some alteration of essential performance is possible. For example, part of the executable image may be overwritten but not all of the executable image.
    - *Complete:* There is a total compromise of system integrity. There is a complete loss of system protection, resulting in the entire system being compromised. Total control over the essential performance of

the device is possible. The attacker can modify any data (including executable) on the target system.

- Availability impact:
  - *None:* There is no impact on the availability of the system.
  - *Partial:* There is reduced performance or interruptions in resource availability, reduced throughput or increased latency in responding to an event, or reduced performance where some essential performance is impact including missed communication periods. An example is a network-based flood attack that permits a limited number of successful connections to an internet service.
  - *Complete:* There is a total shutdown of the targeted device, rendering the device's essential performance nonoperational. The attacker can render the resource completely unavailable. An example is a ransomware attack.
- Collateral damage:
  - *Low:* A successful exploit of this vulnerability may result in slight physical or property damage, including a slight loss of revenue or productivity.
  - *Low-medium:* A successful exploit of this vulnerability may result in moderate physical or property damage, including moderate loss of revenue or productivity.
  - *Medium-high:* A successful exploit of this vulnerability may result in significant physical or property damage, including significant loss of revenue or productivity.
  - *High:* A successful exploit of this vulnerability may result in catastrophic physical harm (such as death) or property destruction, including catastrophic loss of revenue or productivity.

6.4.6 Alternative Approaches to Scoring

Over the past decade, a large variety of scoring rubrics have been created. They are usually created to fit a specific niche need or industry, and all are written from the perspective of a released device, designed to assess threats rather than vulnerabilities. These rubrics usually fall into one of three approaches:

- (A) Multiple subjective factors/attributes, all equally weighted in the final score;
- (B) Multiple subjective factors/attributes, with each attribute weighted separately before being calculated into a final score;
- (C) Either (A) or (B) with the inclusion of a "likelihood" metric.

6.4.6.1 Likelihood

Because the medical device has not yet been created, there is no likelihood of it being exploited. This metric does not pertain to the context in which we are deploying the rubric. Attempts to assign a probabilistic rate of attack or exploit occurrence are futile, impossible to ascertain, except as a best guess regarding conditions that may exist in the future if the device were to be manufactured and sold according to the early-stage design in front of us. It relies on speculation and conjecture, in other words, and is not useful for identifying and incorporating appropriate security mitigations into your design.

Some rubrics try to conflate "likelihood" with other attributes, typically called out as metrics with names like:

- Attacker skill level;
- Attacker motive;
- Opportunity to exploit;
- Size of attack;
- Ease of discovery;
- Ease of exploitation;
- User awareness;
- Intrusion detection.

Most of these share the same basic issues with Likelihood. Not only are they dependent on speculation and guesswork, but they are also focused on threats, not vulnerabilities.

Only one of these makes a reasonable contribution to medical device design scoring, and that is "ease of exploitation." This is still quite subjective, but it does add some meaning to the final score. For example, if exploiting the vulnerability in question requires the active use of elaborate equipment in the same room as the device being attacked, it may be considered relatively more difficult to exploit because of the challenge of transporting and setting up the equipment in close enough proximity without being detected before the attack can be carried out (assuming that attackers cannot achieve their objective by acquiring a copy of the device and taking it to their own workspace to exploit). However, if the exploit can be carried out from anywhere with the click of a button, it may be considered relatively less difficult to carry out.

A good, representative example of how Likelihood and related metrics are typically deployed by these rubrics is the National Institute of Standards and Technology's (NIST) Risk Determination rubric, where "likelihood" and "severity" are both equally influential upon the final score [8]. Under this rubric, the assessor could determine that the vulnerability's "severity" is very high because a patient could die from this vulnerability being exploited. However, in a case where the previous version of the device has been on the market for many years without a known attack against that same design vulnerability, the assessor could rate the "likelihood" as very low. The resultant score would be low, and thus, the scoring rubric would allow the designers to ignore it.

Needless to say, this is extremely problematic because, in cybersecurity, past experience (if indeed such past experience actually exists) has limited usefulness for predicting future events, and could lead to scrutiny and delays in the (still evolving) regulatory market approval process. Either the device will not be approved for the market because the flawed rubric used for the design vulnerability assessment failed to flag the vulnerability and the design was submitted for approval without adequate mitigation, or worse, the flawed rubric will mislead reviewers, who then approve the device for the market, only to see it later become the means by which one or more patients are severely harmed due to this unlikely vulnerability being exploited at some time in the future.

Conversely, if you assume that likelihood is always high, then this metric simplistically magnifies severity and distorts the comparative outcome when identifying vulnerabilities in need of mitigation.

Either way, the fundamental problem with these types of rubrics is revealed: the outcome is easily gamed, intentionally or unintentionally, by the person or team assessing the vulnerability. Even NIST does not know how to define likelihood nonsubjectively [9]:

> The term likelihood, as discussed in this guideline, is not a likelihood in the strict sense of the term; rather, it is a likelihood score. Risk assessors do not define a likelihood function in the statistical sense. Instead, risk assessors assign a score (or likelihood assessment) based on available evidence, experience, and expert judgment. Combinations of factors such as targeting, intent, and capability thus can be used to produce a score representing the likelihood of threat initiation; combinations of factors such as capability and vulnerability severity can be used to produce a score representing the likelihood of adverse impacts, and combinations of these scores can be used to produce an overall likelihood score.

Because there is no objective definition of likelihood, the results will not be consistent or even congruent between iterations, assessors, reviewers, or projects.

6.4.6.2 Other Rubrics

Now that we have made the difficulties about likelihood and related metrics clear, we can more easily assess the various rubrics available for their applicability to design vulnerability scoring. First, let's look at those that are not suitable:

- The FDA's September 2023 Premarket Cybersecurity guidance: Details a simplistic scoring rubric of severity over exploitability for the security risk management system.
- NIST's Risk Determination: Heavy reliance on likelihood invalidates this rubric [10].
- DREAD: Overly subjective; no longer supported.
- CVSS 3.x: This version of CVSS is focused on the general software and IT environment use case and has removed "Collateral Damage" (Severity), which is the single most important factor to consider when scoring medical device vulnerabilities. It is possible to spoof this value by manipulating other attributes in this rubric, but then why use it for medical devices if it is a suboptimal fit [11]?
- NIST's CMSS and CCSS: These are very close variants of CVSS v2, but they introduce likelihood in the form of "general exploit level" and "perceived target value" [12, 13]. These alternatives to likelihood are just as problematic due to the highly subjective nature of these value labels. Any given person performing the scoring may not view one exploit level the same as a person more knowledgeable about the nature of exploits.
- MITRE's Medical Device Rubric: A greatly expanded variant of CVSS v3 created by MITRE, this rubric was designed to be inclusive of multiple users' perspectives, including HDOs and MDMs. This variant introduces a large number of additional questions that must be answered to inform the attribute settings—45 questions in all—which must be assigned to each vulnerability before a base metric can be computed. Although it is a good tool for its intended purpose—recommended by the FDA for postmarket vulnerability scoring, that is, evaluating deployed devices—this rubric is less suitable for design vulnerability scoring because of this additional burden. In the initial assessment of a design, hundreds of vulnerabilities may be identified, and each would need to have 45 answers entered by the assessor before determining a base score [14].

  Some of its other advantages are that it can be used by both MDMs and HDOs for postmarket scoring, as well as its inclusion of patient safety, patient/clinician privacy, and cybersecurity risk from an enterprise vulnerability-management perspective. Furthermore, it provides examples as well as a decision tree to guide a user through the scoring process. However, it is not suitable for the assessment of chained vulnerabilities. The authors like this rubric as it is trying to address all the needs (this has never been attempted by any other rubric) for all the different domains where vulnerability and threat scoring needs to be performed. It was an ambitious effort, but basing it on CVSS v3 harmed it more than helped.

- IVSS: This is an industrial variant of CVSS v3, and has several interesting additions to it, such as "cascading consequences" and "process control consequences." However, at its base level, it is oriented toward a fully designed and released product: Fundamental to its base scoring are attributes such as "report confidence" and "exploit maturity" [15].
- OWASP Risk Rating: This scoping rubric is very simple and is a member of the "likelihood over impact" scoring group. Since it relies on likelihood, it is not suitable for design activities vulnerability scoring [16].
- EPSS: Presented at the Black Hat 2019 conference, this rubric is a predictive variant of CVSS, where the probability of a vulnerability being exploited is considered by basing this on previous past exploits. Analysis of past common vulnerabilities and exposures (CVEs) is brought to bear to help to quantify and exploit probability (likelihood) of each vulnerability. Unfortunately, the vast quantity of CVEs in the database are related to MIS/IT systems, not embedded devices, so many potential vulnerabilities are left unconsidered, such as those requiring physical attack vectors. Also, existing CVEs are skewed towards the simplest of attacks, such as buffer overflows, which causes more complex attacks to be ranked as (relatively) lower risks. Finally, because of restating likelihood as an "exploit probability," this rubric would not be suitable for design vulnerability scoring [17].

Next, let's consider two rubrics that are suitable for design evaluations:

- CVSS v2: As we hope we have already demonstrated, this rubric is an example of a rubric that is simple to perform yet returns a final composite score based on appropriately weighted attributes. This rubric requires a quantity of 7 attributes to be assigned to each vulnerability before a base metric can be computed. "Severity" is supported in the form of collateral damage, which allows for flexibility in supporting impacts to patient safety and efficacy and well as business risks.
- RSS-MD: This rubric is a medical device variant of CVSS, requiring 11 attributes to be assigned to each vulnerability before a base metric can be computed. It is a good rubric to utilize for design vulnerabilities with one minor exception—it has an attribute called "scope of impact,"

which would not be relevant at the time of a medical device's design. Fortunately, the premature inclusion of this attribute can be easily corrected by setting it to "all" for each of the discovered vulnerabilities, which produces a minimal impact on the final calculated score. Because of its close relationship with CVSS, this rubric also shares some of its shortcomings, such as poorly defined "attack vector" settings. This rubric also focuses exclusively on patient safety impacts and is unable to account for other types of negative impacts, such as financial and reputational risks. Its patient focus also excludes the utilization of the medical device under attack as a pivot point into a network with a larger attack surface. However, even considering these limitations, this rubric should still be considered a viable candidate for scoring design time vulnerabilities [18].

6.4.6.3 Final Notes on Scoring Rubrics

Most of these rubrics were not developed specifically with application to medical devices, nor were they designed to score design vulnerabilities during a secure development lifecycle. So, while it is possible to utilize CVSS v2 and RSS-MD for medical device design vulnerability scoring by slightly broadening one's interpretation of some of the scoring attributes, further standardization work needs to be performed before the medical device industry will have a formalized rubric well-suited for this purpose. This work includes developing a rubric specifically for evaluating medical device designs; flexibility and proper weighting to account for vulnerabilities typical of embedded devices, such as portable and home use devices, rather than trying to stretch enterprise software/IT approaches to fit; and robust "severity" metrics that account both for risks to patient (safety and efficacy) as well as potential threats to the MDM's business model (reputation and profitability).

6.5 MDM Development Lifecycle Activities

After this detour to discuss and review vulnerability scoring and its methodologies, we now go back to the lifecycle diagram and development lifecycle (see Figure 6.4). We are ready to discuss applying vulnerability scoring throughout the lifecycle activities, but in different ways depending on whether we are dealing with design versus development versus supply chain vulnerabilities and their respective scoring and mitigations.

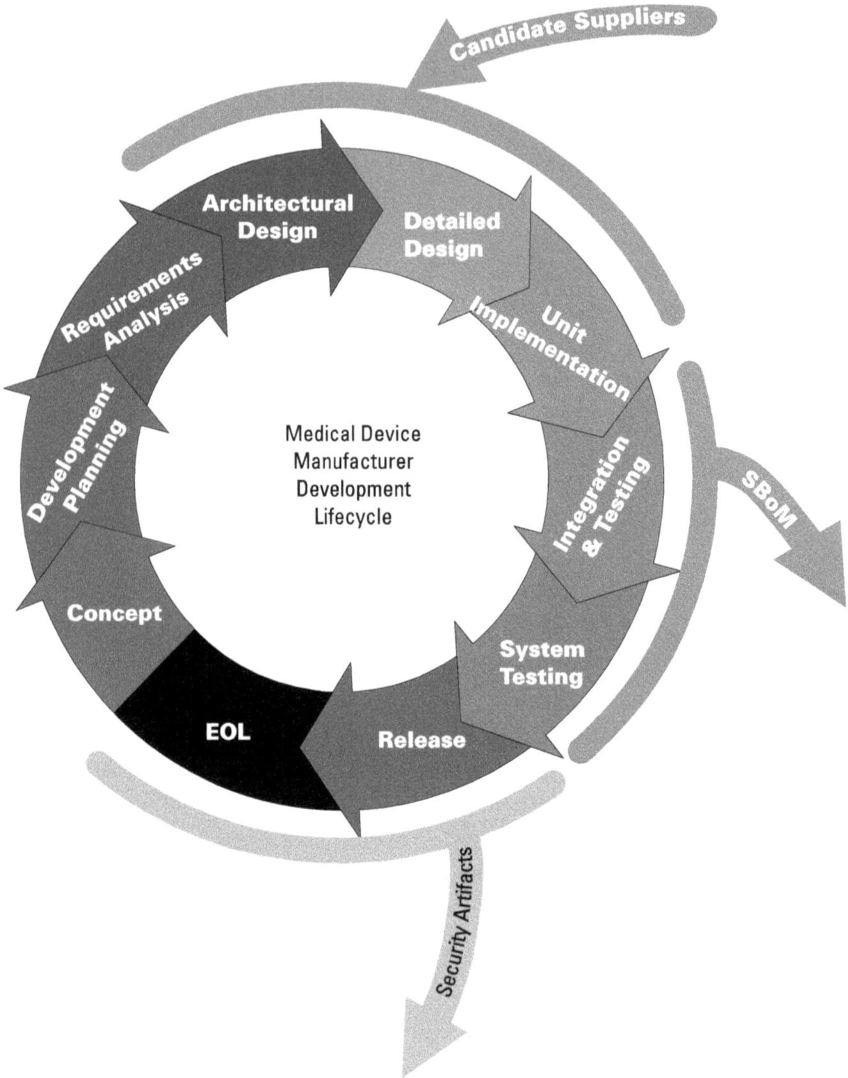

Figure 6.4 Development lifecycle activities.

6.5.1 Concept and Planning Activities

During this early part of the medical device development, the development team needs to consider all of the roles that interact with the medical device and their associated use cases.

Taken as a set, these use cases constrain the security mitigations to consider, as well as the set of potential vulnerabilities that may need to be mitigated.

Examples of these security-relevant concepts and planning scenarios are:

- The user needs the functionality to copy PHI/PII from the medical device to a USB flash drive.
- The medical device needs to allow the user to attach his or her own keyboard via a USB connection.
- The medical device will utilize a Bluetooth low energy connection to any smartphone to communicate information to a cloud service.
- The medical device does not have an advanced user interface to allow for the authentication of a user.

General considerations during the concept activities should include:

- User authentication;
- End-of-life controls;
- Connections to other trusted entities;
- Examples of situations when "break glass" emergency access might be necessary for critical operations, where user-to-machine authentication can be bypassed.

6.5.1.1 Proof-of-Concept Considerations

Early proof-of-concept work can have an undue influence on the technical nature of the final device. To illustrate this, consider a small startup MDM that has limited funding and therefore selects low-cost components during proof-of-concept that cannot be secured. They expect to be able to leverage device-specific software in the final production version of this device. Porting their code to a new platform is expensive and time-consuming and was not accounted for in their cost-projection model. Now the startup faces a sunk-cost scenario that it was not anticipating. Early proof-of-concept systems should utilize components that are capable of being secured, even if they are not required for the proof-of-concept version of the device. Components should include such features as:

- Cryptographic accelerators (e.g., for AES; SHA; HMAC; CMAC);
- Secure boot functions;
- Secure key storage;
- Anti-decapping measures;
- Physically Unclonable Functions (i.e., PUF);
- Virtualized security CPU;
- Side-channel power monitoring mitigations.

While these security features may not necessarily be active in the proof-of-concept build, working with the same components in early builds will pay dividends when transitioning into the development of the real medical device. It can also smooth the way through security aspects of investigative device exemptions and investigative review boards.

6.5.1.2 Incremental Improvements and Secure Development

A common approach for MDMs is to make minor changes to an existing medical device to add a new feature or enhancement. This method of minor upgrades to a device has been an inexpensive way to attempt to keep an older device relevant and marketable while reducing engineering costs and traditionally keeping the regulatory hurdle low. However, great care should be taken before deciding to continue with this approach. Not only are 510(k) and "substantially equivalent" device submissions coming under increasing scrutiny for cybersecurity features specifically, but existing medical devices may not have been designed with electronic components, operating systems, or application software that is capable of being secured against today's sophisticated threats.

The topic of legacy designs is a very difficult one because there are no easy fixes. The best course is to identify unsecurable designs early and allow their implementations in the field to reach their end of life and be replaced with a newer platform that is capable of being secured. Until it becomes business as usual, this approach will not often be a popular topic to discuss with senior management during the next few transitional years. They have likely participated in this model of continual improvement for a very long time, and they may be reluctant to let it go. However, not only is this a more responsible approach necessary for reducing exposures to patient and business risk, we anticipate that regulators and standard bodies will increasingly drive it to become normative.

6.5.2 Development Planning

Often, cybersecurity professionals from other industries are astonished at the lack of security-related planning incorporated into most medical device development projects. All too often, security practices in embedded device development lifecycles look like this:

1. Someone in the development team makes a list of all the security concerns that they can think of. The persons carrying out this activity may or may not be security professionals with a working awareness of the contemporary threat landscape, and they may or may not use rigorous, well-vetted, industry-accepted techniques and rubrics to identify and classify these concerns.

2. These concerns are included in the project as product security requirements.
3. Subsequently, security and security activities are ignored until the first Release Candidate is ready. At this point, a third-party penetration tester is hired to attack the device and provide the MDM with a report of all the security mistakes that were made. This approach typically delivers an overwhelming list of findings and it is way too late in the product development cycle to address them with reasonable effort and cost.
4. Because market pressure is already building, many (or most, or all) of the device's vulnerabilities continue to be ignored, and the product is released.

Security should be treated like any other activity carried out during development. Every other part of the development lifecycle is planned for, scheduled, tracked, and documented. Security should be no different and should be planned for at an early stage. This holistic practice allows for coordination between teams, accurate tracking of project progress, realistic assessment of costs and resource needs, and smooth integration of subsystems for an overall high-quality product.

With cybersecurity aspects of medical devices becoming more and more regulated, it is no longer permissible for MDMs to take such a lax approach to security planning. A well-formulated security plan (further discussed in Chapter 7) that breaks down the phases of the development project, lays out the security activities to be performed in each phase, provides for their traceability and verification, is indispensable to securing approval in key markets worldwide, including in the United States, the EU, France, Canada, Japan, Australia, and China, and is a core prerequisite for producing a product that is secure and can be maintained.

6.5.2.1 Security Goals

The Security Plan can also be a good location to express the security goals of the project. These are not requirements, but rather overall high-level goals for virtually any type of development project. High-level goals for other considerations in development, such as the "voice of customer" or "voice of business," are usually expressed in a separate document, and this can be done with security as well ("voice of security"). However, security's high-level goals are usually few, and therefore a voice of security document would be quite short. Locating them in the security risk management plan also aligns with the latest FDA requirements for content in the security risk management plan. More about planning and the content of the security risk management plan can be found in Chapter 7.

6.5.3 Requirements Analysis

At first glance, it may seem that cybersecurity requirements for medical devices are obvious. Certainly, the traditional CIA trio of Confidentiality, Integrity, and Availability seem like good starting points (and they are); but the need for more granular levels of detail quickly becomes evident.

Depending upon the level of abstractions implemented in your document tree, the highest levels (typically system requirements) may stop with the inclusion of the CIA triad, which could then be traced back to as the fountainhead for lower-level requirements, such as those defined in device-specific requirements documents and interface control documents.

Further guidance for structuring requirements documentation is found in Chapter 7; this section is meant to assist with the task of sorting out the wheat from the chaff in defining the content of cybersecurity requirements. The first step in achieving that is to remove all references to "threats." As previously discussed, threat-based requirements are those designed to counter specific threats ("exploits") or even specific categories of threats, when instead, the focus should be placed on vulnerabilities.

Vulnerabilities can be created or introduced in three ways during development: by Design, during Implementation, or through TPSCs. The traditional approach to securing a medical device is to break these areas down into the well-known security triad: Confidentiality, Integrity, and Availability. Examples of these include:

- Confidentiality:
  - Of the patient's sensor readings;
  - Of the medical device settings;
  - Of the patient's PII;
  - Of the firmware upgrade image used to upgrade the device while in the field;
  - Of the MDM's intellectual property;
  - Of the command-and-control functionality exposed via communications media;
  - Of user or network credentials;
  - Of secret materials such as cryptographic keys;
  - Of network or GPS data that would disclose a patient's geolocation.
- Integrity:
  - Of the essential performance of the medical device;
  - Of the patient's sensor readings;
  - Of the patient's medical device settings;
  - Of the patient's PII;

  - Of the firmware upgrade image used to upgrade the device while in the field;
  - Of the command-and-control functionality exposed via communications media;
  - Of device-specific data like calibration settings or safety limits.
- Availability:
  - Of the essential performance of the medical device;
  - Of the patient's sensor readings in the expected time frame;
  - Of the patient's medical device settings;
  - Of the command-and-control functionality accessed through communications media.

In the healthcare industry, great emphasis has traditionally been placed on the confidentiality of data, thanks to regulations like the HIPAA Breach Notification Law in the United States or the GDPR in Europe. However, there is also an implicit ranking of the CIA triad from the MDM's perspective. MDM security activities have tended to prioritize ICA, where integrity is considered paramount (because it often represents the most obvious business risk to the MDM), while the availability of the device is not recognized as quite so important. This can be seen in the handling of many types of alarms in existing medical devices and proposed designs, where normal treatment is suspended or reduced to a safe mode or fallback mode of operation until the alarm can be cleared by an authorized user. Meanwhile, HDOs may be more interested in being assured of the availability of the device's essential clinical functions and the integrity of both the device and any connected devices within the HDO's network, with confidentiality seen as a slightly lower priority in the face of immediate care delivery needs, AIC. Given these slightly different emphases in priorities, determining detailed cybersecurity requirements necessitates careful evaluation based on the specific nature of the medical device, specifically with reference both to the use profiles and to device architecture.

At its simplest level of categorization, there are two main device architecture types: the embedded PC and the custom hardware with microcontroller.

For the embedded PC, detailed design objectives may include some or all of the following:

- The commercial off-the-shelf operating system shall be hardened to remove all unused functionality, including:
  - Disabling USB ports or, if not possible, disable all but the anticipated functionality supported (e.g., disable the USB autorun function);
  - Disabling all unused network ports and protocols;

- Disabling all unused functionality and processes;
- Inhibiting all hotkeys;
- Automatic encryption of the mass storage device (hard drive or SSD);
- Disabling automatic updating of the operating system.

Note that while hardening baselines such as those provided by the operating system (OS) manufacturer or the Department of Defense (DOD) are good standards for controlling this effort initially, some minor modifications of a given baseline may be necessary to support specific device/system use cases, such as:

- The operating system shall be placed into a kiosk mode or equivalent such that only the medical device application can be executed.
- The BIOS setup shall be protected from alteration.

Specific environmental functionality shall be supported. Considerations should include functions such as:

- Inclusion in an environment's active directory;
- Automatic identification to the intended device install site's asset management system.
- Installation of third-party host intrusion detection system (HIDS) or host intrusion prevention system (HIPS), often referred to as "allow listing." A third-party Host IDS/IPS offers the advantage of avoiding device-hosted signature databases that need to be updated and other disadvantages of traditional anti-malware technology (see Chapter 11 for a more detailed discussion).
- Installation of commercial anti-malware (if appropriate for the device type and use case). Note that including commercial anti-malware tools introduces a new set of considerations, including a degree of unpredictable timing behavior, the need to update the virus signature database, the possibility of false-positive detection, and the possibility of exposing the device directly to the internet or local intranet to retrieve signature updates. Depending on the configuration, other challenges may also require consideration, such as handling how and where a detected virus notification will occur, especially if the OS is in the kiosk mode during normal operation.
- The application shall support secured functionality to grant authorized users access to underlying operating system functionality not typically

granted to a normal user, such as joining a domain or connecting to a Wi-Fi access point.

- End-of-life functionality shall be supported. An example of such functionality is the deletion of all data at rest, as is typically provided by a factory reset.
- Updating the operating system and updating the application under operator or service personnel control shall be supported.

For Custom Hardware with Microcontroller or commercially available single-board computer (SBC) designs, where we are not trying to restrict a flexible commercial operating system into a single-purpose machine, detailed requirements may include some or all of the following:

- Secure coding conventions shall be defined and followed for each of the computer languages utilized in the system.
- Secure coding conventions shall be enforced via static analysis of the generated code.
- Secure boot functionality shall be implemented in your microcontroller to ensure the execution of unadulterated code.
- The protection of confidentiality, integrity, and availability shall be applied as widely as possible throughout the design and implementation of software and hardware for the device and system.
- The implemented security shall not be brittle, meaning that it should allow and/or detect outlying conditions and events present in non-normative conditions without negative impacts on security and safety.
- Security mitigations shall be interlocking with other security mitigations where violations to one have detectable consequences in the other (also known as defense in depth).
- Nonvolatile memory shall be partitioned to ensure that the security mitigations and intellectual property can be contained on the same die as the CPU (as opposed to external nonvolatile memory).
- CPU-provided memory protection shall be utilized in the microcontroller to partition processes and any potential data storage into separate areas, as well as to designate which memory addresses can contain executable code.
- Exception handling shall be configured not only to detect failures but also to log them, notify the user, and continue to deliver the essential performance of the medical device.

- Processes shall never be blocked due to resource exhaustion (e.g., from dynamically allocated memory). Resources shall be statically allocated to processes.
- General-purpose operating systems shall be avoided in favor of smaller schedulers or bare-metal round-robin execution (all extra functionality in operating systems increases the possible attack surface).
- Due to the limited scope of default, general-purpose mitigations, the use of default and/or standard security mitigations implemented in communications mediums shall not be considered adequate, except as part of a defense-in-depth strategy along with other mitigations.
- Modes of operation shall be considered in the system, especially where nonclinical functionality is performed. The usual cases for this are production-line configuration and testing support in a device/system. While manufacturing functionality may need to be active in a device/system that is currently in the process of being manufactured, this functionality should be removed or at least disabled in a device or system once it leaves the manufacturing environment.
- The system design shall ensure that, in the case of a compromise of a single device, the information learned cannot be used to facilitate additional attacks against other similar systems, especially larger-scope attacks. Typically, such exposures are the result of default credentials, a shared secret present in all devices, or hardcoded credentials.
- Hardware root-of-trust functionality shall be preferentially implemented over software-only solutions. This includes key and shared secrets storage as well as encryption/decryption acceleration.
- There shall be encryption of all critical data. This includes data in motion, data at rest, and firmware updates.
- Encryption shall, at a minimum, utilize AES symmetric encryption of not less than 128 bits in width with a preference of 256 bits in width (greater encryption widths may be required in the future as advances in technology may, for example, reduce the time needed for brute force attacks).
- Data integrity shall be ensured via digital signatures, cryptographically strong hashes, or Message Authentication Codes, such as HMACs and CMACs.
- Data authorization shall be ensured via access control lists or similar structures.

- There shall be authorization control of devices in the system's communication network.
- Integrity-secured logging shall be used to ensure the integrity of the content in the log, as well as the integrity of the order of the records added to the log.
- Keys and shared secrets shall be stored in a secure manner (i.e., not in plaintext nor in easily accessible memory).
- Key provisioning during production or initial go-live shall not expose secret materials, especially important if third-party production is intended.
- Keys and shared secrets shall be device-unique and ephemeral.
- Apps targeting commercial operating systems (e.g., Windows, iOS, Android) shall be obfuscated.
- "Freshness" of received data shall be monitored for significant delays and, if present, reported to the user.
- Handling of shared secrets such as encryption and authentication keys shall be performed in a nonpredictable bit or byte manner to prevent the possibility of power monitoring side-channel attacks that would expose these values.
- Comparisons of values utilized for security mitigations shall be performed in a time-agnostic manner, such that comparison success or failure occupies identical periods of time.
- Intrinsics such as memcmp, memcpy, or strcmp shall not be utilized for handling of any shared secret.
- Compiler optimizations shall be disabled around the code that is used as an alternative to intrinsics such as memcmp, memcpy, or strcmp.
- In-the-field firmware updates shall be supported.
- Firmware update images shall be protected from exposure of image contents (i.e., by encryption).
- Firmware update images shall be protected from undetectable adulteration of the image contents and authentication of origin (i.e., by integrity checks via keyed cryptographic hashes).
- Firmware update images shall be protected from rollback attacks (i.e., by allowing advances in version numbers only).
- Firmware updates shall be authenticated (i.e., cryptographically signed) to enable automatic verification of authenticity prior to installation.

- Interruptions to essential performance shall, at a minimum, be reported to the user.

Additional requirements and detailed designs will be created during the design activities, based on the results of the vulnerability assessment.

6.5.3.1 Safe Harbor Versus Full Encryption

The HIPAA Safe Harbor provision is part of the HIPAA Privacy Rule. It applies generally to HDOs and to any MDMs whose products or services include handling of PHI. However, the HDO may contractually require HIPAA-compatible (or stronger) data protection to enable its compliance with the law and lower its risk exposure.

When data is shared outside of a covered entity, Safe Harbor provides a method of deidentification data by removing specific types of information to prevent the reidentification of the PHI. Note that taking this approach means conforming to a legal position and is not the same as truly deidentifying sensitive data. To achieve this legal position, the following types of data need to be removed from the PHI dataset to be considered deidentified by HIPAA [19]:

1. Names;
2. Geographic subdivisions smaller than a state (except for the initial three digits of the ZIP code under certain conditions);
3. Dates (except year) for dates that are directly related to an individual and all birth dates;
4. Telephone numbers;
5. Vehicle identifiers and serial numbers, including license plate numbers;
6. Fax numbers;
7. Device identifiers and serial numbers;
8. Email addresses;
9. Web Universal Resource Locators (URLs);
10. Social Security numbers;
11. Internet Protocol (IP) addresses;
12. Medical record numbers;
13. Biometric identifiers, including fingerprints and voice prints;
14. Health plan beneficiary numbers;
15. Full-face photographs and any comparable images;
16. Account numbers;

17. Any other unique identifying number, characteristic, or code, except as permitted;
18. Certificate or license numbers: biometric data, including fingerprints, voice prints, retinal images, gait, or unique cardiac identifiers.

While the Safe Harbor provision was created to inform covered entities and not specifically for medical devices, it is commonly used to help to inform MDMs of what data elements should be avoided.

Every MDM should be aware that there have been multiple examples (far too many to list) [20–22] of anonymized data being reidentified, going even so far as to track U.S. Secret Service agents [23]. Also, where there used to just be HIPAA, now there are many laws for protecting information and the privacy of the patient (for example, on the state level (e.g., California and Massachusetts)), and many of these laws have serious financial consequences for violating them. Privacy regulations are further discussed in Chapter 3. Lastly, HDOs are now increasingly including language in contracts that specify that all data needs to be encrypted, whether at rest (i.e., in a database, buffer) or in motion (i.e., when being transmitted via wired or wireless communications). It is therefore recommended that encryption for all data be strongly considered for your medical devices, along with strong cryptographic primitives (hashes) to ensure the integrity of the data as well.

6.5.4 Architectural Design and Detailed Design

Identifying, assessing, and managing vulnerabilities during the design activities of a project can be difficult. Little in the way of clear guidance has been made available, and few academic studies have been performed on identifying and preventing design vulnerabilities. Even the concept of what engineers should seek to identify and mitigate during the design activities (vulnerabilities, not threats) is not widely understood.

This is one of the largest differences between MDMs and normal MIS/IT security. IT security teams use tools and techniques to detect intrusion attempts and breaches (threats), whereas the MDM uses tools and techniques to detect vulnerabilities and prioritize their mitigation so to reduce the risk of future threats exploiting these vulnerabilities. As discussed in detail before, this is a subtle and often misunderstood difference, with significant impacts on how product security is practiced by an MDM.

6.5.4.1 Threat Modeling

Utilizing threat tools (which is a confusing name) at design time to expose areas of a proposed design that need mitigating controls is exactly the process detailed in this section. The challenge inherent in this approach is that development engineers must apply and adapt tools and techniques developed and used by MIS/IT security defenders in the field. These are different domains that need slightly different treatments when applied during the design of the medical device. At the most basic level, it is the MDM's goal to remove potential cybersecurity exposures that, if exploited, would result in negative impacts on a patient (risk of harm) or business (risks to business model). However, after a product has been released into the field, threat modeling can inform experts in incident response and planning as well as postmarket risk management. At that point, threat modeling may be able to provide guidance as to where changes in the device utilization, configuration, or environment could be made to reduce the chances of a successful cybersecurity attack.

What is needed during a medical device's design activities is a formal approach that minimizes the participation of a cybersecurity SME in the process, as well as a nonambiguous method to logically define the system being designed that will constrain and define the list of potential vulnerabilities, as derived by decomposition of the defined system. This list of potential vulnerabilities then needs to be screened for false-positives and the remainder prioritized in terms of the use profile for the particular medical device system. The prioritization process also needs to differentiate risks to users (device safety and efficacy) from business risks (threats to the business model) that the proposed design could pose to its MDM if taken as is into full production. Finally, this Design Vulnerability Detection process needs to be as automated as possible, to allow for speed of design, redesign, and reuse in other development projects.

The closest currently available tool for achieving these goals is the STRIDE Threat Modeling Tool from Microsoft. STRIDE stands for:

- Spoofing identity;
- Tampering;
- Repudiation;
- Information disclosure;
- DoS;
- Elevation of privilege.

In the original, IT-focused definition of STRIDE, its six areas of concern were specifically defined to address IT risks. For example, spoofing identity was directed towards spoofing a user's identity. When adapted for medical devices, we have to be assured not only of the user's identity (including a variety of user roles like patient, clinical, technical, or remote users) but also of the identity of any other connecting device (i.e., machine to machine—M2M).

Extending the definition of these terms works well with the existing STRIDE model, so here are the redefinitions of those same six areas for use with an embedded medical device system:

- Spoofing identity:
  - Utilizing another user's authentication;
  - Utilizing a different device or IT system to communicate into the system (e.g., by executing a replay attack).
- Tampering:
  - Malicious or unauthorized modification of data at rest;
  - Malicious or unauthorized modification of data in motion;
  - Malicious or unauthorized modification of program executable.
- Repudiation: the ability to deny performing an action;
- Information disclosure:
  - Exposure of information to an unauthorized user;
  - Exposure of data at rest;
  - Exposure of data in motion;
  - Exposure of secret materials such as cryptographic keys;
  - Exposure of program executable.
- DoS:
  - Negative impacts on medical device system availability and reliability;
  - Preventing valid users from authenticating;
  - Blocking or delaying communication mediums carrying data in motion.
- Elevation of privilege:
  - Unprivileged user gains privileged user's access (a subcase of spoofing).
  - Unauthorized connection into the medical device system.

These are then applied to the set of primitives, as shown in Table 6.2. The example mitigations for each of the STRIDE classifications are:

- Spoofing identity:
  - Perform user authentication, strong (multifactor) if justified by the use case.
  - Perform device-to-device authentication.
  - Prevent exposure of keys used with cryptographic hashes.

Table 6.2
Applying Threat Modeling

	Process	Data Stores	External Entities	Data Flows
S	✓		✓	
T	✓	✓		✓
R	✓	✓	✓	
I	✓	✓		✓
D	✓	✓		✓
E	✓		✓	

  - Protect secrets utilized in authentication (i.e., point-to-point connections).
  - Utilize highly ephemeral secrets in authentication.
  - Utilize third-party authentication, such as PKI.
- Tampering:
  - Perform authorization.
  - Prevent exposure of keys used with cryptographic hashes.
  - Message authentication codes.
  - Digital signatures.
  - Secure boot.
- Repudiation:
  - Digital signatures;
  - Audit trails;
  - Cryptographic hash chains;
  - Forensically significant logging;
  - Logging of key events, including security events;
- Information disclosure:
  - Perform authentication/authorization.
  - Perform encryption.
  - Utilize highly ephemeral secrets, such as encryption keys.
- Denial of service:
  - Perform authentication/authorization.
  - Filtering of communications.
  - Throttling of communications.
  - Monitor quality of service of communications (i.e., check for "stale" data).

- Ensure proper bandwidth for the essential performance of the medical device.
- Elevation of privilege:
  - Provide multiple methods of authentication/authorization.
  - No single layer of authentication (i.e., avoid the binary concept of authentication).

In the intended utilization of STRIDE, an engineer would lay out the Data Flow Diagram (DFD) of newly developed software using the Microsoft Threat Modeling Tool (TMT). An example is provided in Figure 6.5.

The TMT would then automatically apply the STRIDE rules to this DFD and decompose the diagram into a list of potential vulnerabilities, output into a spreadsheet.

At this point, a few items to note. First, STRIDE was initially designed to be a software-only tool and thus assumes that the software was written before STRIDE would be performed. In other words, it was not initially well-suited for design activity work. Second, the list of potential vulnerabilities is as complete as the DFD entered, meaning that the obvious level of completeness displayed in the DFD will be the same as that represented in the list. Third, the DFD

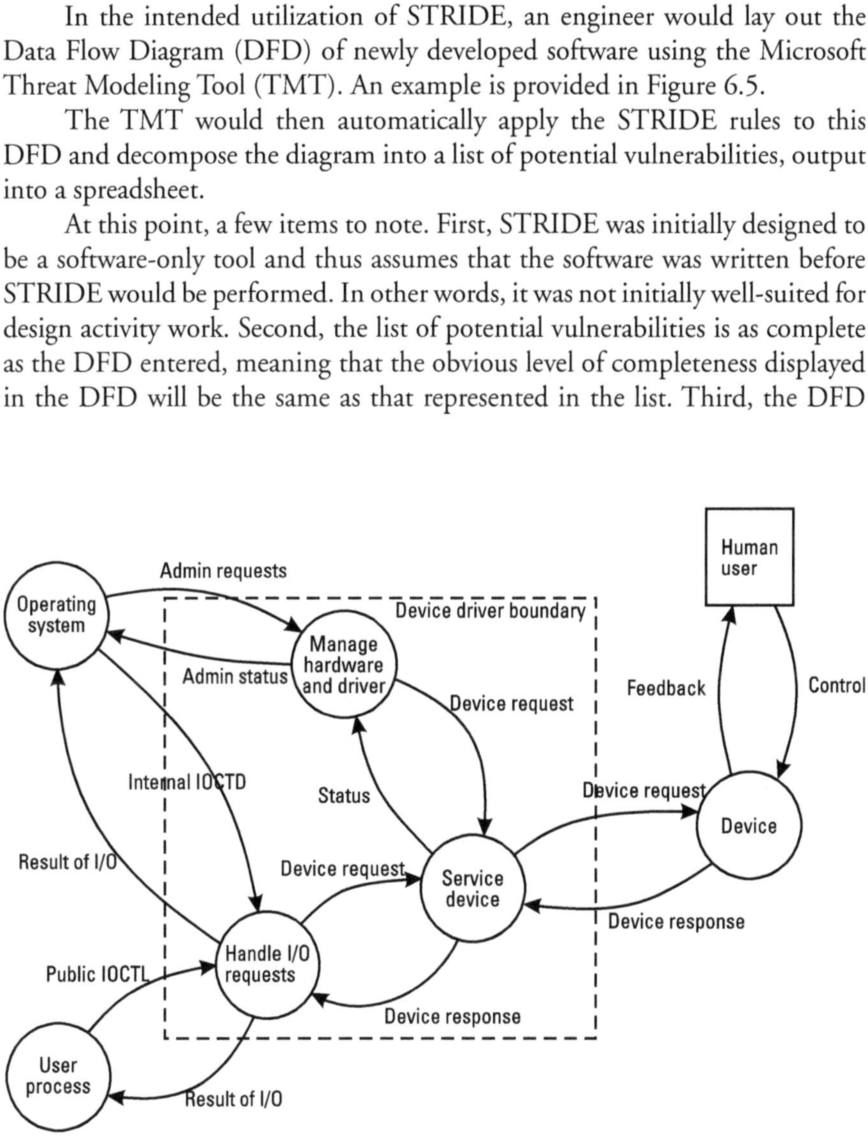

Figure 6.5 Sample TMT output using software data flow.

is formally decomposed into the list of potential vulnerabilities automatically, without regard for the security skill set of the engineer.

However this software-focused tool is very flexible and, with some small alterations, can be used during the design activities to create a list of potential vulnerabilities in the system-level design being proposed. Figure 6.6 provides a second example from the Microsoft TMT, this time representing a system-level topology with logical data flows.

At this system-level representation, a list of potential vulnerabilities is automatically created from the data flow diagram. The TMT accepts as input the overall system topology and any particular device details, including schematics of major components, data stores on smartphones, and the nature of utilized communication media. However, due to the abstract nature of STRIDE (e.g., it does not know the difference between a JTAG connection versus an ethernet connection), the list of potential vulnerabilities usually has about 30% false-positives and duplicates that will need to be removed through manual inspection.

6.5.4.2 Informal Approaches to Threat Modeling

Although we are promoting STRIDE as our preferred formal method of decomposing a design into design vulnerabilities, there are other methods in use, such as the informal decomposition of a design. These include whiteboarding sessions and threat trees. This section addresses some of these inferior approaches; reading it will not make you a better security engineer, but it might help you to avoid a few pitfalls.

Try this mental exercise. Consider a development project example, and imagine yourself in front of a whiteboard. Can you list all changes to be made to the software to prevent ransomware (a threat) from affecting your product? How do you know you have captured all of the necessary changes? Can you think of one more? A dozen more? When do you know you have a complete list?

Similarly, attack trees as commonly used are not helpful during product design. In addition to the reasons discussed under threat modeling, attack trees can also introduce a false sense of the attack surface. As an example, consider Figure 6.7, a simple use case of opening a locked door.

Is this a comprehensive threat model? It looks nice, and it seems that someone put some time into considering the possible attacks. Again, because it is an informal approach to the decomposition of a design, how do the leaf nodes lead to mitigations? Consider the "burst through door" leaf node—it leaves a lot of ambiguity as to the nature of the attack. Does it indicate the attacker is kicking the door in, using a battering ram, or charging through it with a tank? These have very different mitigations needed to protect against each separate type of attack.

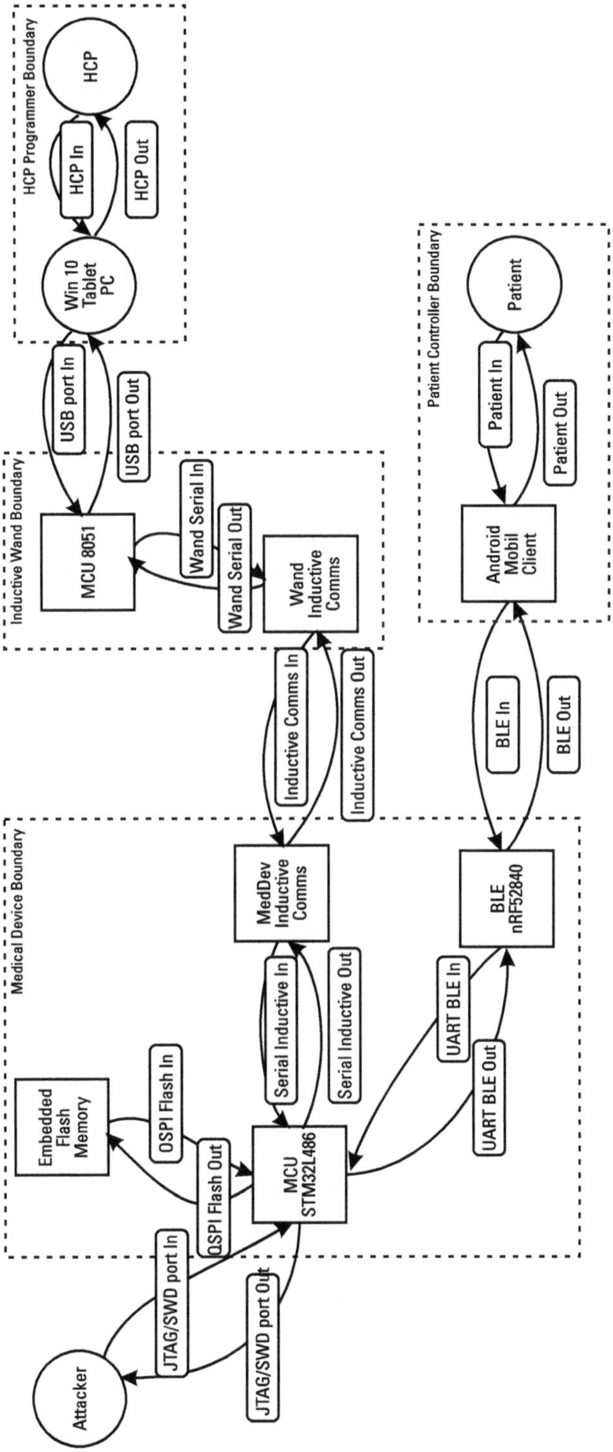

Figure 6.6 Sample TMT output using system-level topology with logical data flows.

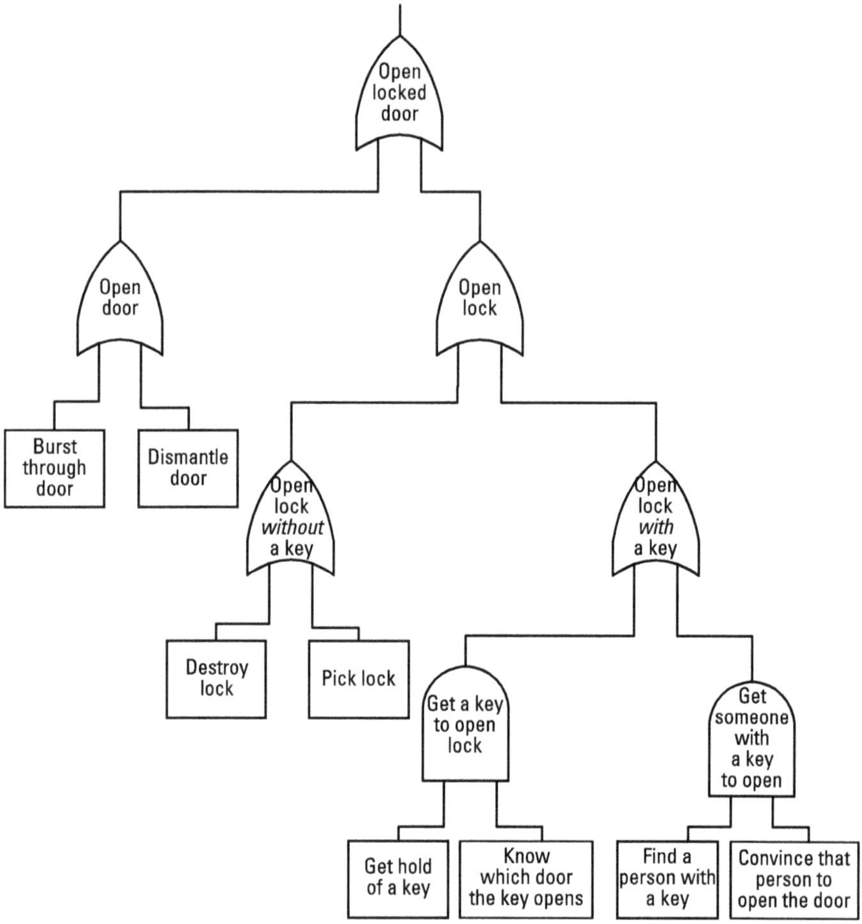

Figure 6.7 Locked-door attack tree. (Note that, for anyone unfamiliar with threat trees, the nodes with the arched undersides are "OR" nodes. The nodes with the flat bottoms are "AND" nodes. The boxes are all terminal leaf nodes.)

Also, how would we know whether the model is complete? The attacker could block the latch from engaging (the technique used during the Watergate break-in), socially engineer a situation where the door does not get locked in the first place, take a photograph of the key and reproduce the key from the photograph, or attack the lock with foil impressioning. Some of these are obvious omissions (such as blocking the latch), but others (reproduction from a photograph, foil impressioning) are attacks that would probably only be anticipated by security experts (or readers of crime novels).

Unfortunately, the majority of software engineers have had no training in security, so popular approaches which propose that informal threat modeling can consist of "whiteboarding," "brainstorming," "threat trees," and "trust boundaries" end up being performed by people with little to no knowledge, experience, or training in security. This yields results that can cause extra work to be performed, or worse, work being performed in the wrong areas of a device while needed work goes overlooked.

To be clear, informal threat modeling is useful when employed by cybersecurity subject matter experts (SMEs) to clarify and document their thinking on a given area of security. Unless being used by such an expert, this tool does not assist a staff software engineer who is in the process of developing a medical device.

As with all the tools and techniques presented in this book, these represent the current state at the time of this book's writing. We anticipate that many, if not all, of these, will be deprecated in the coming years; threat trees are one example where this will likely occur sooner rather than later. The excellent work that was recently conducted in the EU by the TREsPASS team, which demonstrated unique methods of data visualization for large data sets and utilized "attack tree linearization" and "attack clouds," holds promise for other tools that may build on their approach [24].

6.5.5 Unit Implementation, Integration, and Testing

Four major cybersecurity testing and assessment activities should be performed concurrently with implementation:

- Static application security testing (SAST);
- Fuzzing test of key communication media;
- Unit testing boundary analysis;
- Penetration testing.

Performing all of these tests concurrently with development ensures more accurate monitoring of project progress, easier implementation of fixes (we have just completed working on the software being tested, so we are still familiar with the code), and more complete coverage of mitigations for discovered vulnerabilities.

- *SAST:* Static analysis tools analyze software source code syntax and enforce secure coding standards. Various code security guidelines are publicly available, such as SEI CERT C and MISRA 2012 for a variety of computer languages. SAST should be performed often on the codebase to avoid a single huge effort needed to keep the codebase compliant. One of the best ways to accomplish this is to schedule it directly into the build cycle. Enforcing adherence to coding standards is good practice because it prevents poorly implemented software from becoming a vulnerability that may later be leveraged by an exploit. SAST is further discussed in Chapter 7.

- *Fuzzing:* Fuzz testing is a negative software testing methodology that feeds a program, device, or interface with malformed, unexpected input data in order to find defects. This malformed data can be randomly generated or can be based on experience and use data sets known to be problematic in a given scenario or for a given interface.

  Fuzzing should be performed on all communication media implemented in the system. These communication media can be wired or wireless and need not transport key or critical data to be suitable candidates for fuzz testing. Typical mediums to fuzz include Ethernet, Wi-Fi, Bluetooth low energy, USB, serial ports, proprietary low power RF, MedRadio, and Inductive.

  Fuzzing can take multiple forms, such as randomly altering packet structure and timing, or it can also attack the contents of the packet itself, such as the fields and values being received.

  Fuzz testing should be performed for as long a period of time as is practical to the project. Keep in mind that time is the measurement of the success of a device under test, so many millions of iterations should be allowed during the testing regime.

- *Boundary analysis:* Boundary value analysis is a black box type of testing, which stress-tests the supplied values (e.g., parameters) given to a function versus the intended range of values. As an example, consider a function that expects a passed signed integer value of 8 bits to have a value of –100 to +100. As such, there should be at least 6 defined test cases, exercising:
  - –100 and +100;
  - –101 and +99;
  - –99 and +101.
- *Penetration testing:* The goal of penetration testing is to get an independent indication of vulnerabilities that may exist in the system so that mitigating controls can be identified. Penetration testing is a negative black box form of testing where a security expert attacks the system in an unplanned and unstructured manner to identify any obvious vulnerabilities, as well as to provide an indication of the difficulty in exploiting said vulnerabilities. The amount of time needed by the security expert to exploit a vulnerability should be considered as valuable information in deciding if the vulnerability requires mitigation.

Penetration testing should not be left until late in the development lifecycle, as the testing can take significant time to perform, will produce an overwhelming number of findings, and the results of testing (detected vulnerabilities) become increasingly expensive to mitigate late into development with significant impact on cost and schedule. Instead, penetration testing should be performed concurrently with implementation activities, scoped to specifically defined aspects of the system as they are implemented.

As an example, the ethernet-based command and control structure of the device may have reached the status of "feature complete" while the rest of the device is still under development. So, penetration testing efforts would only attempt to compromise this ethernet interface, ignoring the other parts of the device. This results in multiple penetration tests being performed on separate sections of the device. Obviously, planning and scheduling are an important part of this type of testing.

Lastly, it is important that penetration testing is performed by an experienced entity that has no vested interest in the outcome of the test (e.g., a bonus depending on the timely release of the product). Pentesting is a task that requires a unique skill set and needs continual improvement. That is why it is generally advised to contract pentesting with an external entity. In-house testing has been so abused by manufacturers that the FDA in their latest guidance requires the inclusion of evidence in a premarket submission of the "independence and technical expertise of testers" [25], so be prepared to capture artifacts of independence and proper training for cybersecurity testing.

6.5.6 System Testing (V&V)

In a properly structured and executed secure development process, few security-related activities are performed during V&V testing. The only cybersecurity-related V&V testing should be for design vulnerabilities discovered during the cybersecurity risk management process (including threat modeling, software components, unresolved anomalies) that impact safety and efficacy, with subsequent confirmation by the risk management process.

Even if mitigations were implemented for these vulnerabilities, there should still be V&V testing coverage to ensure the safety of the device.

As a part of the risk management process, vulnerability scoring can be performed again during this phase. The scoring rubric utilized should be similar or ideally the same as the rubric used during the design activities. Traceability also plays an important part here, with tracing to be performed from cybersecurity requirements to test results, along with identified risks and implemented controls to test results.

6.6 Transfer to Manufacturing and Related Security Considerations

Congratulations! Your new product has passed all the hurdles of development, including all the security-related activities and regulatory approval. You now need to move the software elements into manufacturing.

The first cybersecurity-related hurdle to this transition is: Did your build system get compromised and subsequently infect your software deliverable? Are you about to manufacture and ship thousands of devices pre-infected with malware [26]?

The correct approach to preventing this is to utilize a tightly controlled build system maintained exclusively for this purpose. Preferably, this build system is offline and passes all infrastructure security checks, including virus detection, Host IDS, Network IDS, and IPS. The software deliverables should all be signed or have hashtag values for each of the deliverables.

6.6.1 Three Different Transfer Models

At this point, the options diverge into three possible approaches:

1. Have nonvolatile components commissioned and programmed with the deliverable at a value-added reseller (VAR) who has implemented secure handling of components.
2. Have the deliverables transferred into a secure programming environment in your company's internal manufacturing facility. There, deliverable integrity shall be confirmed on each programming iteration as well, and endpoint protection tools installed on these programming systems will block attempts to exfiltrate the deliverables out of the programming tools as would be performed during an insider attack. (Competitors may attempt to pay your employees to steal your intellectual property; therefore, freedom of movement for your software deliverables must be restricted.)
3. Have the deliverables transferred into a secure programming environment in a contract manufacturing (CM) facility. There, deliverable integrity shall be confirmed on each programming iteration as well, and endpoint protection tools installed on these programming systems will block attempts to exfiltrate the deliverables out of the programming tools. A certificate management system should also be put into place to allow remote issuance of certificates to be programmed into the devices. This management system is supported by auditing activities against the CM. As an example, if the certificate manager is commanded by the MDM to issue 1,000 certificates and when auditing is performed 950 units where shipped, there had better be 50 units

awaiting destruction (or rework) by the auditing team. This prevents the CM from creating "off-the-books" devices for their own profit.

6.6.2 Production

Transferring design to manufacturing and maintaining security during production is a critical task. Additional considerations need to be made to secure products during the production process, postmarket servicing, and rework on returned products. During the design activities of the development lifecycle, the needs of the production-line requirements need to be considered, such as the anticipated volume of units that will be built each year, the number of lines to be supported, the geophysical location of the lines, the time required to manufacture each unit (takt time), the communication medium used to connect to devices during the production process (e.g., for testing or software installation), or the number of software applications needed on the device (including testing routines needed for production as well as final production code), to name a few.

Further considerations related to securing the manufacturing process include:

- If you are uploading new software images to a unit being built and the communications medium is slow (such as BLE), do you have the time to perform each upload while maintaining secure processes?
- Will unique identifiers be written to each device, or is the identifier based on another variant value, such as the serial number? How is device software versioning (down to a security-relevant level) managed?
- Does the device include cryptographic security technology and, if so, how are unique symmetric key or unique asymmetric key pairs written to or generated by the device so that they are not at risk of exposure or tampering during provisioning?
- How will these cryptographic values be managed? Will they be local or hosted? How will they be managed across possible multiple production locations?
- The use of a back-end server means there has to be an IP connection to the production line, which incurs an additional risk of hacking a production-line workstation, manufacturing equipment, or even products in production.
- Is there some out-of-band (OOB) medium being utilized, such as a 2-dimensional (2-D) barcode that needs to be printed and associated with the correct unit being manufactured, and if so, how is that matched with the device's security-relevant software version?

- Are the programmable components being fully programmed by a VAR before they reach the production line? Are they completely blank components? Are they being programmed by a VAR with just the bootload and software update agent, for loading on the production line?
- At what step in the production line is the JTAG ("debugging port") being disabled?
- At the end of the line, are the units being placed into a low power mode for storage? Is there a risk that they could be awoken by an attacker to deplete the battery life while in storage?
- If there is just one software image loaded into the programmable components, which includes the production line's testing routines, how are these routines permanently disabled at the end of the production line?
- If separate software is installed for testing or calibration functions, how is it removed or the function disabled at the end of the production line?

These types of considerations should be evaluated jointly by the cybersecurity design support and the production teams. Joint evaluation should cover the range of ensuring secure manufacturing infrastructure and processes, secure engineering-to-manufacturing transfer, and security features designed into the device to support these activities. The production team will be in a better position to evaluate the production impact of cybersecurity controls and design mitigations being proposed to be performed by, or on behalf of, production. In situations where personnel from the specific production organization should not be involved in a joint review—for example, because the main production will be handled by a contract manufacturer who should not be exposed to the design process—the MDM should involve an in-house or independent production engineer in the cybersecurity planning process, followed by ongoing audits of the contract manufacturer to ensure that all security processes are being performed and documented.

6.6.3 Production-Line Functionality Left Enabled in a Shipped Device

One of the most common cybersecurity mistakes MDMs make is to leave production-line functionality (features that support testing or calibration) enabled in a shipped medical device. This has resulted in many attacks that have been documented by security researchers, and, in several cases, this has led to a great deal of bad publicity for the MDM.

There are two failures that result in this vulnerability:

1. Leaving the production functionality enabled in a shipped device. The production line does have support needs for nonstandard operating conditions. These needs can include calibrating sensor inputs and driving the device to extreme limits of delivery or mechanical position with the removal of all alarms and safeties. This type of unrestricted functionality, while necessary in a production environment, can have a devastating effect on a patient, and is often trivial to exploit in a fielded device. Ease of exploit is frequently compounded by a total lack of authentication for accessing production-line functionality—either because nobody even considered implementing authentication, or because the additional time requirement for authenticating to each device during production was deemed too costly.

   The bottom line is as simple as this: If any functionality is not related to the normal essential performance of the medical device, then that functionality should be disabled before shipping.

   Fortunately, mitigating this issue is quite straightforward, and there are multiple ways to address it. One option is developing dedicated production software that is loaded into the device exclusively for certain tests and is overwritten upon successful completion of them. However, reprogramming the device on the production line may not be desirable due to the negative effect on the throughput of the production line, as reprogramming a device can be a time-consuming operation.

   Another option is to include production-line functionality in the final fielded firmware, but to disable it once that functionality is no longer needed. At this point, a production line station would issue a command that permanently disables the production-related functionality. Note that, to be effective, this must be a one-way command with no facility to reverse the disabling of the production-related functionality.

   As with the previous option, the production functionality would be included in the final fielded firmware and remain enabled, however, this time with cryptographically significant authentication and authorization required before any functionality can be activated.

2. The other process failure in this scenario is the lack of consideration and direct support for the production line in the design of the device. All production-related and reworking functionality needs to be considered and directly supported in the initial design of the device, not as an afterthought. A common example of this bad practice is leaving the JTAG port enabled in a fielded device.

6.6.4 Factory Service and Rework

Some thought should also be given to the remanufacturing and factory-level service of the medical device, including repair or upgrade of devices that are still in storage and have not been shipped yet, repair or upgrade that is too complex to perform in the field, and forensic analysis of a returned device that was exposed to a cyber incident (see Chapters 8 and 9). The design will need to support such functionality as:

- Uniquely secured (i.e., not a universal password) functionality to download logs and other forensically relevant data from a device (as JTAG is disabled).
- Uniquely secured functionality to return a device to factory default values and erase all PII and PHI that are stored in the medical device.
- Uniquely secured functionality to return a device to a low power mode for storage.
- Uniquely secured functionality to return a device to either the production line test functionality or a separate service level test functionality.
- Utilizing firmware updating functionality for devices being serviced (further discussed in Chapter 8. The methodology described is also applicable to the factory, with some adjustments). Alternatively, design new secure functionality to accomplish the update.
- Uniquely secured functionality to place a device into a configuration mode, such as supplying Wi-Fi credentials when a medical device is being installed in a given working environment.

Again, it is recommended that the MDM involve production engineering during design so that their experience can be used to help to design the secure pathways needed for service and rework.

6.6.5 Securing Production Infrastructure

Another critical area that requires attention is the security posture of the production environment itself. Responsibility for ensuring protection of the production environment typically lies with the enterprise security team or a dedicated operational security team within. From a cybersecurity perspective, there are several objectives to be met:

- Protect the production environment from cybersecurity compromise to avoid costly shutdowns and prevent costly recalls or regulatory notifications.

- Ensure the integrity of the finished product so that: (a) only approved configurations are shipped, and (b) the product itself is not compromised by malware or backdoors that could be exploited by an adversary.
- Correct and timely management and phase-in of product changes that occur in the normal lifecycle of a product or may occur in response to and as part of the mitigation of a newly discovered vulnerability.
- Protect products from security compromise through the hardware or software supply chain.

All of these are related and require a holistic approach to ensure that production security and integrity are met and maintained. If only one of the above is absent, the finished product is at heightened risk of a costly cyber compromise. A cybersecurity risk assessment and management approach specific to the production environment is an advisable approach.

There are differences depending on the type of product. A hardware medical device with embedded software is different from a pure software medical device (SaMD). Furthermore, although not a topic of this book, production security and integrity considerations should also be applied to nonsoftware products in the biotech or pharma industries as a production-line security compromise could result in shutdown or product quality impact. For example, in 2017, the NotPetya ransomware outbreak at a leading pharmaceutical manufacturer resulted in steep financial losses as well as shortages of drugs and vaccines and, to make matters worse, resulted in a legal battle with the cyber insurance (that had classified NotPetya as an Act of War and therefore tried to exclude it from coverage) to obtain coverage [27]. Consequently, many insurers have rewritten their policies or raised the premiums for their cyber insurance plans [28].

As in all areas of cybersecurity, a mature approach requires that we address "people, process, and technology." Production staff needs to be properly trained on not only specific secure-build processes but also on general cybersecurity hygiene and best practices. They should not, for example, use a production-line PC to check their personal email over lunch (this and similar activities should also be prevented by technical means); neither should they, although understandable, bring in a USB data carrier preloaded with their family or vacation photos to use as a wallpaper or screen saver.

The key to ensuring that staff can execute in a way that does not compromise security is properly defined and assigned roles and responsibilities. Who is responsible for which areas of security (e.g., deploying and managing security technology, installing new production equipment, performing data transfer in and out of the production environment)? Sustaining engineering has a key responsibility in addressing security-related issues and tasks as they arise during the production process.

Based on individuals' roles and responsibilities, regularly scheduled training refreshers as well as updates on emergent security topics of relevance should be provided and documented. Furthermore, staff need to have the ability (and confidence) to detect potential cyber risks and compromise and take appropriate action.

From a process perspective (summarized in Figure 6.8), the areas that need to be addressed are:

- Receiving engineering transfer of new products or updates.
- Processes to secure license, key, or certificate lifecycle management and deployment during production (further discussed in Chapter 12). Insufficient lifecycle management processes or security of the license, key, or certificate infrastructure can result in devices with compromised security (when using cryptography-based security technologies) or financial losses (loss of revenue due to compromised licenses).
- Managing change during the production process (e.g., phasing in updates as a result of a newly discovered vulnerability, which leads to engineering change and requires mitigation before the finished product ships). Note that, for the purpose of product change management during production, a manufacturer may define a "point of no return," after which changes will no longer be implemented at the factory and will re-

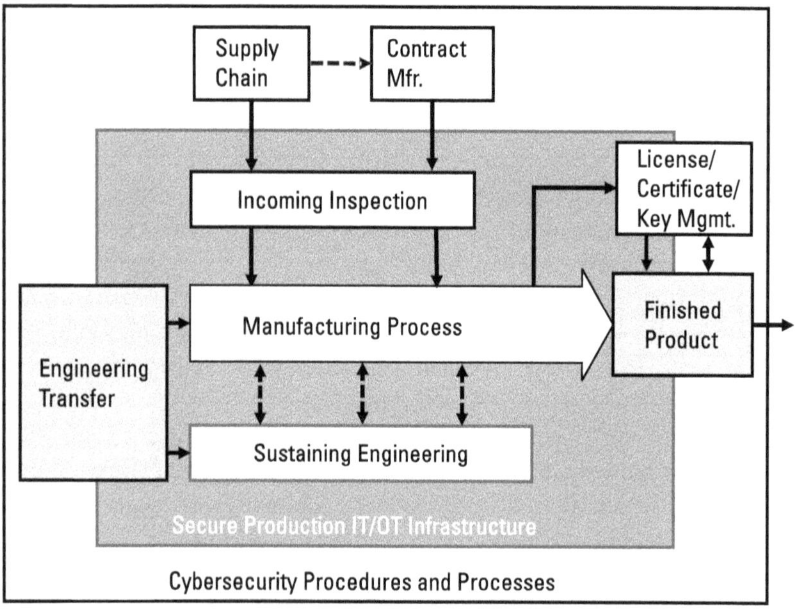

Figure 6.8 Securing production infrastructure.

quire a field update. This will, obviously, depend on the type of product, production process variables, and the type of change.

- Production V&V (engineering V&V asks "Did we design the right product?" whereas production V&V asks "Are we building the right product?"). Includes test runs and audits in cooperation with customers (to test the entire chain including shipment, delivery, installation, documentation, and license or key deployment).
- Production environment security policies and procedures (how to keep the production environment secure).
- Specific policies and procedures addressing staff onboarding, termination, and training, as well as supportive processes such as incident response.

From a technology and infrastructure perspective, we need to assume that the production environment is a mix of IT (workstations, servers, storage) and OT (SCADA and ICS devices, robots) integrated on a dedicated production network. Serendipitously, many of the security challenges in the OT environment are similar to that of the medical device ecosystem itself. For example, we often find an older infrastructure that, from a security perspective, is difficult to maintain and protect. A well-planned and well-executed defense-in-depth approach is the answer, ensuring security across all control points, complemented by network architecture and network-based security to compensate for the remaining vulnerabilities that cannot be addressed by commercial security technology. Applicable solutions are:

- Endpoint security for traditional IT devices in the production environment (workstations, servers, smart devices). However, due to the complexity of today's endpoint security products, they are typically not suitable for nontraditional endpoints with limited system resources and the need for highly predictable behavior.
- Host intrusion detection and prevention systems as an alternative security approach for IT endpoints (more complex to implement but lower on system resources and resulting in a more secure posture) as well as OT/IoT endpoints (may require cooperation with the manufacturer).
- Network-based security technologies ranging from traditional firewalls, (network) intrusion detection and prevention system ((N)IDS/(N)IPS), security gateways, and anomaly detection solutions can all be applied to improve the overall security posture of an enterprise at the network level.

- Data loss prevention (DLP) to monitor and prevent loss of confidential data such as source code or other intellectual property.
- Security management, including security infrastructure management console, security operations, and incident detection and response capabilities.
- Other security technologies should be applied as appropriate, for example, network segmentation, virtual private networks (VPN), and multifactor authentication (MFA) for remote access as well as on an as-needed basis, data-at-rest, and data-in-motion encryption.

In today's security environment of targeted, sophisticated, and stealthy attacks conducted by highly capable cyber criminals and adversarial nation states, none of the above by itself will provide sufficient security. Defense in depth requires the combination and cooperation of most if not all of the suggested technologies.

Furthermore, in combination with the above technologies the production environment requires a secure network architecture that is properly segmented (using traditional VLAN or more recent software-defined network (SDN) approaches) and minimizes integration points (which, in turn, minimizes security exposure) with other enterprise IT systems. For example, the production network may need to have integration with other enterprise systems (although some may only need to be temporary) for engineering-to-production transfer, or enterprise resource planning/management (ERP/M) systems such as scheduling, inventory, materials, and order management. However, unnecessary integrations should be avoided, for example, between the production network and general business systems, like email. Certainly, a production network should not be connected to a coffee machine [29].

Some additional areas that typically require attention, yet may be overlooked in the production environment, are:

- Incident response, typically includes the key steps of planning, practice, identification, triage, containment, eradication, recovery, and follow-up. Good forensic data collection is critical to, for example, understanding the extent and duration of a compromise and determining if the last 24 hours of production were affected, the last week, or longer. Another important aspect of incident response is the communication with and responsibilities of stakeholders external to the production organization: legal, regulatory, executive management, and, if warranted, customers, and users. Incident response is further discussed in Chapter 9, specifically concerning the medical devices themselves. Some adaptation and flexibility may be required to extend product incident response principles to address incidents in the production environment. A consensus standard on incident response for medical device production environments is currently being developed by the HSCC.

- Physical security of the production environment to prevent malicious and nonmalicious tampering with infrastructure and products. For example, the last thing you want is for an eager engineer to enter the production environment with a critical software update or test tool on a USB stick.
- Data carrier handling. Data carriers, such as USB thumb drives, are extremely useful but are also known to transfer malware into even air-gapped or heavily isolated networks. Data carriers may be used for internal purposes (e.g., software or documentation transfer) or external purposes (e.g., maintenance of production equipment). Ideally, any production environment should be able to function without requiring sneaker nets, but there may be practical limitations and exceptions. For that purpose, strict procedures should be in place to prevent the contamination of the production environment. This should include strict handling requirements, hardware limitations, and prescanning of data carriers on stand-alone scan stations.
- Integrity and security maintenance of the production environment requires the timely deployment of updates and patches and as needed, cooperation with the vendor of the equipment in use (SCADA, ICS, robots).

The above describes the most generic production environment as an example. Special considerations apply depending on the specific situation as it may include a geographically distributed production infrastructure, inclusion of contract manufacturers, and supply chain of off-the-shelf components, as well as access requirements for suppliers, vendors, or other partners.

It is essential to properly manage any third party that interacts with the production environment, such as the ICS service provider or a supplier of hardware or software components. As part of the incoming inspection, checks of security critical properties should be applied (e.g., assessment of component version and compatibility, or performing malware scans). Components should have the ability to be inspected for their integrity (e.g., through the use of checksums or hashes for software).

In addition, suppliers and contract manufacturers should be contractually required to comply with established security requirements and procedures, and compliance should be confirmed by audit.

Unfortunately, we are aware of several cases where manufacturers have been compromised by their supply chain or where manufacturers delivered devices to HDOs that had malware installed on them "out of the factory." This highlights the importance of a secure and well-managed production environment to ensure the integrity and quality of the finished product.

References

[1] TrapX Labs, "Anatomy of an Attack: MedJack (Medical Device Hijack)," May 7, 2015, https://www.cs.tufts.edu/comp/116/archive/fall2018/smeggitt.pdf.

[2] Department of Health & Human Services (DHHS) Centers for Medicare & Medicaid Services (CMS), *Pub. 100-07 State Operations, Transmittal 103*, February 21, 2014, https://www.cms.gov/Regulations-and-Guidance/Guidance/Transmittals/Downloads/R103SOMA.pdf.

[3] Mujezinovic, D., "AIDS Trojan: The Story Behind the First Ever Ransomware Attack," *Make Use Of*, November 28, 2021, https://www.makeuseof.com/aids-trojan-the-first-ransomware-attack-in-history/.

[4] Brunau, C., "Common Types of Ransomware," *Datto*, August 22, 2019, https://www.datto.com/blog/common-types-of-ransomware.

[5] Center for Internet Security, "Security Primer—Ransomware," 2020, https://www.cisecurity.org/white-papers/security-primer-ransomware/,

[6] Whitwam, R., "Australian Police Warn of Ransomware USB Drives Showing Up in Mailboxes," *ExtremeTech*, September 22, 2016, https://www.extremetech.com/computing/236157-australian-police-warn-of-ransomware-usb-drives-showing-up-in-mailboxes.

[7] https://www.first.org/cvss/v2/guide.

[8] NIST Joint Task Force, Computer Security Division, "800-30 Rev. 1 Guide for Conducting Risk Assessments," September 2012, https://nvlpubs.nist.gov/nistpubs/legacy/sp/nistspecialpublication800-30r1.pdf.

[9] NIST Joint Task Force Transformation Initiative, "SP 800-30r1: Guide for Conducting Risk Assessments," September 2012, https://csrc.nist.gov/publications/detail/sp/800-30/rev-1/final.

[10] NIST Joint Task Force Transformation Initiative, "SP 800-30r1: Guide for Conducting Risk Assessments," September, 2012, p. 81, https://csrc.nist.gov/publications/detail/sp/800-30/rev-1/final.

[11] FIRST CVSS Special Interest Group, "Common Vulnerability Scoring System v3.1 Specification Document," June 2019, https://www.first.org/cvss/v3.1/specification-document.

[12] LeMay, E., K. Scarfone, and P. Mell, "The Common Misuse Scoring System (CMSS): Metrics for Software Feature Misuse Vulnerabilities," July 2012, https://csrc.nist.gov/publications/detail/nistir/7864/final.

[13] Scarfone, K., and P. Mell, "The Common Configuration Scoring System (CCSS): Metrics for Software Security Configuration Vulnerabilities," December 2010, https://csrc.nist.gov/publications/detail/nistir/7502/final.

[14] Chase, M., and S. Christey Coley, "Rubric for Applying CVSS to Medical Devices," September 2019, https://www.mitre.org/publications/technical-papers/rubric-for-applying-cvss-to-medical-devices.

[15] Bodungen, C., "Industrial Vulnerability Scoring System (IVSS)," https://securingics.com/IVSS/IVSS.html.

[16] The OWASP Foundation, "The OWASP Risk Assessment Framework," https://www.owasp.org/index.php/OWASP_Risk_Rating_Methodology.

[17] Jacobs, J., et al., "Exploit Prediction Scoring System (EPSS)," August 31, 2019, https://arxiv.org/abs/1908.04856.

[18] QED Secure Solutions, "Risk Scoring System for Medical Devices (RSS-MD)," https://www.riskscoringsystem.com/medical/techspecmedical.pdf.

[19] U.S. Department of Health and Human Services, "Guidance Regarding Methods for De-identification of Protected Health Information in Accordance with the Health Insurance Portability and Accountability Act (HIPAA) Privacy Rule," https://www.hhs.gov/hipaa/for-professionals/privacy/special-topics/de-identification/index.html.

[20] Bode, K,, "Researchers Find 'Anonymized' Data Is Even Less Anonymous Than We Thought," *Vice*, February 3, 2020, https://www.vice.com/en_us/article/dygy8k/researchers-find-anonymized-data-is-even-less-anonymous-than-we-thought.

[21] Ehrenkranz, M., "Researchers Reveal That Anonymized Data Is Easy to Reverse Engineer," *Gizmodo*, July 23, 2019, https://gizmodo.com/researchers-reveal-that-anonymized-data-is-easy-to-reve-1836629166.

[22] Campbell-Dollaghan, K., "Sorry, Your Data Can Still Be Identified Even If It's Anonymized," *Fast Company*, December 10, 2018, https://www.fastcompany.com/90278465/sorry-your-data-can-still-be-identified-even-its-anonymized.

[23] Thompson, S., and C. Warzel, "How to Track President Trump," *The New York Times*, December 20, 2019, https://www.nytimes.com/interactive/2019/12/20/opinion/location-data-national-security.html.

[24] https://visualisation.trespass-project.eu/.

[25] FDA, "Cybersecurity in Medical Devices: Quality System Considerations and Content of Premarket Submissions Guidance for Industry and Food and Drug Administration Staff," September 27, 2023, https://www.fda.gov/media/119933/download.

[26] Perlroth, N., "Microsoft Finds PCs That Ship Pre-Infected," *The New York Times*, September 13, 2012, https://archive.nytimes.com/bits.blogs.nytimes.com/2012/09/13/microsoft-finds-pcs-that-ship-pre-infected/.

[27] Vittorio, A., "Merck's $1.4 Billion Insurance Win Splits Cyber from 'Act of War'," *Bloomberg Law*, January 19, 2022, https://news.bloomberglaw.com/privacy-and-data-security/mercks-1-4-billion-insurance-win-splits-cyber-from-act-of-war.

[28] Jones, D., "Changing Cyber Insurance Guidance from Lloyd's Reflects a Market in Turmoil," *Cybersecurity Dive,* August 29, 2022, https://www.cybersecuritydive.com/news/lloyds-cyber-insurance-exclusions/630535/.

[29] u/C10H15N1, "How the Coffee Machine Took Down a Factory's Control Room," *r/talesfromtechsupport,* 2017, https://www.reddit.com/r/talesfromtechsupport/comments/6ovy0h/how_the_coffeemachine_took_down_a_factories/.

7

Documentation and Artifacts

During the development lifecycle of a medical device, a large amount of documentation needs to be created, reviewed, approved, controlled, and managed. This is not only a good engineering practice, but it is also required by most regulators, for example, as found in the U.S. 21 CFR Part 820 Subpart D [1] and ISO 13485:2016 [2]. While there are excellent electronic tools available to perform requirements management (e.g., Jama Connect), it is still fairly common to see small MDMs using standard office productivity suites (e.g., Microsoft Office) as their tools of choice to manage requirements. It does not matter if an MDM uses a "paper-based" approach or a commercial tool, the content represented in this chapter is equally valid for both the simple and more advanced requirements management systems.

Over the course of the past 10 years, we have seen those MDMs who were early adopters of secure medical device development practices completely change how security artifacts are managed in their organizations. These artifacts represent the output of all the security activities, including plans, reports, worksheets, test results, and summaries. In the past, these artifacts were extremely tightly controlled, with very limited distribution, even encrypted and gated with "eyes only" types of access. During the past decade, however, we have learned that little is gained by such secrecy, while much is lost in the productivity and growth of the organization's teams and overall security posture, as well as the customers' opinion of the MDM.

Today, it is common practice to share security artifacts throughout the MDM's organization and even with third-party contractors. Due to changes in regulatory guidance, much of the content found in these artifacts is now available to regulatory agencies, as well as HDOs and other end users. We have increasingly realized that keeping security measures secret generally results in an overall worse security posture, not better. By the same token, sound security measures do not become less secure if stakeholders are aware of them. Locked doors remain locked, even if the door manufacturer publishes a statement declaring "this door is locked with a deadbolt." We are now seeing this principle codified in common lifecycle management practices.

7.1 Overview of Secure Development Deliverables

While the following content elements need to be incorporated into the device's design history file (DHF), the vehicle that contains each of these elements certainly does not need to adhere to the structure below. However, this structure does work and aligns with typical development activities and flows.

In a sense, these artifacts are an output of the Development Lifecycle and its supporting processes. When defining and implementing structured processes (as described in other chapters), the MDM must anticipate the need for processes to produce artifacts that facilitate auditability, documentation, and transparency.

The following list of artifacts are all created during the development lifecycle (Chapter 6) and may be used for engineering purposes, test planning and execution, regulatory submission, customer documentation and support, and postmarket management and to inform future development (e.g., next-generation devices). While all of these elements must be entered into the DHF, as with any other set of DHF artifacts not all will necessarily be required for all activities. Artifacts can include:

- Security risk management plan;
- Design and supply chain vulnerability assessment;
- System security architecture;
- Interface control document (ICD);
- Testing reports;
- SBOM (includes software end of life and end of support information);
- System security report;
- Labeling, including technical service manuals and Instructions for use (IFU) and supporting infrastructure requirements;
- Instructions on device security event/incident response, fail-safe mode(s), and restoration (as appropriate for device capabilities and use case, further discussed in Chapter 9);

- Maintenance instructions, including information on how to obtain and install updates and patches (further discussed in Chapter 8).

It is assumed that each of these deliverables contains the usual boilerplate sections, such as purpose, scope, glossary, references, approval, and a traceability matrix where applicable (to demonstrate traceability from security requirements and identified risks to implemented controls and test results).

Additionally, a risk assessment also needs to be performed for the device as a whole, and subsequently on any newly discovered vulnerabilities, to determine if and to what degree they could result in patient harm. While such vulnerabilities would be discovered during the secure development lifecycle, the assessment, triage, supporting documentation, and testing that may result from such discovery are performed by the risk management organization, not by engineers or security experts. Good references for security process and risk management include AAMI TIR 57, AAMI TIR 97, AAMI SW96, and the Joint Security Plan version 2 (JSP2) from HSCC.

7.2 Security Risk Management Plan

This artifact is referred to by several different names, both in the FDA's September 2023 Premarket Guidance and in eStar, including alternate designations such as the "Cybersecurity Management Plan." Regardless of what it is called, this artifact defines, for a given project, all the security-related activities required for development. As of the most recent finalized guidance from the FDA [3], it must also cover how the manufacturer intends to support its postmarket activities. Such a plan may be defined via corporate standards that cover some or all of these items but must additionally detail any modifications or additions for a specific project. The activities defined include: what is being performed; what part of the system under development is being exercised; and during which phase of the total product life cycle the activity is being performed.

While each activity is documented, the manner of grouping the content together into reports can be flexible.

The security risk management plan should include:

- *Security goals:* Defines the voice of security's goals, similar to voice of customer or voice of business. These are not requirements (requirements should never be expressed in a plan), but instead high-level objectives from which requirements can be derived in subsequent documents.
- Justification of a standardized secure development framework as the basis for the secure development activities in the given project. Recognized

standards include IEC 81001-5-1 Health Software—Activities in the Product Life Cycle and NIST SP 800-218 Secure Software Development Framework (SSDF). IEC 81001-5-1 is specific to medical device software while SSDF is a more generic software development framework. Each of the activities should be specific in the plan to assist with the assigned roles and responsibilities in the next section of the plan.

- *Cybersecurity roles and responsibilities:* For each of the activities described in the secure development process, designate which role(s) are responsible, accountable, consulted, and informed. This is most easily achieved with a traditional RACI diagram.
- *Definition of TPSC activities:* For all TPSC, there needs to be descriptions of their related activities, such as:
  - How SBOMs are created and distributed to end users.
  - How SBOM vulnerability monitoring is performed, and the periodicity of monitoring.
  - How CISA's KEV Catalog monitoring is performed, and the periodicity of monitoring.
  - How discovered vulnerabilities are processed.
  - Description of the process for addressing when a software component has reached the end of life.
  - If the manufacturer does not have access to source code, then describe how the manufacturer can protect the source code in case the component's vendor goes out of business (e.g., utilizing an escrow to hold the source code in the event that the vendor is no longer available).
- *Patches and updates:* Description of how updating and patching of devices in the field is performed. Updating is when a trigger event occurs (adding features, mitigating vulnerabilities) that creates a new version of the device software; patching is a periodic event (e.g., quarterly) providing software updates from third-party software vendors.
- *Intended operative environment:* Describe the intended operating environment of the medical device.
- Describe the postmarket SBOM management processes (including sources, methods, and frequency) for:
  - Monitoring for new vulnerabilities;
  - Processing new vulnerabilities (describe how they will be identified, scored, and prioritized);
  - Mitigation (including communications and update distribution);

  - Criteria and procedure for processing a vulnerability under CAPA, if needed;
  - Coordinated vulnerability disclosure (CVD) process, including references to any internal or external standards used to guide or govern this process.
- Design vulnerability assessment and asset threat modeling;
- Process threat modeling (supply chain, manufacturing, deployment, interoperation with any other devices, maintenance, updating, and decommissioning of the device);
- System security architecture (results of all threat modeling and security architecture views);
- Development testing (fuzzing or malformed inputs, penetration, static analysis, known vulnerability scanning, anomaly testing);
- Periodic postmarket testing performed every 6 to 12 months until reaching the end of support for this device (note that there have been indications from the FDA that they are expecting continued cybersecurity activities even after the end of support has been reached):
  - Malformed inputs or fuzz testing;
  - Attack surface analysis;
  - Anomaly testing;
  - Closed box testing of known vulnerabilities;
  - Integrity testing of executables or binaries used in manufacturing (including software composition analysis of the binary);
  - Static analyses of the source code;
  - CISA's KEV Catalog scanning;
  - Penetration testing.
- Security risk management report summarizing the results for all development cybersecurity testing activities and final assessment at the end of development.
- Description of the cybersecurity content in device labeling, including:
  - Recommended cybersecurity controls;
  - Diagrams of cybersecurity implementations;
  - Communication mediums being utilized;
  - Any infrastructure requirements;
  - Human readable SBOM;
  - Software update processes.

7.3 Design Vulnerability Assessment

This includes adaptation of threat modeling tools for system modeling [4], itemizing vulnerabilities, screening for false-positives, and scoring the remainder. Further details on these activities and their inputs and outputs are provided in Chapter 6.

7.4 Risk Management Report

Developed during the early design phase, this artifact is the result of several other activities, broadly categorized as:

- Formal decomposition of the system and scoring of the vulnerabilities discovered, as described in Section 7.3 and in detail in Chapter 6.
- Investigating mitigations to be applied to the highest-scoring vulnerabilities.

(The artifacts created during decomposition and scoring may be kept for the DHF but are not typically needed for submission, as the results of these activities will be summarized in this architecture document.)

The risk management report is one of the most important artifacts, as it guides the design and implementation of security in the medical device. Developing it early will minimize impacts on the device design and development process, as well as minimizing end-user burden for security controls.

The architecture document should contain:

- Overview of the system topology used for decomposition;
- Cybersecurity hygiene best practices [5];
- A description of the decomposition process, for example, a description of STRIDE;
- A description of the scoring rubric, for example, CVSS v2;
- A section containing all the high-scoring vulnerabilities; each vulnerability should itemize:
    - The vulnerability identification number from the threat modeling tool;
    - The environment of the vulnerability (the area in the system diagram where the vulnerability exists);
    - Name of the vulnerability;
    - Category (e.g., which letter of the STRIDE methodology?);

- Description of the vulnerability;
- Indication of potential impacts to safety and efficacy;
- The pre-mitigation vulnerability score;
- Trace to mitigation(s) that address this vulnerability.

- A section containing all the recommended mitigations. These mitigations should be highly detailed technical descriptions, including diagrams, algorithms, and libraries to utilize and should not be limited to software-based mitigations alone. It includes:
  - Hardware-based mitigations, such as the selection of a microcontroller that supports hardware acceleration of cryptographic operations;
  - Guidelines for hardening operating systems and other commercial software, as applicable;
  - Physical security controls including side-channel protections;
  - Supporting security infrastructure (e.g., PKI, patch distribution, signature updates) and applicable constraints.
- State any assumptions about the intended environment of use.
- A rationale for the various threat modeling methodologies utilized.
- Process threat modeling, including:
  - Supply chain;
  - Manufacturing;
  - Deployment;
  - Interoperation;
  - Maintenance/update activities;
  - Decommission activities.
- Security architecture views provide the FDA reviewer with easily understood diagrams along with descriptive text on cybersecurity in four areas:
  - Global System View;
  - Multipatient Harm View;
  - Updateability/Patching View;
  - Security Use Case Views.

7.5 Interface Control Document

While technically not a security document and one that will not be provided to the FDA via eStar, an interface control document (ICD) is a highly relevant and valuable technical document that covers all aspects of each communications interface called for in the device design, including the technical aspects of security and security mitigations for a given communications medium. Sequence diagrams, algorithms, and timings are itemized in this artifact.

Tracing is commonly performed from this document to the system security architecture and detailed design documents. This can be an extremely important artifact when addressing questions from the FDA related to a premarket submission.

Security-relevant sections in the ICD include:

- Overview;
- Description of the communications medium and standard;
- Assumptions, constraints, or risks pertaining to:
  - Environment of operation;
  - Unsupported environments;
  - Expected environmental interferences;
  - Expected performance;
  - Communication quality of service (QoS);
  - Inherent risks of this communication medium;
  - Robustness (e.g., can it withstand port scanning?).
- Use cases;
- General interface requirements;
- Security Features:
  - Authentication;
  - Authorization;
  - Encryption;
  - Integrity;
  - Anti-replay.
- Detailed interface requirements: all interactions detailed to the lowest level of abstraction.

7.6 Cybersecurity Risk Assessment

This assessment further describes the results of cybersecurity activities performed during design, including:

- Capture of risks and controls identified from the threat model;
- Vulnerabilities assessed according to the device and the environment;

- Rubrics used for scoring the vulnerabilities pre-mitigation and post-mitigation;
- Associated risk acceptance criteria;
- Method for transferring security risks into the safety risk assessment;
- Residual risk assessment after security vulnerabilities and controls are applied;
- Assessment of vulnerabilities identified during cybersecurity testing;
- Description of the risk-scoring rubric utilized.

7.7 SBOM Support Report

This report documents the level of support and the expected end of life for each component in the SBOM. Most of this information will not be known or obtainable, so a section must be included that details how, in the future, your organization will manage components that are no longer available or no longer being maintained by the OEM.

7.8 Software Component Risk Management Report

This report describes the organization's response to identified risks in software components, substantiating the following:

- Vulnerabilities found in the CISA KEV catalog are eliminated.
- Risk assessments for each known component vulnerability are included.
- Security mitigating controls are utilized for each component vulnerability.

7.9 Unresolved Anomalies Risk Management Report

For each anomaly (i.e., "bug") identified in the product, a security risk assessment needs to be performed and the results need to be documented in this report.

7.10 Cybersecurity Metrics Report

There are three metrics that the FDA is requiring MDMs to maintain on an ongoing basis:

- Percentage of identified vulnerabilities that are patched;
- The time periods from vulnerability identification to remediation;
- The time periods needed for update rollouts.

7.11 Cybersecurity Controls Report

This report addresses all of the cybersecurity mitigations utilized in the product. The structure of this document is also specified to include the following sections, and each section must detail the named mitigations (e.g., cryptographic algorithms, cryptographic key width, sequence diagrams of security-related communications). The required sections are:

- Authentication controls;
- Authorization controls;
- Cryptography controls;
- Code, data, and execution integrity controls;
- Confidentiality controls;
- Event detection and logging controls;
- Resiliency and recovery controls;
- Firmware and software update controls.

7.12 Cybersecurity Testing Report

This is an amalgam of separate internal security reports for all of the cybersecurity testing that has been performed. These include the following tests.

7.12.1 Fuzz Testing Report

Fuzz testing (using automation to deliberately feed invalid, unexpected, and random data into the system or software interfaces in an attempt to provoke a crash or other unanticipated, possibly exploitable, behavior and log the results) should be performed as early as possible during the implementation phase of a new medical device development effort. The fuzz testing report should include:

- A description of what interface is being fuzzed and to what level the fuzzing will be performed;
- Target device under test (DUT) identification and version number;

- Name, version, and settings of the tool used;
- Name of the fuzz tester;
- The date the fuzzing began;
- Description of any specific setup needs such as previous packets, settings, configuration, or environment;
- Iteration count of the total fuzz testing execution;
- Packet contents of failed fuzzing events;
- Description of the crash or interruption, including any reported assertions, or processor exceptions, and the possible results of such an event. However, at no time should assertions about the impacts on safety and efficacy be made by the fuzz tester or report writer.
- If applicable, citation of relevant CWE ID numbers [6].

7.12.2 Malformed User Inputs

These are similar to fuzz testing; however, this is more manual and directed toward user input as opposed to machine-to-machine communications. The testing process usually requires writing specific code snippets to simulate and exercise the user's input.

7.12.3 Attack Surface Analysis

Attack surface analysis is about mapping out what parts of a system needs to be reviewed for extra mitigation controls to be put in place. The point of this activity is to understand the risk areas in your product. This is very similar to asset threat modeling and global security architecture views.

7.12.4 Vulnerability Scanning

This process employs commercial tools to exercise communication media for known vulnerabilities. This is almost always constrained to IT types of communications, such as Wi-Fi and Ethernet connections.

7.12.5 Software Content Analysis of Executables

The content of executable images should be closely tracked to prevent unintentional malware corruption from occurring. The use of SCA tools and cryptographic hashes of known good executables are used to perform and maintain this level of integrity.

7.12.6 Static Analysis Report

Static analysis is a method of computer program review and debugging that is performed by examining the code without executing the program. The process provides an understanding of the code structure and helps to ensure that the code adheres to industry standards. By running the source code through a set of predetermined criteria, static analysis tools expose various types of flaws in the code. Static analysis tools function independently of compilers, linkers, and hardware and can analyze the most commonly utilized languages, C and C++.

Commercial static analysis tools do not yet replace traditional (human) code review entirely, but they make the code review process more efficient and productive. The next generation of static analysis tools does show promise of improved automation of the code review process, and with further development, complete review automation could become possible in the future.

Coding rules and guidelines exist to ensure that software is:

- Safe: Usable without causing harm. This is especially critical in medical devices where code supporting functions that would typically be innocuous could produce a ripple effect impacting patient safety.
- Secure: Its vulnerabilities are mitigated. This is the primary focus of SAST.
- Reliable: Functions as it should, every time.
- Testable: Can be evaluated.
- Maintainable: As the codebase grows, we want to ensure that it is designed for maintainability. Coding rules and code analysis are established for many purposes, including ensuring software quality, but it is important to ensure that security objectives are included and met. We saw an example of unmaintainable code with the Toyota acceleration issue. When Toyota analyzed their dysfunctional code, they found that it had become unmanageably complex over years of continuous development with insufficient checks and oversight [7].
- Portable: Functionality is the same across various environments and systems.

Coding standards are a collection of coding rules, guidelines, and best practices to improve the safety, quality, and security of the implementation. These may also include guidelines on coding style. Using established coding standards:

- Ensures compliance with the International Standards Organization (ISO), which is required for most major medical device markets.
- Guarantees consistent code quality, no matter when or who writes the code.
- Secures the software right from project start.
- Reduces costs by speeding up time to market through reduced variability and uncertainty, as well as by decreasing overhead for postmarket support.

Static application security testing (SAST) is a subset of static analysis where the emphasis is on security-related vulnerabilities. By applying the appropriate security-related standards and rule set, weak or substandard code with potentially exploitable security vulnerabilities can be identified, triaged, and addressed. The most common standard and rule set utilized is the CERT C and C++ by SEI 2016, which presents hundreds of examples that illustrate coding practices widely acknowledged to represent potential cybersecurity concerns side-by-side with compliant solutions [8].

Ideally, SAST should be performed often as possible. Given the current state of continuous integration (CI) tools, it is possible to integrate the rule sets and static analysis tools into the build process and have an intermediate and informal result set reported to the developers on an established cadence (e.g., nightly builds, weekly builds).

SAST is expected by worldwide regulators. It is also a best practice to ensure that the source code is properly secured, both from a technical and business aspect. We see many examples of security exploits due to improperly tested code that exact a heavy cost on industry and end-users.

DHF records from SAST activities include the following:

- *Rule set:* To conduct proper SAST, the development and operations teams should select the appropriate standards and guidelines (i.e., CERT C) and then down-select the necessary rule sets. Note that it is possible and valid to mute specific checks within the rulesets as long as there is a rationale and justification provided in the rule set record, which should be preserved in the DHF.
- *Initial assessment results:* Using the rule sets, an initial run of SAST is used to identify all the violations. The pre-mitigation result should be recorded in the DHF. Note that it is possible and valid to tweak previously selected and previously muted checks as long as there is a rationale and justification provided. Also remember to update the rule set records appropriately.

- *Post-mitigation results:* The pre-mitigation results are used to triage the identified violations. These fall into three categories:
  - True-positives that must be addressed; these require changes to the code, refactoring, or even possible design.
  - True-positives that can be suppressed; these typically apply to low-risk violations that do not provide the necessary risk or benefit. These are referred to as residual risk elements.
  - False-positives that are either suppressed or muted; violations are identified but do not apply, in which case they can be suppressed or the check itself can be muted (this is only if the violation is globally false).

  After all the violations have been triaged and addressed, the post-mitigation results are also recorded in the DHF. Note that the triage and address of violation should not be a one-type exercise and should be executed and repeated as often as practical throughout the development lifecycle process, with records captured for each instance.
- *Final SAST Report:* The SAST artifact needed for submission is the Final SAST Report. This report collects and summarizes all three preceding activities. It provides the rationale and justification for the standards and rule sets selected and muted, the pre-mitigation results, and post-mitigation results, including any suppressions and residual risk elements.

7.12.7 Penetration Testing Report

Penetration testing is a (white hat) "hacker-for-hire" service where a security expert attacks the medical device under development. Penetration testing can be performed at two different times in the development lifecycle, concurrent with the implementation phase and/or at the end of all development activities. If performed concurrently with implementation, then penetration testing can consist of several iterations, each scoped to specific areas of the medical device. This is sometimes referred to as μPen or micro-penetration testing. This is the preferred approach to penetration testing, because results are delivered while the implementation phase is still in progress and require fewer resources to address compared to one-time, post-development penetration testing, which delivers results after all development activities have concluded.

The penetrating testing report should include:

- A description of the penetration testing effort;
- The scope of the penetration testing (full system or specific subsystems in an μPen testing scenario);

- Target device under test identification and version number;
- Name of the penetration tester;
- Date of the penetration testing;
- Equipment utilized in the penetration testing;
- Description of any specific setup needs such as communications, settings, configuration, or environment;
- Description of the results of the penetration effort, including the possible implications of such an attack (however, at no time should assertions about the impacts on safety and efficacy be made by the penetration tester or report writer);
- If applicable, citation of relevant CWE ID numbers.

Ideally, penetration testing should be performed by an independent entity, that is, a contracted third party or an internal entity that has not shared objectives or business interests. The FDA stipulated that [3]: "device manufacturers should indicate in the test reports by whom the testing was performed (e.g., independent internal testers, external testers) and what level of independence those responsible for testing devices have from the developers responsible for designing devices. In some cases, it may be necessary to use third parties to ensure an appropriate level of independence between the two groups, such that vulnerabilities or other issues revealed during testing are appropriately addressed. For any third-party test reports, manufacturers should provide the original third-party report. For all testing, manufacturers should provide their assessment of any findings including rationales for not implementing or deferring any findings to future releases."

7.13 SBOMs

Knowing what we are getting when we buy something is a foundational concept of the modern marketplace. Every piece of candy one buys from the supermarket, for example, comes with a list of ingredients. Most of the time, we may not care about the specifics of what is enrobed in creamy chocolate. We may not care, that is, until someone in our life has a serious food sensitivity. In this case, it is imperative that the data about these ingredients is available, in an accessible format. Why is it that we expect more transparency from a Twinkie than from the software on which our medical devices depend?

Building on the model of a bill of materials, in manufacturing, an SBOM amounts to a list of ingredients that make up software. Just as every manufacturing process begins by determining the necessary underlying parts, a good software development process should also explicitly track what components are used. Producing an SBOM is a reflection of the MDM's security maturity, and sharing an SBOM extends that security commitment to the customer to enable them to understand the composition of the product. This also gives the end-user insight into the potential risks of discovered vulnerabilities that may be disclosed for items in the SBOM.

Software is seldom written from scratch in today's digital economy. Instead, we use existing libraries and modules to take advantage of the efficient functionality that others have developed, either in the form of open-source software libraries or licensed commercial software from others.

This, in turn, introduces risk. Unlike custom software, third-party components are more widely deployed and are therefore often more known to malicious actors and are more targeted. When vulnerabilities are discovered in a third-party component, an SBOM will help all involved understand whether or not a specific device is at risk.

7.13.1 Elements of an SBOM

An SBOM is a nested inventory of third-party software components as well as the MDM's application software. A recent international initiative is underway to better define SBOMs as identifying and listing software components, information about those components, and the supply chain relationships between them [9]. The SBOM enables visibility and clarity into the complexity of the underlying components that make up modern software. Information about these components can be mapped to other sources of data, which, in turn, supports better decision-making.

The basic unit of an SBOM is the software component. A component means a TPSC or any identifiable subdependencies that make up that TPSC, such as a software library, that originate or could be used outside the primary organization. Components may contain more subcomponents, which, in turn, may be made of more components. This level of specificity allows us to get more granular than, say, a package. However, it does not require the complexity and public detail of, say, a particular function.

Components should be "sufficiently uniquely identifiable," as should the SBOM at its highest level. There are several objectives for identifying components. First, two SBOMs referring to the same component should be able to be certain that they are truly referring to the same component. As such, a cryptographic hash should be used to uniquely identify components. Second, the component identified must be able to be mapped to other sources of data about the software component, such as to a vulnerability database, or a set of data about software licenses.

These components are organized in a dependency tree, documenting subcomponents that are included in a component. Alice's application may depend on Bob's Browser and Bingo's Buffer. Bob's Browser, in turn, may include Carol's Compression Engine. Ideally, this set of dependencies would include all subcomponents, until primitives are identified that have no further dependencies. In reality, obtaining this level of transparency may not always be possible. The necessary depth will be a function of market demand and of the cost of obtaining that information.

7.13.2 SBOM Formats

As a relatively new security concept in many industries, including healthcare, there is no single broadly accepted mechanism for defining or handling SBOMs. Two major data formats exist that can be used for defining TPSC data in machine-readable formats: CycloneDX [10, 11] and SPDX [12]. Each of these can capture minimum elements of as SBOM as defined by the NTIA for each component. Both of these standards can be expressed as XML or JSON, although JSON is more frequently used. The format selected should follow the relative advantages of each standard as they apply to each organization [13, 14].

There is an entire ecosystem surrounding SBOMs, which was further discussed in Chapter 5.

7.14 Security Risk Management Report

This report is created at the end of all development and cybersecurity-related activities. The purpose of this report is to summarize for relevant stakeholders the outcome of all security activities performed during the development of the medical device. The System Security Report contains:

- A reiteration of the cybersecurity risk determination and justification found in the Security Plan artifact;
- Like the Security Plan, this Report details each of the performed activities with summarized results, including:
  - The number of vulnerabilities discovered;
  - The number of vulnerabilities mitigated;
  - The number of residual vulnerabilities, including a description and a risk-benefit analysis justifying the residual status for each one.
- Risk analysis, mitigations, and design considerations for cybersecurity risks;
- A traceability matrix of security risks to security controls;

- Traceability to the verification reports for documented security controls;
- When and how security updates or patches will be provided;
- The steps taken to ensure devices are delivered malware-free;
- Summarize the risk evaluation methods and processes;
- Summarize the residual risk conclusion from the security risk assessment;
- Summarize the risk mitigation activities undertaken;
- Traceability between the threat model, cybersecurity risk assessment, SBOM, and testing documentation.

MDM senior management must review and approve this artifact before it can be considered complete. This demonstrates senior management's awareness of the risks and potential attack surface of the new medical device.

7.15 Labeling

According to the FDA's "Cybersecurity in Medical Devices: Quality System Considerations and Content of Premarket Submissions Guidance for Industry and Food and Drug Administration Staff," finalized September 27, 2023 [3]:

- "Labeling that does not include sufficient information to explain how to securely configure or update the device may limit the ability of end users to appropriately manage and protect the medical device system."
- "...cybersecurity information discussed in this guidance is important for the safe and effective use of devices and should be included in device labeling..."
- "The device manufacturer should also provide users with whatever information they may need in the device labeling to allow them to manage risks associated with the software components, including known vulnerabilities, configuration specifications, and other relevant security and risk management considerations."

The term "labeling" includes package, part, and document labels; technical manuals; instructions for use; promotional material including websites; and statements made by the MDM. Specifically, with regards to cybersecurity and labeling, the agency stated that [15, 16]: "Inadequate cybersecurity information in the device labeling may cause a device to be misbranded under section 502(f) of the FD&C Act if its labeling does not bear adequate directions for use or under section 502(j) of the FD&C Act because it is dangerous to health when used in the manner recommended or suggested in the labeling, among other possible violations. For cyber devices, failure to comply with any requirement under section 524B(b)(2) (relating to ensuring device cybersecurity) is considered a prohibited act under section 301(q) of the FD&C Act."

FDA has listed the following as "examples of information that may be included in labeling to communicate relevant security information to users":

- Device instructions and product specifications related to recommended cybersecurity controls appropriate for the intended use environment (e.g., anti-malware software, use of a firewall, password requirements).
- Sufficiently detailed diagrams for users that allow recommended cybersecurity controls to be implemented.
- A list of network ports and other interfaces that are expected to receive and/or send data. This list should include a description of port functionality and indicate whether the ports are incoming, outgoing, or both, along with approved destination end-points.
- Specific guidance to users regarding supporting infrastructure requirements so that the device can operate as intended (e.g., minimum networking requirements, supported encryption interfaces). Where appropriate, such guidance should include technical instructions to permit secure network deployment and servicing, and instructions for users on how to respond upon detection of a cybersecurity vulnerability or incident.
- An SBOM as specified in Section V.A.4. or in accordance with an industry-accepted format to effectively manage their assets, to understand the potential impact of identified vulnerabilities to the medical device system, and to deploy countermeasures to maintain the device's safety and effectiveness. Manufacturers should provide or make available SBOM information to users on a continuous basis. If an online portal is used, manufacturers should ensure that users have up-to-date links that contain accurate information. The SBOM should be in a machine-readable format.
- A description of systematic procedures for users to download version-identifiable, manufacturer-authorized software and firmware, including a description of how users will know when software is available.
- A description of how the design enables the device to respond when anomalous conditions are detected (i.e., security events). This should include notification to the user and logging of relevant information. Se-

curity event types could be configuration changes, network anomalies, login attempts, or anomalous traffic (e.g., send requests to unknown entities).

- A high-level description of the device features that protect critical functionality (e.g., backup mode, disabling ports, or communications).
- A description of backup and restore features and procedures to restore authenticated configurations.
- A description of methods for retention and recovery of device configuration by an authenticated authorized user.
- A description of the secure configuration of shipped devices, instructions for user-configurable changes, and identification of user-configurable changes that could increase security risk for the medical device system. Secure configurations may include end point protections such as anti-malware, firewall or firewall rules, allow lists, deny lists, security event parameters, logging parameters, and physical security detection, and resetting of credentials.
- Where appropriate for the intended use environment, a description of how forensic evidence is captured, including but not limited to any log files kept for a security event. Log file descriptions should include how, where, and in what format the log file is located, stored, recycled, and archived and how it could be consumed by automated analysis software (e.g., Intrusion Detection System (IDS) or Security Information and Event Management (SIEM)).
- Information, if known or anticipated, concerning device cybersecurity (including components) end of support and end of life. At the end of support, a manufacturer may no longer be able to reasonably provide security patches or software updates. If the device remains in service following the end of support, the manufacturer should have a pre-established and pre-communicated process for transferring the risks highlighting that the cybersecurity risks for end-users can be expected to increase over time.
- Information on securely decommissioning devices by sanitizing the product of sensitive, confidential, and proprietary data and software [3].

The FDA has stated that [3]: "a revision-controlled, Manufacturer Disclosure Statement for Medical Device Security (MDS^2) and Customer Security Documentation as outlined in the HSCC Joint Security Plan (JSP) may address a number of the above recommendations."

For more information regarding FDA's policy on labeling, see the FDA Guidance Search Tool [17] as well as IEC TR 80001-2-2, IEC TR 80001-2-8, and IEC TR 80001-2-9.

7.16 Marketing and Sales Supporting Materials

7.16.1 Security Considerations in the Sales Process

Traditionally, MDMs have been conservative compared to other industries. The nature of medical devices and their regulatory status have mostly discouraged MDMs from publicly making unsupportable claims to boost sales and from pursuing risky product development activities. This conservative approach has also extended into the field of product security claims. Some MDMs have been ahead of the industry in, for example, having a formal coordinated disclosures process, where external third parties can report vulnerabilities that are then managed through a formal triage process and mitigated. Yet it is less common for an MDM to make public claims about the positive aspects of their product security for fear of becoming a hacking target.

This reticence in bragging about following good security practices and delivering secure products means that product security has not historically played a role in marketing or sales efforts. However, changing conditions have opened the doors for MDMs to leverage product security as a sales advantage in direct communications with potential customers, without making public claims that might not be supportable or might single them out as a target.

As a result of increasing security and safety concerns, medical device customers, such as HDOs, are now including security requirements in their purchasing contracts and typically expect that the MDM supplies three separate deliverable artifacts as part of the sales and acquisition process. These are:

- Description of device software composition (i.e., an SBOM);
- Description of device security properties (e.g., MDS^2);
- Security commitments (technical and process) in contracts between the HDO and the MDM (e.g., MCS2).

HDOs, based on guidance provided by the FDA, may stipulate delivery of additional artifacts that help them assess a device's security posture. These may include:

- Documentation of security risk analysis, mitigations, and design considerations (including list of cybersecurity risks considered and cybersecurity controls applied).

- Traceability matrix that links cybersecurity controls to the identified risks as well as evidence of control effectiveness through testing (e.g., test reports or certification).
- A plan for cybersecurity postmarket management, including how the MDM intends to provide software updates and patches.
- Documentation of software integrity controls (i.e., assurance that the software remains free of compromise and that the device is protected from the installation of malware).
- Instructions and specifications related to management and maintenance of cybersecurity controls (e.g., anti-malware, firewalls) as well as instructions for secure deployment, handling, and integration practices.
- Instructions related to operation of the device within an environment of enterprise security tools (see Chapter 11), including integration of the device with security management systems (e.g., vulnerability scanners, risk and compliance management tools, SIEM, SOC, or external security management services).

7.16.2 SBOMs in Sales

Both human and machine-readable SBOM formats are a useful sales tool, as they indicate to an HDO that the MDM is taking their cybersecurity responsibility seriously and performing best practices in the Total Product Lifecycle. As HDOs are increasingly requiring security documentation, including SBOMs, as part of their pre-procurement assessment as well as for the purpose of security maintenance, the availability of machine-readable SBOMs for use within the HDO's asset and risk management systems will serve as a huge incentive for acquiring medical devices from vendors that provide compliant security documentation.

7.16.3 MDS^2

The "Manufacturer Disclosure Statement for Medical Device Security" (MDS^2) was initially developed in 2008 through the cooperation of the National Electrical Manufacturers Association (NEMA) and the Health Information and Management Systems Society (HIMSS) Medical Device Security Task Force, in collaboration with multiple industry associations, government agencies, and other stakeholders. It provided a basic overview of a device's security posture in 41 questions, allowing medical device manufacturers to describe to their customers the basic security and privacy properties of the medical device they are purchasing. This included parameters like operating system and version, type of network connections, the ability of the operator to install anti-malware, or what and how much PHI is stored on the device and whether it is transient or permanent.

Simply put, the goal of the MDS^2 is to assist the HDO's professionals responsible for security-risk assessment and management of the medical device security on their networks. The MDS^2 becomes an important sales tool when shared in advance of procurement, allowing an HDO to make an informed decision about its purchase.

Filling out the MDS^2 is purely voluntary and typically needs to be specifically requested by the purchasing entity. However, a few manufacturers have made it their practice to publish the MDS^2 by linking to it on their web page or customer portal.

A major update to the MDS^2 was issued in 2013 (adding a further 81 questions), and, in October 2019, a new revision of the MDS^2 was been released by the NEMA. The scope of the MDS^2 was increased dramatically, to 216 questions.

The new MDS^2 has 23 sections that now include:

- Document information (version, date);
- Manufacturer organizational information (including security contact, ISAO membership, vulnerability disclosure process);
- Basic device security properties (operating system, network diagram, hosting of SaMD components);
- PII management (type, storage, transmission, maintenance, duration, deletion);
- Device security capabilities (aligned with ISO/IEC 80001-2-2):
    - Automatic logoff (ALOF);
    - Audit controls (AUDT);
    - Authorization (AUTH);
    - Cybersecurity product updates (CSUP);
    - Healthcare data deidentification (DIDT);
    - Data backup and disaster recovery (DTBK);
    - Emergency access (EMRG);
    - Health data integrity and authenticity (IGAU);
    - Malware detection/protection (MLDP);
    - Node authentication (NAUT);
    - Connectivity capabilities (CONN);
    - Person authentication (PAUT);
    - Physical locks (PLOK);

    - Roadmap for third-party components in device life cycle (RDMP);
    - SBOM;
    - System and application hardening (SAHD);
    - Security guidance (SGUID);
    - Health data storage confidentiality (STCF);
    - Transmission confidentiality (TXIG);
    - Remote service (RMOT).
- Other security considerations.

This new MDS^2 improves the granularity of the information and provides compatibility information that may facilitate the integration of medical devices within the larger enterprise. HDOs have always been able to use the MDS^2 data to inform purchasing decisions; however, this latest version provides more detailed data for the finer levels of decision-making, such as flagging the availability of an SBOM from the MDM or anti-malware integration. Moreover, its alignment with ISO/IEC 80001-2-2 allows for integrated information flow with the HDO's risk management process compliant with the standard. However, the MDS^2 does not specify service processes, support commitments, or other activities that are better addressed via contracts.

Given its obvious utility and the number of benefits provided to the HDO, we anticipate that MDS^2 will be leveraged more thoroughly in the future for both pre-procurement and post-procurement activities and increasingly adopted as a requirement for product evaluation and management. In the meantime, the MDS^2 provides a strong sales tool for MDMs wishing to highlight their secure development practices and preparedness for postmarket cybersecurity support. General adoption also makes things easier for the MDM, as all their customers expect the same type of security information in the same format.

7.16.4 Cybersecurity in Purchasing Contracts

While contracts have been and will continue to be the vehicle that defines and constrains the business relationship between the HDO and the MDM, detailing such items as price, service, and warranty, there are now additional elements that are more commonly being added to contracts by HDOs in the form of cybersecurity assurances and ongoing services to be provided by the MDM. These new terms are being utilized to define and formalize the cybersecurity relationship between the HDO and the MDM from the early stages of the purchasing and contracting process. These efforts are far from having established standards, but some of the publicly available examples [18] and voluntary standards and guidelines (e.g., as provided by ENISA and others) [19, 20] do paint an overall picture of where the industry is heading. We can expect these contract standards to be continually updated.

Some common threads run between most of these artifacts, and, by reviewing these, we can start to level-set certain expectations for the language that we expect to be more frequently included in new contracts. Having such common and increasingly aligned understanding of cybersecurity requirements helps both parties: HDOs can be more confident that they have defined the right level of security, and MDMs benefit from consistent requirements across customers. To that end, the Health Sector Coordinating Council has released a valuable resource titled "Model Contract-Language for Medtech Cybersecurity (MC^2)," available for free online [21].

The following topics are now commonly found in contracts between the HDO and the MDM, and include technical and process requirements:

- Secure software development lifecycle (SSDLC):
    - The MDM commits to following a current best-practices secure development program.
    - There is documentation to substantiate this secure lifecycle.
    - The MDM follows industry best practices for internal staff training and awareness.
    - The MDM agrees to auditing by third-party experts.
    - Conformance and compliance with existing laws and regulatory guidelines.
- Examples of security requirements:
    - Fail-safe modes of operation to safely maintain essential clinical performance during and after an incident.
    - Features to allow restoration of device functionality and configuration data.
    - Hardened design following secure by design and secure by default principles [22].
- Security updates and patches:
    - Setting up specific time periods (measured in days) for the MDM to notify the HDO of any discovered vulnerabilities.
    - Setting up specific time periods (measured in weeks) for the MDM to provide a patch to mitigate the vulnerability; the exact time period may depend on vulnerability risk level, the higher the risk the sooner the update should be provided.
    - All patches will be cryptographically secured and authenticated (i.e., signed or hashed).

    - Patches to TPSC (anti-malware, operating system) delivered on a mutually determined cadence.
- Security events at the HDO:
    - Point of contact at the MDM for event notification and support requests;
    - Support and response to the events in a specific period of time (measured in days);
    - Defined support processes for identification and remediation efforts;
    - Support with the investigation of security issues and active incidents;
    - Support with forensic data capture, retention, and analysis.
- Security events at the MDM affecting the HDO's business or patient data (e.g., an incident impacting services that an MDM may provide to an HDO as a HIPAA business associate):
    - Notify the HDO within a specific time period (measured in weeks).
    - Respond to the events in a specific period of time (measured in days).
    - Support with the investigation of security issues and active incidents.
- Liability allocation and remedies for events:
    - MDM shall maintain cybersecurity and liability (including cybersecurity) insurance.
    - Assignment of liability for costs incurred by the HDO for events and incidents that affect cybersecurity;
    - Retesting assets attached to the HDO's network;
    - Loss of productivity;
    - Application of mitigation patches and updates;
    - Liability of MDM for related damages.
- SBOM obligations:
    - MDMs agree to provide and maintain an up-to-date and accurate SBOM.
    - TPSCs are traced to any CVEs that are reported in vulnerability databases, such as NVD.
    - SBOM is routinely assessed for any newly disclosed vulnerabilities.
    - SBOM is maintained to match current product versions.
- Sensitive data protection and encryption:
    - All communications of sensitive or critical data between devices or devices and other systems is cryptographically protected.
    - All sensitive or critical data at rest is cryptographically protected.
- Authentication:

- User to device;
- Device to network;
- Device to device;
- Device to back end (local or hosted);
- Enforcement of secure password requirements for login credentials;
- Support for 2-factor authentication, especially for a high privilege or remote access;

- Logging:
  - Create log entries for authentication;
  - Create log entries for data access;
  - Create log entries for changes to sensitive or critical data;
  - Create log entries for security events and incidents.
- Rights for the HDO to conduct penetration testing.
- Rights for public disclosure of a discovered vulnerability under a responsible disclosure.

By nature, business arrangements have to remain flexible. Some of the language that we currently see being suggested for use in contracts will need to be revised, either to be more realistic ("the device will have no vulnerabilities" is extremely unrealistic), or to specify clearly which party is going to shoulder which ongoing cost of meeting the obligations set forth by regulation and/or in these contracts.

References

[1] "Title 21: Food and Drugs, PART 820—Quality System Regulation, Subpart D—Document Controls," U.S. Code of Federal Regulations, April 13, 2020, https://www.ecfr.gov/cgi-bin/text-idx?SID=df2431a9085035a24294a4709b05253c&mc=true&node=sp21.8.820.d&rgn=div6

[2] International Standards Organization, "ISO 13485:2016 Medical Devices—Quality Management Systems—Requirements for Regulatory Purposes," March 2016, https://www.iso.org/standard/59752.html.

[3] FDA, "Cybersecurity in Medical Devices: Quality System Considerations and Content of Premarket Submissions Guidance for Industry and Food and Drug Administration Staff," September 27, 2023, https://www.fda.gov/media/119933/download.

[4] Shostack, A., *Threat Modeling: Designing for Security*, New York: Wiley, 2014.

[5] Cybersecurity Forum, "CyberSecurity FAQ – What Is Cyber Hygiene," https://cybersecurityforum.com/cybersecurity-faq/what-is-cyber-hygiene.html.

[6] https://cwe.mitre.org/.

[7] Safety Research & Strategies, Inc., "Toyota Unintended Acceleration and the Big Bowl of 'Spaghetti' Code," November 7, 2013, https://www.safetyresearch.net/blog/articles/toyota-unintended-acceleration-and-big-bowl-%E2%80%9Cspaghetti%E2%80%9D-code.

[8] SEI CERT, "SEI CERT C Coding Standard," 2016, https://resources.sei.cmu.edu/downloads/secure-coding/assets/sei-cert-c-coding-standard-2016-v01.pdf.

[9] NTIA Framing Working Group, "Framing Software Component Transparency: Establishing a Common Software Bill of Material (SBOM)," November 12, 2019, https://www.ntia.gov/files/ntia/publications/framingsbom_20191112.pdf.

[10] NTIA Open Working Group, "Survey of Existing SBOM Formats and Standards," October 25, 2019, https://www.ntia.gov/files/ntia/publications/ntia_sbom_formats_and_standards_whitepaper_-_version_20191025.pdf.

[11] https://cyclonedx.org/.

[12] https://spdx.dev/.

[13] Lear, E., R. Droms, and D. Romascanu, "Manufacturer Usage Description Specification," Internet Engineering Task Force, March 2019, https://tools.ietf.org/html/rfc8520.

[14] https://www.mudmaker.org/.

[15] FDA, "Device Labeling," October 23, 2020, https://www.fda.gov/medical-devices/overview-device-regulation/device-labeling.

[16] "Consolidated Text: Regulation (EU) 2017/745 of the European Parliament and of the Council of 5 April 2017 on Medical Devices, Amending Directive 2001/83/EC, Regulation (EC) No 178/2002 and Regulation (EC) No 1223/2009 and Repealing Council Directives 90/385/EEC and 93/42/EEC (Text with EEA Relevance)," March 3, 2023, https://eur-lex.europa.eu/eli/reg/2017/745/2023-03-20.

[17] https://www.fda.gov/regulatory-information/search-fda-guidance-documents/.

[18] https://iuhealth.org/about-our-system/vendor-relations.

[19] ENISA, "Procurement Guidelines for Cybersecurity in Hospitals," February 2020, https://www.enisa.europa.eu/publications/good-practices-for-the-security-of-healthcare-services.

[20] HSCC, "Supply Chain Risk Management Guide v2.0," October 2023, https://healthsectorcouncil.org/hic-scrim-v2/.

[21] HSCC, "Model Contract Language for Medtech Cybersecurity MC2," March 2022, https://healthsectorcouncil.org/model-contract-language-for-medtech-cybersecurity-mc2/.

[22] CISA, "Secure-by-Design: Shifting the Balance of Cybersecurity Risk: Principles and Approaches for Secure by Design Software," October 25, 2023, https://www.cisa.gov/resources-tools/resources/secure-by-design.

8

Medical Device Manufacturer Postmarket Lifecycle

8.1 Understanding Postmarket Regulatory and Customer Expectations

The introduction of new medical devices into regional markets is generally regulated either within the country (e.g., United States) or region (e.g., European Union) through local regulators, or by falling back on approvals in markets that are deemed mature. (For example, Israel's medical device registration is based on approval in one of five regions of the Global Harmonization Task Force (GHTF): Australia, Canada, European Union, Japan, or the United States. Similar, New Zealand requires certificates from a European Union notified body, Health Canada, TGA, or U.S. FDA.) For example, in the United States, medical devices were put under the oversight of the FDA when Congress enacted the Medical Device Regulation Act of 1976. At the time of signing, President Gerald Ford stated, "The Medical Device Amendments of 1976 eliminate the deficiencies that accorded FDA 'horse and buggy' authority to deal with 'laser age' problems," observing that [1], "When well designed, well made, and properly used, [medical and therapeutic devices] support and lengthen life. If poorly designed, poorly made, and improperly used, they can threaten and impair it."

In 1976, President Ford would have had no awareness of cybersecurity or the risks that it would someday pose to medical devices, yet his statement applies equally well to today's cybersecurity risks. It remains a testimony to the flexibility of that Amendment and the actions resulting from it that today's FDA is able to protect America's population by setting expectations for medical device security under the umbrella of device safety and effectiveness. International regulators in many countries, as discussed in Chapter 3, have since taken similar action and implemented a comparable approach.

Following the initial FDA guidance, "Content of Premarket Submissions for Management of Cybersecurity in Medical Devices" 2014, the next step taken by the FDA was in December 2016 [2] when it released the "Postmarket Management of Cybersecurity in Medical Devices" Guidance. Following that, proposed ("draft") updates to the 2014 Premarket Submissions Guidance were released in 2018 and 2022, and a finalized version was released in September 2023. Together, these documents explain the postmarket security management responsibilities of MDMs during what we now call the Total Product Lifecycle (TPLC), including postmarket activities. These activities include:

- Monitoring of cybersecurity signals;
- Identification of potential vulnerabilities and assessment of resulting risks;
- A mechanism to make available timely mitigations and remediations (e.g., security patches and software updates) to devices in the field;
- Management and maintenance of device cybersecurity posture, including periodic security testing [2].

Another of the FDA's expectations for postmarket activities is the sharing of vulnerability and other cyber risk information within the larger medical device and health delivery community. However, even 8 years after its release, many MDMs are struggling with understanding and implementing these basic postmarket activities. These MDM activities are time-consuming and stand outside of traditional project-oriented activities. Moreover, the level of security and software expertise required to perform them also puts these activities outside the skill set of the traditional sustaining engineer. This has contributed to difficulties and challenges in introducing these postmarket activities into many existing MDMs' processes.

Another area of contention is the time period called out in the postmarket guidance for providing a security patch to the party responsible for maintaining the vulnerable device. Specifically, for vulnerabilities with uncontrolled risk [2]:

> As soon as possible but no later than 60 days after learning of the vulnerability, the manufacturer fixes the vulnerability, validates the change, and distributes the deployable fix to its customers and user community such that the residual risk is brought down to an acceptable level.

The legislation passed in the "Consolidated Appropriations Act, 2023" (H.R. 2617), also known as The Omnibus Act, has raised the bar for expected security patch response. That law indicates [3]:

> ...availability of postmarket updates and patches to the device at regular cycles for known unacceptable vulnerabilities and out of cycle critical vulnerabilities that could cause uncontrolled risks...

To meet these legal requirements, MDMs must plan for regular security updates (perhaps quarterly, semiannually, or annually) and be prepared in addition to issue and distribute urgent patches for uncontrolled risks out of cycle and at a more aggressive timeline.

Other regulatory expectations of the Omnibus Act include:

- Development of plans to monitor, identify, and address postmarket cybersecurity vulnerabilities and exploits, including the coordinated vulnerability disclosure.
- Design, development, and maintenance processes to ensure that a device is cybersecure.
- Provision of an SBOM that includes commercial, open-source, and off-the-shelf software components [4].

As these regulations have evolved, one organizational entity that increasingly has had to take on cybersecurity tasks is the MDM's service and support organization. This entity has several responsibilities relating to cybersecurity that are essential to ensure patient safety, regulatory compliance, and privacy protections (see Figure 8.1). It is advisable that MDMs perform a separate and dedicated security risk analysis for their service, support, and maintenance activities, and mitigate risks through appropriate processes, security technologies, and training. These should meet the following maintenance and service security objectives: (1) monitor for cybersecurity signals; (2) identify potential vulnerabilities and assess resulting risks; (3) provide a mechanism to make timely patches and updates available to devices in the field; and (4) enable management and maintenance of the device's security posture over its useful life.

This chapter addresses those postmarket objectives and advises MDMs on how to monitor and securely update their devices already in the field, how to avoid recalls, and how to manage the device at the end of life. The considerations and general guidelines offered here are not only common security best practices, but they are also based on practical experience from working for and with several MDMs.

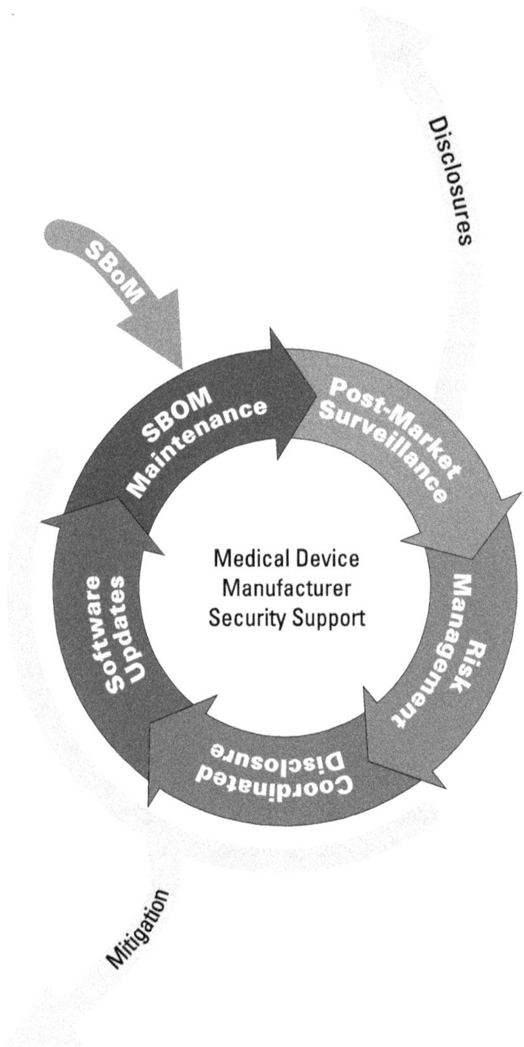

Figure 8.1 MDM postmarket management and support lifecycle.

Postmarket considerations should also include security risks that result from the manufacturer's production or support and maintenance activities. For example, we are aware of several medical device security incidents with significant impact on care delivery where a USB thumb drive used by a traveling service technician introduced malware to medical devices, resulting in the shutdown of medication cabinets or even entire cath labs. Such a USB attack vector is especially concerning as it can bridge into an air-gapped environment (intentionally or accidentally) or can affect network-segmented devices. Another example was provided by security researchers who reported that an MDM's download site was compromised by malware, to which any hospital downloading legitimate device software updates was then exposed [5]. If not performed systematically, circumspectly, and in accordance with best practices, the activities needed to maintain a device's security can themselves become the means of its compromise.

Typically, service accounts provide high-access (or even highest-access) privilege. Compromised service credentials could enable an attacker to completely take over a device. Therefore, it is advisable that: (1) the design of the device includes a strong and well-planned role-based access control model (RBAC) and (2) that local and remote service access accounts should be equipped with multifactor authentication. Depending on device architecture, this could be based on commercial solutions (e.g., for computer-like devices) or may require a software solution specifically designed for the device that provides that multifactor feature (or a comparable security control).

Some HDOs may require local control over usernames and passwords for administrative accounts for devices on their network. This can create a conflict between MDM service personnel and their desire for consistent service access across all devices, and the security needs of the HDO to deny use of devices with service-wide access credentials.

Furthermore, different support scenarios require different approaches to maintain cybersecurity controls. For example, when providing local support (on-site or at-factory), service technicians need to be fully trained in cybersecurity best practices and how to maintain the device's security posture as well as that of its connected IT infrastructure, if applicable. Service tools that they use (e.g., laptops, USB data carriers) need to be verified to be free of compromise, need to be used and handled so as to minimize exposure, and regularly assessed and sanitized per an established procedure. Similarly, remote support activities must not introduce any cybersecurity risks to the HDO IT infrastructure. This includes ensuring that MDM IT infrastructure is free of compromise; secure and encrypted access (e.g., via VPN); and protection of support user accounts and access (e.g., no default or easy-to-guess usernames and passwords; use of multifactor authentication). Specific guidance for remote support activities is provided through ISO 11633-1 and ISO 11633-2 [6]: "Health Informatics — Information Security Management for Remote Maintenance of Medical Devices and Medical Information Systems." At the same time, accessing the HDO IT infrastructure to perform support should be likewise controlled so as not to introduce any compromise back to the MDM.

Needless to say, the device must be designed and manufactured with these maintenance and service security needs in mind in order to provide features that support the above and other use cases (such as providing support to devices in a home health care setting). This includes specific security features (e.g., disable USB autorun), access and authentication (e.g., 2-factor authentication for high-privilege access), PHI protection during the lifecycle (including data destruction when a device is taken out of operation), as well as incident response and recovery/restoration features.

Increasingly, HDOs are including postmarket support, incident response, and device restoration capabilities in their contractual relationship with the MDM. For more information, see the discussion on cybersecurity and contracts in Chapter 7.

The above review of a service organization's security needs is largely based on the typical scenario of hospital-based equipment. However, many medical devices may not fit that pattern, for example, home-based systems (discussed in Chapter 4), or patient-worn and implantable devices. These are typically not serviced by dedicated technical personnel but are often maintained by the patient themselves or by the patient's clinical care provider. Even in these scenarios, the same basic security requirements exist and need to be supported by the device and the device's infrastructure: (a) prevent compromise, (b) maintain security posture, (c) protect sensitive data, and (d) detect and respond to incidents. To enable a nontechnical user to perform these tasks requires special considerations, which will be discussed in more detail throughout this chapter.

8.2 Streamlining Management of Postmarket Security Vulnerabilities

Manufacturers have traditionally assumed that software changes to a device require reapproval by the FDA, disincentivizing the creation of cybersecurity-only patches. However, the FDA in their December 2016 postmarket guidance has narrowed the situations where new premarket approval would be required (i.e., only for certain uncontrolled vulnerabilities [7]). In the *FDA Fact Sheet* on dispelling myths about the agency's role, they explicitly state [8]:

> *Myth: Medical device manufacturers can't update medical devices for cybersecurity.*
>
> *Facts: Medical device manufacturers can always update a medical device for cybersecurity. In fact, the FDA does not typically need to review changes made to medical devices solely to strengthen cybersecurity.*

However, the manufacturer is still obligated to perform and document validation activities before releasing a cybersecurity update, whether a premarket review is required or not [8]:

> *Myth: The FDA is responsible for the validation of software changes made to address cybersecurity vulnerabilities.*

Facts: The medical device manufacturer is responsible for the validation of all software design changes, including computer software changes to address cybersecurity vulnerabilities.

So, the speed at which a manufacturer can turn a patch from a third-party supplier into one for the device is solely dependent on their software validation process efficiency.

Additionally, resource allocation can prevent or impede rapid response. In a typical medical device manufacturer, only a single group or focused team holds knowledge of the software and its testing for any specific device. Typically, that group is located in research and development and works on the next-generation device or feature update. Historically, this has worked well since the only time software updates were required on released devices was when a safety issue was identified, an unusual situation that is managed by disrupting the software group's new product development work. Devices were rarely patched for cybersecurity vulnerabilities because lack of evidence of exploitation meant that they did not trip the manufacturer-defined defect monitoring systems.

With broader connectivity comes the increased risk that unpatched, latent vulnerabilities turn into active exploitations at any time. Clearly, regulatory agencies are changing the expectation for the frequency of cybersecurity updates and patches, as evidenced by the Omnibus Bill discussed earlier. The assessment of exploitability as a measure of likelihood of future occurrence, instead of evidence of a past active exploitation, becomes a more important measure. The effort needed to manage security patching must be planned for in the postmarket device phase.

Manufacturers need to accept that regular patching of fielded devices to correct vulnerabilities in commonly deployed third-party software will be expected and a constant activity. Such software is far more likely to be exploited as it is used in many other high-profile or high-impact systems besides medical devices. The scenario of a cyber adversary seeking out a specific medical device may be deemed unlikely, but a cyber adversary (or piece of malware) seeking out a vulnerability in a piece of third-party software that happens to run on a medical device is highly plausible. A recent study found that the share of malware explicitly targeting vulnerabilities in operational technology (OT) increased by almost 30% from 2021 to 2022 [9].

Proper device architecture decisions and design of test suites can minimize the cost of postmarket patching and allows for the activity to be reassigned from the primary research and development team, thereby minimizing resource conflicts. Regular and periodic patching is a best practice for companies commercializing medical devices with intrinsic responsibility to monitor both safety and performance.

Checks and balances to determine if the manufacturer has implemented adequate security will include regulators, which expect to see evidence of security risk analysis and threat modeling as well as traceability of security requirements and identified risks back to implemented and validated security controls. The absence of this type of documentation or clear weaknesses in their execution will delay or prevent device approval as well as lead to difficult implementation decisions late in the development process.

It is commonly understood that security cannot be bolted on. Discovering security gaps when the device is being prepared for or has been submitted for regulatory approval often leads to attempts to "bolt on" just enough security to get by. This undesirable approach will neither result in a secure device nor be economical in the long term. The other check on poor security decisions will be in the purchasing process when customers review the manufacturer's MDS^2 forms to determine whether the device can be secured appropriately in their environment. Customers require good security and expect that security updates will be made available for their purchased devices. Security updates should therefore be separated from paid feature updates.

8.2.1 Sources of Postmarket Vulnerabilities

As shown in Figure 8.2, postmarket vulnerabilities arise from several sources, and the skill set and processes used to respond to them vary.

8.2.1.1 Vulnerabilities from Architecture or Design Choices

During the architecture development and design processes, certain decisions are made about how the device will be secured. Some examples are:

- What form of identification, authentication, and authorization will be required?
- What encryption technology will be used to protect data at rest and data in motion?
- Does the device have a hardware security element or other protected mechanism to store credentials and validate software?

The process of reaching good decisions during and developing a well-secured architecture and design typically uses a threat modeling approach. The lack of a documented threat model and inexperienced security skills may lead to many gaping security holes even before a single line of code has been written. Later in the development cycle, penetration testing will expose those security holes overlooked or introduced during design and development, leading to schedule and/or product availability disruption.

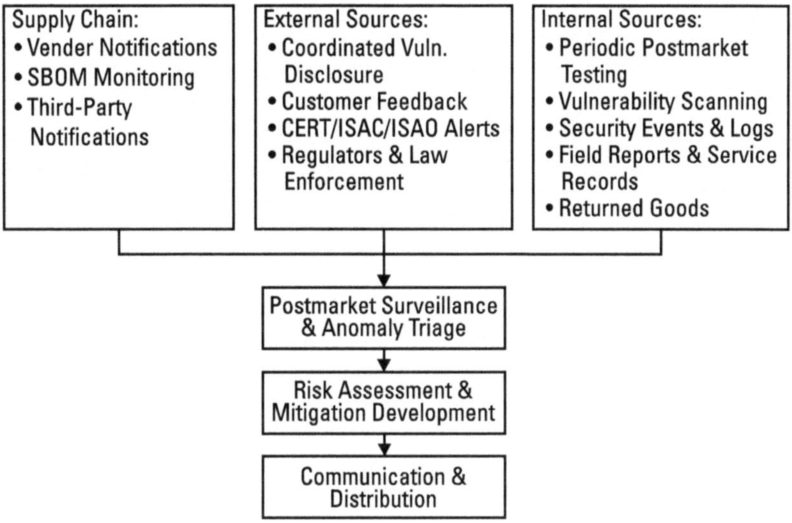

Figure 8.2 Sources of postmarket vulnerabilities.

8.2.1.2 Vulnerabilities from the MDM's Implementation Errors

Many vulnerabilities come from software implementation errors. Buffer overflows, "use after free" and others are examples of insecure coding practices that may be exploited by an attacker to make the device perform outside of its specified behavior.

Elimination of manufacturer implementation vulnerabilities is accomplished by a set of processes and tools, as follows:

- Secure coding standards (as confirmed by SAST) such as CERT C, CERT Java, MISRA;
- Static analysis tool (SAST) with support for the secure coding standard selected;
- Dynamic analysis security testing (DAST);
- Security code reviews;
- Misuse and abuse cases;
- Attack surface analysis;
- Vulnerability chaining test regime;
- Closed box testing of known vulnerability scanning;
- Software composition analysis of binary executable files;
- Malformed and unexpected input testing;
- Fuzz testing;

- Penetration testing.

While these tests have been performed during development, the September 2023 premarket cybersecurity guidance document from the FDA also requires the re-performance of these tests during the postmarket phase until the end of support for the device is reached. The periodicity of reperforming these tests varies from every 3 to 12 months, as determined by the premarket approval reviewer.

Any deployed system of even minimal complexity can have undetected vulnerabilities present. Common venues for uncovering security vulnerabilities in software are:

- A researcher or threat agent explicitly examines the device software to identify exploitable vulnerabilities.
- An updated penetration test by the manufacturer reveals new vulnerabilities.
- An updated static analysis tool recognizes previously unknown vulnerabilities.

Keep in mind that security researchers and threat agents can freely purchase most devices on the open market, especially secondhand devices (on eBay, for example). They can then take all the time that they want, in an environment they control and with the most sophisticated tools available, to work out the device's vulnerabilities. These activities could include extracting software and redistributing it through various channels, including online. It is a mistake to think that obscurity ("with all the devices out there, what are the odds that someone chooses to study or go after mine?") is in any way relevant to the device's security posture. Manufacturers must keep their security testing tools and processes current.

8.2.1.3 Vulnerabilities from Incorporated Third-Party Software

This will be the area of most frequent activity. Third-party software includes the operating system and various libraries and frameworks to simplify the device software development. Because third-party software is broadly used, it is subject to more scrutiny by researchers and threat agents, leading to regular patch cycles to close newly discovered vulnerabilities.

As an example, Microsoft has a monthly "patch Tuesday" where it issues patches for dozens of vulnerabilities across various products. Linux, Java Frameworks, and other commonly used tools have regular publications of newly discovered vulnerabilities with their associated patches. Such patches eliminate implementation errors that allow that software to be manipulated to perform unintended functionality, and the removal of these errors via patch installation should rarely impact the software performing the device's primary functionality.

Similar considerations apply to open-source software, but with additional complications. Management of patches and updates is not synchronized and may happen on an ad hoc basis. Furthermore, the presence of open-source code in any particular project may not be obvious, as it may enter a product via a supplier's hardware or software. In such cases, this challenge becomes a fourth-party problem and will require a high degree of software transparency by all parties (see Chapter 5 for a detailed discussion of supply chain security management).

Open-source software is often deeply embedded (and may have been for many product generations) and may be widely deployed across a large installed base, making any actions—such as determining affected devices and deploying patches—a complex problem affecting a larger number of products. A classic example would be the Log4Shell or Heartbleed vulnerabilities that affected millions of devices across the globe. The specific challenges of such endemic and pervasive vulnerabilities are further discussed in Chapter 5.

8.2.2 How Device Architecture Impacts Vulnerability Management

With an understanding that most postmarket vulnerability management will center on third-party software patching, there are device architecture aspects that can simplify validation testing and patch deployment. Several resources address how to construct highly secure devices, including Microsoft's 2020 report *The Seven Properties of Highly Secured Devices* [10], and the example provided by Synopsis ARC SEM Processors [11].

The following are a few examples of architectural choices that improve a device's security posture and security maintainability.

8.2.2.1 Hardware Root of Trust

When implementing cryptography (see Chapters 6 and 12), a fundamental early architecture/design decision is whether to use a hardware security module to store cryptographic secrets such as keys and certificates. It is difficult to fully secure a device if the crypto material is held in unprotected memory. This design decision is also a basis for providing a secure software updating mechanism (via code signing), which becomes more critical if updates are delivered remotely over the internet. Because it is difficult to add later, making the right design decision ultimately leads to a more secure device.

8.2.2.2 Isolation of Software Components

Isolation, also called "dynamic compartments" in the Microsoft report, is the concept of designing the system with hardware-enforced separation between functions of different criticalities. The fundamental hardware component is a Memory Management Unit (MMU) on the selected processor. MMUs can enforce access rights (e.g., read, write, execute) to software processes and can help to ensure that functions more at risk of being exploited, such as the network interface, can be managed to make it difficult for a successful exploit to break out of that function and affect other parts of the medical device.

Isolation is a key element in defense in depth. If an attacker gains access to one part of the system, it can make it difficult for them to leverage that to control other parts. As an example of an isolation architecture, see "A Reference Architecture for Secure Medical Devices" [12].

8.2.2.3 Interfaces to Physical Systems

By definition, medical devices are cyber-physical systems (CPSs) that interact with the physical world through sensors (e.g., to obtain physiological measurements) and/or actuators and transmitters (e.g., to deliver therapy, such as electrical stimulation and drugs). This involves custom hardware and software, which must be accounted for when evaluating changes to the device's security posture (e.g., a new vulnerability is discovered) and when evaluating changes to the device itself, such as when designing a software update or security mitigation. Patches and updates can have varying impacts on devices, including communication behavior, timing, resource consumption and availability, and user interaction. Timing dependencies that may be required to precisely interact with the physical world may depend on the processor being sufficiently powerful to always finish in time. For example, MDMs must consider and account for how cybersecurity software changes could impact the overall performance of the device.

As discussed in the next section, the crafting of validation tests to focus on the expected behavior of the underlying platform can reduce retest costs around specialized interfaces to physical hardware.

8.2.3 Designing Test Suites for Efficient Validation

Again, with an understanding that most postmarket vulnerability management will center on third-party software patching, the specific organization of validation test suites can impact the amount of testing required before a patch can be released to fielded devices. Test suites that intermingle device functional tests with those that test the underlying platform should be avoided.

Identify the explicit services and performance required of the underlying computing platform and build test suites that verify platform services are performing correctly. Selective functional testing may build confidence that the patch can be issued to the field.

8.2.4 Adopt a Software Sustaining Engineering Model to Speed Up Response

Software sustaining engineering may not be a group that exists at a manufacturer, but with device connectivity comes the need for regular patching of third-party software. The validation of patches and their distribution does not require detailed knowledge of device operations, especially if the test suites are properly structured. Establishing a group specifically charged with third-party patch validation and distribution will streamline the ability to quickly respond as expected, without impacting the resources and delivery deadlines of the new software in development.

8.3 Meeting Postmarket Cybersecurity Objectives

As stated at the beginning of this chapter, an MDM's postmarket cybersecurity objectives fall into four broad categories: (1) monitor for cybersecurity signals, (2) identify potential vulnerabilities and assess resulting risks, (3) provide a mechanism to make timely patches and updates available to devices in the field, and (4) enable management and maintenance of the device's security posture over its useful life. This section will consider each of these objectives in turn.

8.3.1 Monitor for Cybersecurity Signals

The process of postmarket monitoring for cybersecurity signals introduces a new set of asynchronous activities that are not initiated by any other MDM development or support activity. These postmarket surveillance activities include:

- Regular monitoring for disclosed vulnerabilities related to TPSC utilized in devices.
- Security logging, reporting, and analysis for fielded devices as well as security-specific analysis of returned devices.
- Regular monitoring for vulnerabilities in the MDM's application software through software practices such as code review and testing.
- Practicing coordinated vulnerability disclosure: Participating in vulnerability intake, analysis, and mitigation for TPSC and MDM application software.
- Surveillance activities that support market engagement and communication of findings and recommended actions to the MDM's customer base.
- Engaging with and supporting product users and consumers regarding product security.

- Maintaining active involvement in an Information Sharing and Analysis Organization (ISAO) to be informed of current threats.

8.3.1.1 Monitoring TPSC Vulnerabilities

As part of software supply chain risk management activities, this process is linked to SBOM creation and maintenance, as the SBOM serves as the TPSC content authority for any version of the software in a device. On a regular basis, the SBOM's contents are reviewed against disclosed vulnerabilities that may impact the operation of a device or other devices in the device's environment (such as an attack utilizing the device as the entry point into an HDO's network).

The sources of these disclosed vulnerabilities can vary significantly, including the National Vulnerability Database (NVD) [13] (but be aware of possible delays in posting new vulnerabilities), Vulners [14], a Github or other software version control system's repository history, ISAO disclosure, software vendor communications, or popular media. Monitoring some of these sources can be automated, but mostly this is still a manual process.

8.3.1.2 Security Logging, Reporting, and Analysis for Fielded Devices

"You do not know what you do not measure." Security logging is an essential element to understanding what the actual environment is for a fielded medical device. Anomalies in the interactions of the device with its environment may be indications of an attempted security compromise. Failed login attempts, errors during cryptographic key verification, attempts to connect to a disabled port, and other examples should be logged as activities potentially indicating a security event.

There are two consumers of device security information:

1. MDMs who want to understand patterns of security activities in order to improve device design.
2. HDOs who want data to feed into their security information and event management (SIEM) or risk management systems to be able to respond to active attacks. MDMs should consider creating feeds from their devices to the HDO SIEM or similar systems to facilitate that action.

8.3.1.3 Engagement with End-Users

In the past decade, we have seen an increase in MDM engagement with customers and end-users and this is especially true when it comes to supporting cybersecurity. While some of these postmarket activities and communication will be driven by regulators as well as contracts between MDM and HDO, in other cases, it is as simple as addressing product security on a webpage and having a portal where end-users and security researchers can contact the MDM to report product security vulnerabilities, voice concerns, and ask for information about the security of the medical device. Such a communications portal should, at a minimum, take the form of an email address in one or all of the following formats:

- security@[domain name];
- security-alert@[domain name];
- secalert@[domain name].

All emails sent to any of these addresses should have an autoresponder to inform the sender that the email was received. Following that, a human should respond in a timely manner (24 to 48 hours) to address the questions and concerns of the sender, including informing them, if appropriate, that the reported vulnerability is being escalated to a team that will investigate the claim.

Many MDMs also make an effort to ensure the confidentiality of the communication by publishing a public encryption key that can be used to secure these messages. This is often seen as beneficial to both the reporter and the MDM.

As the vulnerability is progressing through the investigation, there should be regular and timely emails to the reporter, informing them of the progress throughout the investigation. If deemed necessary, reporters can be included in determining mitigation strategy and confirming its effectiveness and should be kept informed during mitigation development and deployment. This type of continual engagement with the reporter will help the MDM in case any further information or details of the vulnerability is needed from the reporter during the investigation, as well as inform the reporter that the MDM is taking their claim seriously, which should also discourage the reporter from publicly disclosing the vulnerability prematurely.

All communications should be in a respectful and thankful tone, no matter how incredible or fanciful the claim being presented. In one case of which we are aware, the messenger reported that "their medical device infected their cell phone following a large flash of light emanating from their medical device." Cybersecurity is a scary topic for most people. Combine that fear with medical devices and healthcare, another pair of anxiety-inducing topics, and the outcome can be surprising.

Maintaining positive public relations during the postmarket period depends on reporters feeling that their concerns are heard and are being taken seriously. A reporter's concern does not need to be technologically valid to threaten the MDM's reputation or business model. If the MDM chooses to respond rudely or dismissively, it may provoke the reporter to broadcast the concern and/or response, which can result in a backlash of fear and/or negative sentiment toward the MDM, either of which can damage the brand. There is one anecdotal example of a highly skilled researcher who did not feel that an MDM treated their findings seriously and made it their personal mission to obtain various devices from that MDM, find more security issues, and report them publicly and on an ongoing basis. (The specifics of a formal Coordinated Vulnerability Disclosure (CVD) process are discussed in Section 8.3.2.1.)

8.3.2 Identify Potential Vulnerabilities and Assess Resulting Risks

Once any vulnerability is discovered (either externally or internally to the organization), it needs to be evaluated for its impact on the medical device's operation using methodology similar to design vulnerability analysis (discussed in detail in Chapter 6). It is very common for a TPSC with a vulnerability to be utilized in a device yet have absolutely no impact on the security standing of that device. This situation can exist either because the TPSC's vulnerable element is not functionally utilized by the device, or because the device's design renders the vulnerability unable to be exploited.

As an example, OPENSSL v1.0.1f (the last version susceptible to the Heartbleed vulnerability) has 494 defined macros that control an untold number of the preprocessor's conditional compilation functionality (i.e., "#IFDEF"). This functionality delimits blocks of program statements that are compiled only if the specified condition is true (see Figure 8.3). These directives are frequently nested and can contain combinations of multiple macro names.

The result is billions (if not trillions) of possible binary executables options that may or may not include the Heartbleed vulnerability from this one version of OPENSSL, depending upon the status of 494 macros. It is entirely possible, even likely, that the application software utilizing the OPENSSL library does not call any of the vulnerable functions, which may render an exploit based on them much more difficult to execute.

This kind of variability in the resultant binary output means that each software component with a disclosed vulnerability needs to be evaluated in consideration of how, specifically, that software component is being utilized in the final product before any definitive statement can be made about the impact of the vulnerability on the device.

However, every disclosed vulnerability needs this type of investigation into its impact, and, if it is a risk that can be exploited, then mitigations need to be explored. Mitigation options include:

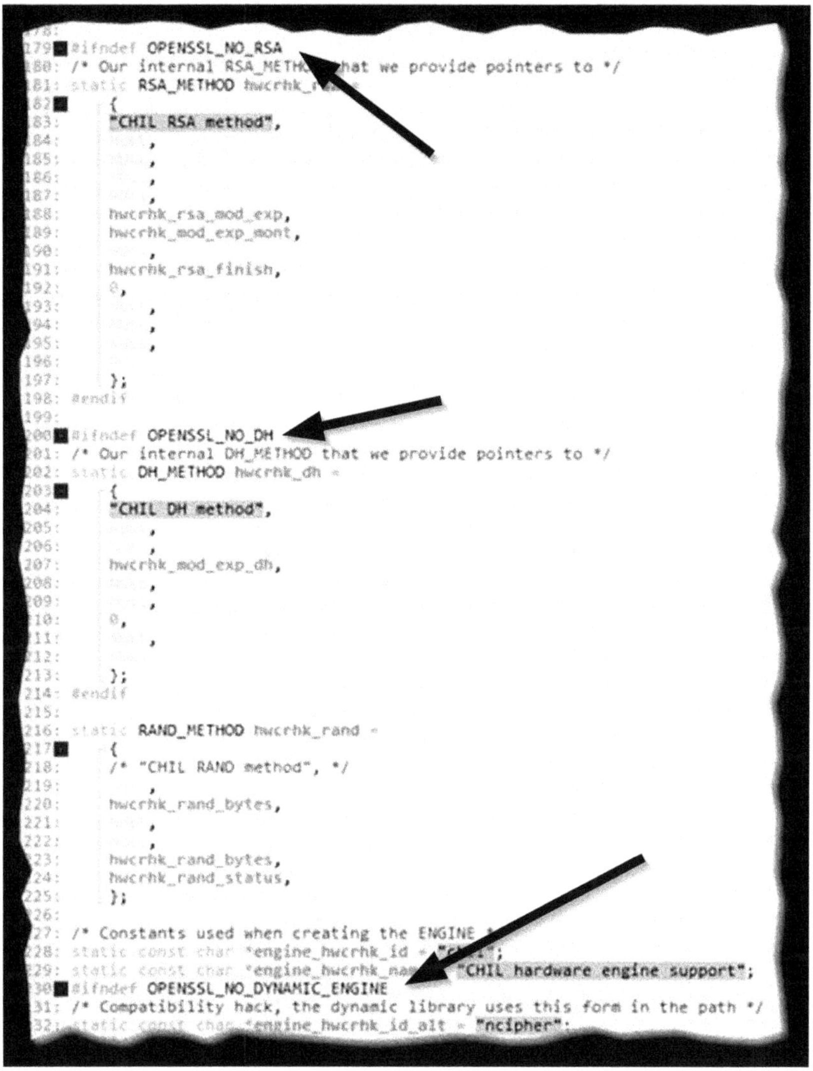

Figure 8.3 Example from OPENSSL.

- A field security patch that includes a new version of the third-party software component that has been mitigated by the supplier;
- A temporary workaround that is communicated to the operator of the medical device;
- A medical device recall.

Once an exploitable vulnerability is discovered, the MDM should perform initiate its defined communication process and disclose the status of this investigation and response process to an ISAO and/or directly to the end-user. Depending upon the nature of the information recipient (HDO, home healthcare, international consumer), the methods for keeping them informed about the risk status will vary.

If a vulnerability is determined to not be exploitable but is a high-profile vulnerability about which customers may be concerned, the actual exploitability in a given implementation can be communicated using the Vulnerability Exploitability eXchange (VEX) format that has been developed alongside SBOM definitions and frameworks [15].

8.3.2.1 Coordinated Vulnerability Disclosures

No matter how security-focused an MDM's development lifecycle is, vulnerabilities will exist in the released device at some point in the future (tomorrow, or in 5 years). This can be caused by a number of factors:

- A flawed process within the secure development lifecycle;
- A missed implementation bug;
- An intentional backdoor/vulnerability created by an MDM insider during development (even for nonmalicious purposes such as debugging);
- Unintended configurations or communications;
- A previously unknown vulnerability in a TPSC (including cryptographic libraries);
- Advances in computing that render existing cryptographic primitives deprecated;
- A completely new type of attack (i.e., AI-assisted side-channel analysis) [16].

All mature design processes assume that there will be vulnerabilities discovered in the device after it has been released to the field. These vulnerabilities will be discovered through a variety of channels, including accidental operation, focused analysis, reverse engineering, intentional testing including subjecting the product to a variety of stress tests, and side-channel analysis.

The person performing this discovery process is referred to as the reporter. The reporter can be anyone, from a casual user, security researcher, hacker, or even someone internal to the vendor or manufacturer who created the product.

When a new vulnerability is discovered, there are several paths the discovery and reporting process can follow:

- *Full disclosure:* Where the reporter publishes the analysis of the vulnerability as quickly as possible, making the analysis information accessible to everyone without delay or restriction.
- *Responsible disclosure:* Where the reporter informs the vendor of the vulnerability first, sharing the analysis with no one else besides the vendor until either of these actions happen:
    - The vendor releases mitigation to address the vulnerability (e.g., remediation via an approved patch) or initiates a voluntary recall (if warranted).
    - An agreed-upon period has elapsed. Note that this period should tolerate the added burden placed on MDMs versus nonmedical reporting scenarios. This additional burden can include many months of testing prior to being able to release new patched software. The primary interest here should be the safety of the patients, not an arbitrary period.
- *Nondisclosure:* Where the vulnerability information is not shared, and this bifurcates into:
    - Disclose analysis only to the vendor, typically under contract, bug bounty, or nondisclosure agreement (NDA).
    - Sold to the intelligence community to remain in place as a secret vulnerability to be used in a future attack.
- *Coordinated vulnerability disclosure (CVD):* Where the reporter informs the coordinator (such as an ISAC or ISAO) and either the coordinator or the reporter and coordinator together inform the vendor. Limited information is released to the public until such time as the vendor has provided a mitigation (e.g., remediation via an approved patch). This has the added benefit of the coordinator being an independent judge who can monitor the vendor's progress in the investigation, mitigation, and distribution of a patch. The coordinator can "encourage" both the reporter and the vendor to act responsibly in addressing the vulnerability in a timely fashion. This also facilitates more effective CVD and discourages its misuse as a mere PR option rather than a legitimate risk mitigation tool.

These methods of disclosure have been the source of conflicts for decades between vendor and reporter. Vendors often fear that any vulnerability disclosure would be harmful to their products and thus their business model, while the reporter often feels that vendors are not addressing their security concerns in a timely fashion, and either misrepresenting the serious nature of the vulnerability or just ignoring the vulnerability claim completely until the reporter has no other alternative than to publicly disclose [17, 18]. In some cases, it appears that both sides are at fault [19]. In other cases, no party appears to be at fault, but a series of poor judgments and miscommunications occur to the detriment of all.

Fortunately, this war between reporters and MDMs seems to be cooling down (while other institutions still seem rooted in the past) [20], in large part due to the increasing adoption of CVD practices, which is viewed by both sides as a compromise position beneficial to both reporter and vendor and is being increasingly recognized and recommended by regulators around the world. With the requirement in the Omnibus Act for MDMs to develop plans to [21]: "monitor, identify, and address postmarket cybersecurity vulnerabilities and exploits, including coordinated vulnerability disclosure," the resistance to this activity should rapidly fade.

There are many excellent standards for implementing CVD (available from NTIA, ISO (specifically, ISO/IEC 21947, ISO/IEC 30111, and ISO/IEC TR 5895), CERT, MDIC, IETF, and Google's Project Zero), as well as good organizations (CERT/CC, DHS CISA, H-ISAC, MedISAO) that will assist the researcher or the MDM in publicizing the vulnerability. For smaller manufacturers that may find standing up to their own CVD program to be challenging, organizations such as MedISAO provide such services [22].

MDMs may find CVDs uncomfortable to perform, feeling that they are tarnishing their good name and ruining their position in the marketplace. Several years ago, when cybersecurity awareness was low and the realities of the threat landscape widely misunderstood (and we were living with much simpler threats), that may have been a valid business concern (although it should be noted that it was never a valid concern from the point of view of protecting patients and end-users). To avoid public exposure, many MDMs would conduct a silent recall and some MDMs may still be reluctant to do otherwise, despite the FDA's efforts to incentivize responsible management of the disclosure. However, many MDMs have taken the lead and implemented excellent disclosure sites. H-ISAC maintains a current set of links to manufacturer sites on their web site [23]. MedISAO provides the opportunity to submit a disclosure to its members via its website [22].

Coordinated disclosure has changed from being a negative public relations event to a positive one, where an MDM can celebrate their proactivity in addressing security issues before patients are harmed. In today's environment, disclosures help to promote and maintain public trust in the MDM brand, which can lead to increased market share for its current and future devices, as well as limiting the MDM's exposure to the perception of negligence. In 2018, the Medical Device Innovation Consortium (MDIC) released a white paper that articulated that a company's legal risks are reduced by implementing a

CVD program. This report also documented best practices from companies that had implemented effective programs [24].

8.3.2.2 Information-Sharing Organizations

The Critical Infrastructure Information ACT of 2002 and Executive Order 13691 established the concept of an ISAO with the objective to promote cybersecurity information sharing among private sector and critical infrastructure industries, including the healthcare and medical device industries, that would otherwise see collaboration on cybersecurity among competitors as prohibited by antitrust laws. ISAOs have been formed across industries [25] for the purpose of gathering, analyzing, communicating, disclosing, and voluntarily disseminating critical cyber information to its members. Full participation in an ISAO is strongly recommended by the FDA and provides certain advantages with regards to cybersecurity reporting. ISAO participation must be documented if the MDM wants the FDA to recognize their participation. The criteria for active participation are as follows:

1. The manufacturer is a member of an ISAO that shares vulnerabilities and threats that impact medical devices.
2. The ISAO has documented policies pertaining to participant agreements, business processes, operating procedures, and privacy protections.
3. The manufacturer shares vulnerability information with the ISAO, including any customer communications pertaining to cybersecurity vulnerabilities.
4. The manufacturer has documented processes for assessing and responding to vulnerability and threat intelligence information received from the ISAO. This information should be traceable to medical device risk assessments, countermeasure solutions, and mitigations [2].

For maximum efficacy, the information must be shared in as close to real time as possible. Participation includes identifying standards and guidelines for robust information-sharing and analysis related to cybersecurity risks, incidents, and best practices.

Both information sharing and analysis centers (ISACs) and ISAOs exist. H-ISAC is the ISAC for the healthcare and public health critical infrastructure sector. Other ISAOs serve subsectors (e.g., MedISAO for MDMs) and are acceptable mechanisms for FDA participation.

The FDA has signed Memoranda of Understanding with the Department of Homeland Security (DHS), H-ISAC, and MedISAO to facilitate information-sharing about potential or confirmed medical device cybersecurity vulnerabilities and threats. To motivate MDMs to join an ISAO, the FDA has lowered its expectations for when the MDM's recall must be reported for a security correction, as shown in Figure 8.4.

The expected key outcomes for belonging to and participating in an ISAO include:

- Improved understanding of vulnerabilities and how they may affect medical devices;
- Improved understanding of security practices;
- Improved solution development for the entire stakeholder community;
- Harmonized best practices for device security information-sharing;
- Improved efficiency to market while simultaneously improving security, safety, and privacy profiles for devices and associated networks.

Clearly, there are many benefits to belonging and participating in an ISAO beyond just meeting regulatory expectations. To summarize, it is in any MDM's best interest to belong to an ISAO.

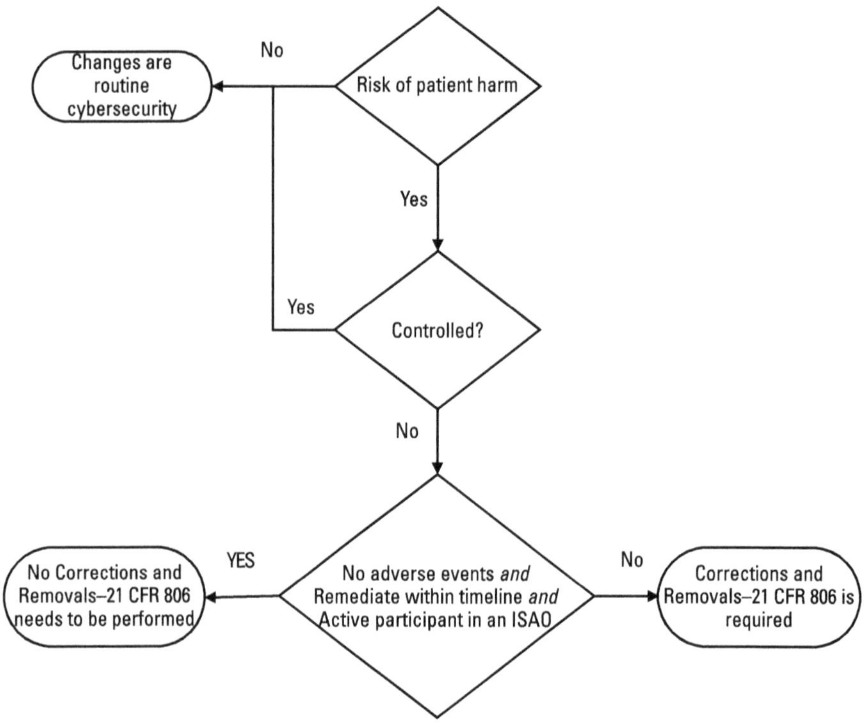

Figure 8.4 Avoiding corrections and removals.

8.3.3 Provide a Mechanism to Update Devices in the Field

"Updating the software in fielded medical devices is easy—just look at all the examples from semiconductor manufacturers. Or use existing Linux package managers to get the job done." This, or something similar, has been the uninformed opinion of nearly every engineer when they first start down the path of updating device software in the field. Most realize rapidly that nothing could be farther from the truth. Creating a secure, reliable, robust system to provide update functionality is, in itself, a major project requiring skills in a whole host of different technologies. This functionality is referred to by several different names, including FOTA and Device Firmware Update (DFU).

MDMs are required by the FDA to be able to provide field updates in a timely manner to address security issues as they are discovered. As discussed previously, under the new authorities provided to the FDA through the Omnibus Act, the agency requires MDMs to deliver postmarket patches on a regular basis and updates as needed, plus have the option to expedite an update if the risk is uncontrolled.

Even if an MDM is currently following the best practices for designing a secure device, that is, even if the device is completely secured to the maximum degree possible at the time of development, the tools and techniques of attackers are constantly evolving and improving. What may be secure today can be insecure tomorrow. For this reason alone, having the capability to update devices in the field is mandatory.

However, field updates create, at a minimum, four potential types of risks:

1. Exposure of intellectual property by reverse-engineering the update image deliverable;
2. Corruption or modification of the update image deliverable;
3. Rollback of device software to a previous update version with exploitable vulnerabilities;
4. Misuse of the update process as part of a supply chain attack.

Mitigations to these implicit vulnerabilities must be included in every field updateable system.

There are many examples of cybersecurity update failures that have occurred in just about every industry, including automotive [26, 27], personal computers [28, 29], smartphones [30], printers [31], medical devices [5], and just about everything else [32, 33]. In some instances, updates have been used to intentionally "brick" the device (i.e., render it completely unable to function) [34]. The need for a controlled, well-engineered update process cannot be overstated.

Starting at the top and going down to individual devices, we begin with an update engine. This is the database of the update images and is responsible for:

- Securely storing the update images.
- Determining the correct update image (if any) for a given model number. This includes determining the device's current hardware and software version, as well as verifying the upgradability of a given device.
- Building the secured deliverable package, which may contain one or more update images and update metadata, and producing an updated SBOM according to the changed software contents.
- Reporting on the version status of any device.
- Reporting on the status of a given update rollout, including depth of progress in the rollout.
- Communicating the update deliverable package to the device/system to be updated, in a device-agnostic manner.
- Logging all actions performed, including additions, deletion of update images, associations between different compatible versions of update images, execution of update rollouts.
- Extensive levels of user authentication and authorization to perform actions, such as multiparty approvals, with MFA-level authentication for each party.

As seen in Figure 8.5, there are a multitude of paths and communication mediums that could be involved in the process. Communication infrastructure along these pathways should be designed to be agnostic about the nature of what is being transmitted to avoid creating unnecessary roadblocks in update delivery and reporting.

In Figure 8.5, it should be noted that although there are a large number of potential communication media, ultimately the update deliverable package arrives at the medical device(s), where each medical device may have several internal components that also may or may not need to be updated together, plus several other medical devices or device accessories that may need updating. This results in a whole constellation of elements that need to be managed as part of the same update process. Coordinating this, especially in a way that allows for errors that may require aborting the update and rolling back to the previous version, is a difficult and complex task (see Figure 8.6).

To avoid a whole class of failures, including those represented in the failure examples previously noted, the target element that is being updated should

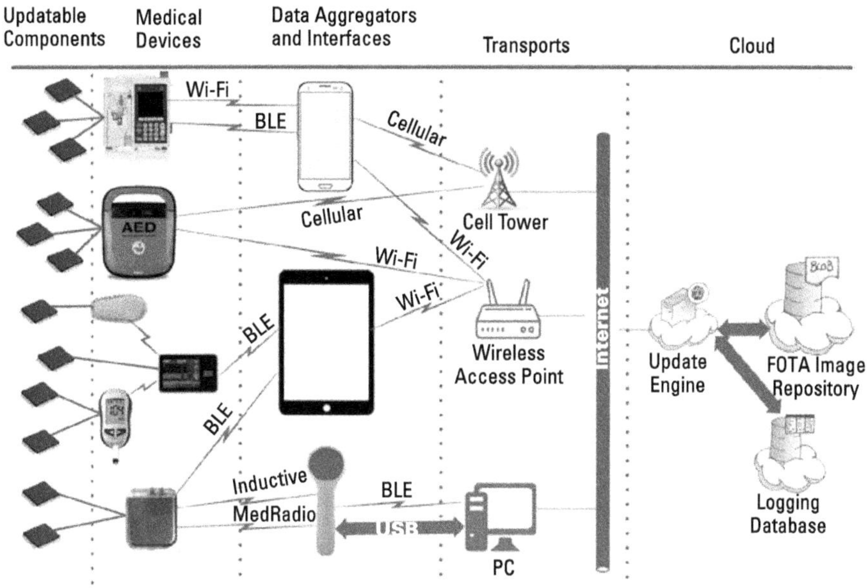

Figure 8.5 Potential paths and mediums involved in field software updates.

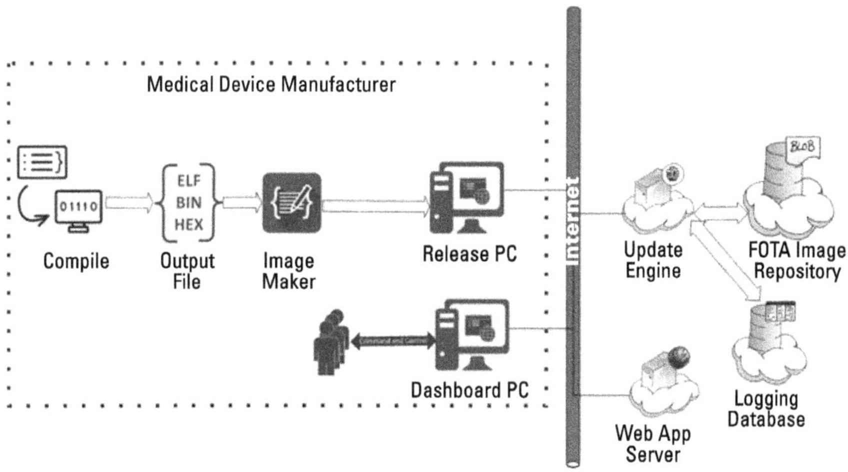

Figure 8.6 Update package assembly and delivery.

have sufficient nonvolatile program memory storage to accommodate two complete and separate images of the executable (Note: this is not just "buffering" in random access memory (RAM); this needs to be a fully executable image in the nonvolatile location in which it is placed.) By taking this approach, two separate versions of the executable can reside in the device and allow for rollback to the previous version if a failure occurs in any part of the update process.

Once a program has compiled and the resultant output file is created, it has to be converted into an update image file via an "image maker" program that adds:

- Versioning information;
- Target hardware information;
- Salt appended to the beginning and end of the compiled output file (this is necessary to avoid encryption of common compiler output formats and patterns that could be utilized to attack the encryption; salt is a nonsecret value used in a cryptographic process to ensure that the results of computations for one instance cannot be reused by an attacker);
- Cryptographic protection;
- Signature or hash.

This data structure is now the update image file. It is added into the FOTA Image repository and traced to any other update image files with which it may need to interact. As communication protocols and content changes, incompatibilities between update targets can occur if one target is updated and another target is left with an older incompatible version of its software.

There are several fundamental approaches to the update process:

1. The MDM is in complete control and has unfettered communication access to the device to be updated; this is most common with home healthcare uses and at some hospitals. With this approach, the device can be updated in the background with no user interaction or approval.
2. Similar to the previous approach, however, in this case, the medical device informs the end-user that an update is available and the update process needs to be confirmed and authorized by a user physically collocated with the medical device before the update can be performed by the MDM. This is a very common approach in hospitals as it allows the hospital to control the process and possibly sequester the medical device(s) to be updated.
3. The medical device is not permitted by the hospital to access the MDM's servers and as such the medical device cannot inform the end user of a pending possible update. Some alternative method (service portal, mail, email, phone call, ISAO notification) can be utilized where the MDM contacts the hospital and informs them of a pending update. At that point, the hospital can sequester the medical device(s),

establish a means of open communications between the MDM and the sequestered device(s), and the end user can approve the update process allowing it to proceed.

4. Similar to the previous approach, the medical device is not permitted by the hospital to access the MDM's servers and as such the medical device cannot inform the end user of a pending possible update. Some alternative method (mail, email, phone call, ISAO notification) can be utilized where the MDM contacts the hospital and informs them of a pending update. At that time an appointment can be scheduled where an MDM representative can be physically present in the hospital to perform the update locally with specialized tools under the supervision of the hospital staff.

This process can be quite involved, depending on the device's use case and other risk factors. It may entail on-site personnel performing post-update testing as part of installation qualification to ensure the device is operating as expected following the update and before it is reintroduced to the HDO's care deliver rotation. An end-user may or may not need to be physically onsite with the device to approve the update process. This is due to the wide variety of medical device types, such as life support devices that cannot be interrupted in any manner while in active use and so must be approved by on-site personnel for safety, versus small sensor devices that have no user interface on the device at all.

Traditionally, very insecure approaches have been utilized in this process, such as shipping hospitals USB flash drives (or other data carriers) containing an update to be applied to all of the devices in the hospital. This approach results in the following issues and concerns:

- It is difficult for the MDM to maintain records of devices that have been updated and when they were updated for record keeping and reporting to regulatory agencies.
- Flash drive contents can be reverse-engineered to expose other medical device vulnerabilities.
- Flash drives can be a source of cross-contamination of malware from device to device.
- None of the typical security controls can be implemented in a flash drive, such as authentication, authorization, and per device unique encryption keys.
- The medical device would need an exposed USB port, which, given the underlying operating system, could allow for easy exploitation of the medical device.

- Speed of the update process is slow compared to a more online approach, especially if the devices are under active attack. If the FDA believes that the update is taking too long they may issues a mandatory Class I recall.
- Control over the update process lies with the customer or operator of the device and is (largely) out of the control of the manufacturer.

8.3.3.1 Steps to Perform a Field Update

The following example assumes a deployment scenario similar to #1 or #2 above. In this situation, a controlled series of predetermined steps should be followed to perform a device update:

1. Communications is established with the target device.
2. The target device is notified of the presence of a newer version.
3. Depending on the device type and use case, the target device may or may not require end-user or operator approval.
4. The target device reports back all of the hardware configurations and all of the software versions currently in use in its system to the cloud update engine. This may include multiple central processing units (CPUs) per device and multiple devices and accessories, depending on the system.
5. Using the versioning information for each system and the tracing information provided when each update image was added to the FOTA repository, the cloud update engine builds a secured deliverable package (which may consist of multiple compatible update images). This secure deliverable package is uniquely encrypted for each device and instance of the process, as well as signed or key-hashed to ensure the integrity and authenticity of its origin.
6. This secured deliverable package is then communicated back along the path of communications. "Data in motion" security mitigations are not assumed or required to be in place, due to the secure nature of the deliverable package.
7. From this step forward, any errors or detection of compromised data is the reason to terminate the entire process and perform rollbacks to the previous version software if necessary.
8. Upon receipt of the secured deliverable package by the medical device, the integrity and origin of the package are confirmed.
9. The package is decrypted and the constituent update image files are communicated to the correct component or device in the system.

10. As each target component receives its update image, the integrity of the image is confirmed and then decrypted.
11. The appended salt is removed and discarded.
12. The version information is compared to the current version. The update image version must be greater than the current version for the update to proceed.
13. The plaintext version of the update image is written to its intended operating location in program memory.
14. Any pre-update activity that needs to be performed would be performed now, including any backout state to be utilized for recovery, if needed.
15. Each target device and component would confirm the integrity of the stored update image. This does not necessarily have to be a cryptographic operation and could be as simple as a cyclic redundancy check (CRC).
16. Utilizing whatever communication media are in place between these components and devices, each target indicates its integrity and readiness to proceed.
17. If any components do not report that they are ready to proceed, the entire update process on all devices and all components are halted. The failure report—or failure to report, whichever is applicable—is communicated back to the cloud update engine for logging.
18. If all components are ready, then the cloud update engine is informed that the update process was a success (and the success report is logged).
19. If all components indicate their readiness, then each component performs a reboot and allows the boot manager code to determine the most recent version of intact software to which to transfer execution.
20. In the update image for each component, there may have been post-update activities to be performed, such as cleaning up any temporary data, converting persistent data structures' contents, or indexing addresses. If present, these activities are now performed.
21. The system has successfully completed the update process.

The success or failure of each update attempt in the field is logged for reporting activities performed at the MDM and may be needed for review by the FDA or the DHS to confirm the progress of the update process against the number of devices in the field. This attrition rate should be balanced against patient risk and if the rate of updating the software is found to be too slow [35] or too problematic [36], additional methods may have to be pursued, such as additional lines of communication to the consumers of the medical device, and even the potential for a recall.

In all cases, the safety of the patient is paramount and must be the primary consideration when structuring the update process. A good example of this would be an implantable pacemaker where a vulnerability has been discovered, but no evidence of active or past exploits has been identified. This situation would require a balanced decision between the technical risk of the update and the security risk of the update being exploited. Since the update would require the patient to go into the physician's office, where emergency care could be provided if needed, the overall time required to roll out the updates will be much longer and, rightfully, some devices may never be updated.

8.3.4 Enable Management and Maintenance of the Device's Security Posture

As part of the MDM's maintenance lifecycle (see Figure 8.1), the manufacturer will, from time to time, provide security updates as part of their risk mitigation process, such as in response to a newly discovered vulnerability or threat. The HDO may have a formal, contractual relationship with the MDM to provide maintenance service to their devices, which would include cybersecurity updates such as configuration changes or patches to be deployed as a special action (if high priority) or as part of the regular maintenance schedule (if low priority).

In case the MDM is not contracted to perform this type of update, service personnel may still be asked to assist customers with planning, managing, and executing what could be quite complex field update projects. In either case, the maintenance activity needs to be coordinated with hospital operations and care delivery priorities and may require complex change management. Change-supporting activities may include planning, supply of update, installation, testing, and training.

8.3.4.1 Providing Security Alerts to Customers

Alerting customers about observed cybersecurity risks, or actual events and incidents pertaining to a medical device, can be based on a formal relationship and as part of a managed service contract under which the service provider collects and analyzes event logs and alerts. Another possibility is that, during the time of service, a technician may observe indicators (functional or performance compromise, or event log data) that a device has been compromised or that a security event is taking place.

No matter what the scenario is under which a manufacturer becomes aware of a security issue, having clearly identified contact points and defined channels of communication between manufacturer and customer is critical for timely and effective response. This can be even more challenging to identify if devices are sold through distributors or to leasing companies. Ideally, at the time of purchasing the parties should establish a Responsibility Agreement (e.g., as per ISO/TR 80001-2-6:2014) that includes a definition of how the parties will communicate on security issues and identification of contact points.

Because the manufacturers have visibility across their entire installed base, they are able to gather and analyze event data, observing trends that should be shared with the customer base as part of expected postmarket activities.

See Chapter 9 for a detailed discussion of incident response and related communications.

8.3.4.2 Providing Customer Support

MDM functions may include analysis of ongoing incident data and support of impact assessment, recovery of device functionality and/or data, the final restoration of the device to its original state, and supporting collection and analysis of forensic data. In the typical case, the HDO will have the main responsibility for incident handling and recovery and, typically, the MDM's service organization will provide a supporting role. Incident response activities are discussed in more detail in Chapter 9.

8.4 Product Recalls

In the United States, a recall is a method of removing or correcting devices in the field that are in violation of the FDA risk regulations. Failure to correct a violation could lead to legal enforcement.

There are three types of recall:

1. Voluntary recall under 21 CFR 7;
2. Mandatory device recall under 21 CFR 810;
3. Corrections and removals under 21 CFR 806.

The vast majority of recalls are voluntary and are initiated by the MDM. Technically, a recall is [37]: "a firm's voluntary removal or correction of a marketed product that the FDA considers to be in violation of the FD&C Act and against which FDA would initiate legal action." Recall activity may need to be supported by requirements for medical device reporting for device user facilities, manufacturers, importers, and distributors in the case of deaths and serious injuries (see 21 CFR 803).

All recalls are classified according to the relative degree of health hazard posed by the continued use of the medical device as follows [38]:

I. A situation in which there is a reasonable probability that the use of, or exposure to, a violative product will cause serious adverse health consequences or death.

II. A situation in which use of, or exposure to, a violative product may cause temporary or medically reversible adverse health consequences or where the probability of serious adverse health consequences is remote.

III. A situation in which use of, or exposure to, a violative product is not likely to cause adverse health consequences.

Recalls are one of the more challenging events that can occur for an MDM. Voluntary or mandatory recalls are both bad for customers, public relations, future regulatory activities, and the business. Recalls cost MDMs an average estimated $2 million per recall, and costs for a single incident can be as high as $600 million, not including associated impact on the MDM's stock price and market share losses that potentially push into double-digit percentages [39]. Recalls can also cause stagnation in development as engineering resources are diverted to fix the problems rather than create new products, thus allowing competitors opportunity to gain on the MDM's market share. Clearly, recalls are bad and, in many cases, can be avoided by implementing a robust software-updating system. Furthermore, as previously discussed in this chapter, correction and removal recall reporting due to cybersecurity risks are affected by membership in an ISAO.

The vast majority of medical device recalls are not cybersecurity-related. More often than not, cybersecurity issues lead to a vulnerability alert and, in severe cases, may even trigger a warning letter or safety communication. Even so, for the 5-year period from January 2017 through December 2022, the FDA lists 60 cybersecurity-based recalls, both voluntary and involuntary [40].

The best way to avoid having to perform a cybersecurity-related recall is to design medical devices with secure field-update capability so that vulnerabilities can be mitigated without exercising the option of a recall.

8.5 Managing End of Life and End of Support

End of life (EOL) is a somewhat subjective concept. When examined from the HDO's perspective, it applies to the individual physical device and can mean:

- Decommissioning the device to destruction, waste disposal, or electronic recycling;
- Ownership transfer of the device to another HDO (through sale or donation) or back to a vendor;
- Return of a leased or loaned device.

From the perspective of the MDM, EOL applies to an entire device type or version and can mean:

- Ownership transfer of the device's product line to another MDM;
- End of the device's warranty period;
- End of the device's support period;
- EOL for a component or TPSC utilized in the device that necessitates redesign of the device;
- End of the device being marketed;
- End of the device being manufactured;
- Expected life of a device (see C.F.R 21 part 803.3).

For MDMs, the generalized steps under the EOL umbrella are End of Production; End of Sales; End of Guaranteed Support; and End of Support. A good definition of how various end stages can be defined and should be communicated between MDM and HDO is provided by HSCC's "Health Industry Cybersecurity – Managing Legacy Technology Security (HIC-MaLTS)" [41]. MDMs must communicate any of these events to the customers well in advance of the End of Support date; typically, 3 years is considered a reasonable minimum period of notification. From a cybersecurity perspective this results in three major problems:

1. Following End of Support (EOS) from the MDM, there will be no further support or vulnerability patching, so risk-based decisions have to be made to determine whether continued use of the medical device by the HDO is warranted and the device can be operated with sufficiently low risk.
2. For any of the HDO's types of EOL, a common theme emerges. Before the medical device leaves the hands of the HDO, any sensitive data being stored in the device needs to be erased. This includes PHI, PII, and any configuration settings, including security-related data such as device, network, or user credentials.
3. Following a transfer of ownership of the medical device to a new HDO, the MDM may not be able to locate the new owner to notify of discovered vulnerabilities or to provide security patches to the device.

These three problems have possible solutions, with varying degrees of effectiveness:

1. The MDM should along with the end of support notification could also provide technical descriptions, as to how the device could be placed into a more secure configuration, or environment, such as disabling communication medium, segmenting the network that is connected to the device. The MDM should take this as an opportunity to support the HDO by offering newer updated device options in an attempt to maintain the business relationship with an existing customer.
2. A "restore to factory default" functionality should be a mandatory feature in any device that stores or buffers any patient or other sensitive data. This should include features (and instructions) that destroy said data. This functionality should be gated by some form of authentication, to prevent patient data from being destroyed in a normally functioning system that is not at EOL.
3. Unfortunately, it can be very difficult for the MDM to track geographical location changes and new owners of a medical device (especially when selling though as distributor network) to inform them of cybersecurity events, and software upgrades. This is one more reason why MDMs need a product security portal on their web presence, where customers can come to them to seek out security information about their newly acquired, old medical device. However, if the device implements a periodic check to an update server, those queries may be used as an IP Beacon to identify where those devices are located on the internet.

An additional complication may arise when the device's cybersecurity EOL happens before its general EOL (e.g., spare parts may still be available but a device main software component, such as the operating system, is no longer supported). Further, EOL considerations may include contractual changes as well as activities such as formal license and risk transfer from the MDM to the HDO.

8.6 Conclusion

The postmarket phase has seen a lot of expansion in cybersecurity activities and responsibilities over the past 10 years, and this trend is sure to continue with additional requirements for every aspect of critical infrastructure, including MDMs. The job description of "sustaining engineer" has largely disappeared in medical device companies, but this role may need to be resurrected or reimagined as a "sustaining security engineer" due to all of the ongoing postmarket cybersecurity activities now expected by regulators, customers, and patients.

All of these improvements in postmarket cybersecurity add to the burdens placed upon the MDM. What has yet to be determined is who will bear the costs of all of these new activities.

References

[1] Ford, G., "Statement on Signing the Medical Device Amendments of 1976," May 28, 1976, https://www.presidency.ucsb.edu/documents/statement-signing-the-medical-device-amendments-1976.

[2] FDA, "Postmarket Management of Cybersecurity in Medical Devices; Guidance for Industry and Food and Drug Administration Staff," FDA Guidance Documents, December 28, 2016. https://www.fda.gov/regulatory-information/search-fda-guidance-documents/postmarket-management-cybersecurity-medical-devices.

[3] U.S. Congress, H.R. 2617 - Consolidated Appropriations Act, 2023, December 29, 2022, https://www.congress.gov/bill/117th-congress/house-bill/2617.

[4] Hammerand, J., "FDA Launches Cybersecurity Requirements for Cyber Device Reviews," *Medical Design & Outsourcing*, April 3, 2023, https://www.medicaldesignandoutsourcing.com/fda-cybersecurity-requirements-for-cyber-device-reviews-rta-rejection/.

[5] Fu, K., "Click Here to Download Your AVEA Ventilator Software Update. Trust Me," Archimedes Center for Medical Device Security, June 8, 2012, https://secure-medicine.blogspot.com/2012/06/click-here-to-download-your-avea.html.

[6] International Standards Organization, "ISO/TS 11633-1:2019 Health Informatics — Information Security Management for Remote Maintenance of Medical Devices and Medical Information Systems — Part 1: Requirements and Risk Analysis," August 2019, https://www.iso.org/standard/69336.html.

[7] Hoyme, K., "Streamlining Management of Post-Market Vulnerabilities," *AAMI CyberInsights*, March 16, 2022, https://array.aami.org/content/news/streamlining-management-of-postmarket-security-vulnerabilities.

[8] FDA Fact Sheet, "The FDA's Role in Medical Device Cybersecurity: Dispelling Myths and Understanding Facts," https://www.fda.gov/media/103696/download.

[9] Palo Alto Networks Unit 42 Research Team, *2023 Unit 42 Network Threat Trends Research Report*, June 2023, https://start.paloaltonetworks.com/unit-42-network-threat-trends-report-malware-2023.html.

[10] Microsoft, *The Seven Properties of Highly Secured Devices*, 2nd ed., November 2020, https://www.microsoft.com/en-us/research/publication/seven-properties-2nd-edition/.

[11] Synopsis, "DesignWare ARC SEM Security Processor Family," 2023, https://www.synopsys.com/designware-ip/processor-solutions/arc-sem-family.html.

[12] Harp, S., T. Carpenter, and J. Hatcliff, "A Reference Architecture for Secure Medical Devices," AAMI/Array, October 2018, https://array.aami.org/doi/10.2345/0899-8205-52.5.357

[13] https://nvd.nist.gov/.

[14] https://vulners.com/feed.

[15] CISA, "Minimum Requirements for Vulnerability Exploitability eXchange (VEX)," April 2023, https://www.cisa.gov/resources-tools/resources/minimum-requirements-vulnerability-exploitability-exchange-vex.

[16] Bursztein, E., and J. -M. Picod, "A Hacker Guide to Deep Learning Based Side Channel Attacks," *Defcon 27*, Las Vegas, NV, August 8–11, 2019, https://elie.net/talk/a-hackerguide-to-deep-learning-based-side-channel-attacks/.

[17] Donohue, B., "Critical Security Vulnerabilities in Drug Injection Pumps," *Kaspersky Daily*, May 12, 2015, https://usa.kaspersky.com/blog/drug-pump-security-bugs/5331/.

[18] Kan, M., "Medical Device Security Disclosure Ignites an Ethics Firestorm," *Computerworld*, August 29, 2016, https://www.computerworld.com/article/3113385/medical-device-security-ignites-an-ethics-firestorm.html.

[19] Krebs, B., "Iowa Prosecutors Drop Charges Against Men Hired to Test Their Security," *KrebsOnSecurity*, January 31, 2020, https://krebsonsecurity.com/2020/01/iowa-prosecutors-drop-charges-against-men-hired-to-test-their-security/.

[20] Wakabayashi, D., "He Won Praise for Halting a Global Cyberattack. Then He Was Arrested," *The New York Times*, August 3, 2017, https://www.nytimes.com/2017/08/03/technology/cybersecurity-researcher-hailed-as-hero-is-accused-of-creating-malware.html.

[21] Hammerand, J., "FDA Launches Cybersecurity Requirements for Cyber Device Reviews," *Medical Design & Outsourcing*, April 3, 2023, https://www.medicaldesignandoutsourcing.com/fda-cybersecurity-requirements-for-cyber-device-reviews-rta-rejection/.

[22] MedISAO, "Vulnerability Reporting," https://members.medisao.com/vulnerability_disclosure/

[23] H-ISAC, "Medical Device Manufacturing Security," 2023, https://h-isac.org/mdm-security/.

[24] MDIC, "Medical Device Cybersecurity Report: Advancing Coordinated Vulnerability Disclosure," October 2018, https://mdic.org/resource/medical-device-cybersecurity-report/.

[25] ISAO Standards Organization – Info Sharing Groups, https://www.isao.org/new-information-sharing-groups/.

[26] Howard, B., "It Won't Be the Last Time: Over-the-Air Glitch Bricks Lexus Infotainment," *ExtremeTech*, June 14, 2016, https://www.extremetech.com/extreme/229877-it-wont-be-the-last-time-over-the-air-glitch-bricks-lexus-infotainment.

[27] Greenberg, A., "Tesla Responds to Chinese Hack with a Major Security Upgrade," *Wired*, September 27, 2016, https://www.wired.com/2016/09/tesla-responds-chinese-hack-major-security-upgrade/.

[28] Morris, L., "A Dell Update Bricked My Computer," *Fstoppers*, March 2, 2019, https://fstoppers.com/originals/dell-update-bricked-my-computer-344764.

[29] Lovejoy, B., "Limited Reports of Catalina Installation Bricking Some Macs Via EFI Firmware," *9TO5Mac*, October 24, 2019, https://9to5mac.com/2019/10/24/efi-firmware/.

[30] Davis, J., "3 Ways to Fix iOS 12/12.3/13 Update Bricked My iPhone," *Wondershare*, December 26, 2019, https://drfone.wondershare.com/ios-12/update-bricked-iphone.html.

[31] Cui, A., M. Costello, and S. Stolfo, "When Firmware Modifications Attack: A Case Study of Embedded Exploitation," *NDSS Symposium 2013*, San Diego, CA, April 23, 2013, http://ids.cs.columbia.edu/sites/default/files/ndss-2013.pdf.

[32] Brignall, M., "Samsung TV Owners Furious After Software Update Leaves Sets Unusable," *The Guardian*, August 24, 2017, https://www.theguardian.com/technology/2017/aug/24/samsung-tv-buyers-furious-after-software-update-leaves-sets-unusable.

[33] Palmer, A., "Not-So-Smart Sneakers: Nike 'Working to Fix' Software Update That Bricks Its $350 Self-Tying Trainers, Leaving Them Unable to Tighten," *The Daily Mail*, February 21, 2019, https://www.dailymail.co.uk/sciencetech/article-6731599/Not-smart-sneakers-Software-update-bricks-Nikes-350-self-tying-trainers.html.

[34] Kieler, A., "Samsung Software Update Will Deliberately 'Brick' Remaining Galaxy Note 7 Phones," *Consumer Reports*, December 9, 2016, https://www.consumerreports.org/consumerist/samsung-software-update-will-deliberately-brick-remaining-galaxy-note-7-phones/.

[35] Saxon, L., et al., "Factors Influencing the Decision to Proceed to Firmware Upgrades to Implanted Pacemakers for Cybersecurity Risk Mitigation," *Circulation*, Vol. 138, September 18, 2018, pp. 1274–1276, https://www.ahajournals.org/doi/pdf/10.1161/CIRCULATIONAHA.118.034781.

[36] Lee, J., et al., "Pacemaker Firmware Update and Interrogation Malfunction," *Heart Rhythm Case Reports*, Vol. 5, No. 4, April 2019, pp. 213–216, https://www.heartrhythmcasereports.com/article/S2214-0271(19)30001-6/pdf.

[37] FDA, "Recalls, Corrections and Removals (Devices)," April 25, 2019, https://www.fda.gov/medical-devices/postmarket-requirements-devices/recalls-corrections-and-removals-devices.

[38] FDA, "Medical Device Recalls," September 26, 2018, https://www.fda.gov/medical-devices/medical-device-safety/medical-device-recalls.

[39] Fuhr, T., K. George, and J. Pai, "The Business Case for Medical Device Quality," McKinsey Center for Government, October 2013, https://www.mckinsey.com/~/media/McKinsey/dotcom/client_service/Public%20Sector/Regulatory%20excellence/The_business_case_for_medical_device_quality.ashx.

[40] FDA's "Medical Device Recalls Database," https://www.accessdata.fda.gov/scripts/cdrh/cfdocs/cfRES/res.cfm.

[41] HSCC Cybersecurity Working Group, "Managing Legacy Technology Security," March 2, 2023, https://healthsectorcouncil.org/legacy-tech-security/.

9

Incident Response Planning

Incident response planning is a critical element of every postmarket management program. Unfortunately, with today's sophisticated threats, it is not a matter of *if* a cybersecurity event occurs but *when*. Therefore, it is of critical importance that an organization is prepared in advance to minimize the impacts on patient safety, data and equipment loss, and operational and financial impact, as well as optimize recovery capabilities to minimize incident spread and duration and enable capture of valuable forensic evidence. Responding to incidents is not merely a technical task, but also includes several important nontechnical steps, including sound decision-making and ensuring operational continuity as well as multiple rounds of communication with various internal and external stakeholders.

When looking at medical devices, there are two main categories of incidents. One is a general, reportable event that resulted in or could lead to patient harm (but is not necessarily related to cybersecurity) and may necessitate regulatory notification and action, such as a recall. The other is a cybersecurity incident, which will be the focus of this chapter, that needs to be handled appropriately. If such a cybersecurity incident resulted in or could lead to patient harm, it will also require regulatory corrective action including a recall and/or a software update to fielded devices. In that sense, a cybersecurity incident response process should have the ability to escalate to a formal corrective action process.

Sometimes the term *incident* is more broadly used to include, for example, the discovery of a new vulnerability or the emergence of a new threat. Although we acknowledge the broader use of the term, in this chapter, we apply the more narrow definition of an actual occurrence as the result of a compromise. Thus, for our purposes a cybersecurity incident is the result of a threat successfully exploiting a vulnerability, resulting in or having the capability to compromise an (tangible or intangible) asset. Incidents can be detected by observing evidence of adversarial activity (sometimes referred to as Indicator of Compromise (IoC)), which could lead to an (actual or potential) compromise of a device or the device's ecosystem (e.g., its network, a cloud service, or connected accessory). Any compromise of the device's integrity, availability, or confidentiality is considered an incident, even if the attempt did not or has not yet interrupted or degraded a device's essential functions. The NIST defined an "incident" as [1]: "An occurrence that actually or potentially jeopardizes the confidentiality, integrity, or availability of a system or the information the system processes, stores, or transmits or that constitutes a violation or imminent threat of violation of security policies, security procedures, or acceptable use policies."

Simply put, an incident is any occurrence that could have an actual or potential security impact.

As with other cybersecurity terms, there is, unfortunately, a degree of discrepancy in the use of traditional safety terms as compared to traditional security terms. For example, consider how broadly the EU's Medical Device Regulation defines "incidents," of which "cybersecurity incidents" are one type:

> Any malfunction or deterioration in the characteristics or performance of a device made available on the market, including use-error due to ergonomic features, as well as any inadequacy in the information supplied by the manufacturer and any undesirable side-effect.

9.1 Incident Response Scenarios

There are several different scenarios that should be covered in an incident response procedure (or a set of individual procedures for each scenario). The traditional division of responsibilities is that the MDM is responsible for designing and delivering a sufficiently secure device to an HDO and ensures continued safety and effectiveness through postmarket management, for example, updates and patches (see Chapter 8). However, the responsibility for operational safety (including security) lies with the HDO as the operator and responsible party, including managing the technical, operational, and legal aspects of an incident. (This is true even in home health care scenarios—the HDO employing the device prescriber and/or the physician supervising the patient's use of the device are ultimately responsible for operational safety, even when certain responsibilities are contractually delegated to a third-party healthcare provider.) For example, in the United States, the Centers for Medicare & Medicaid Services (CMS) State Operations Manual defined that [2]:

> In order to ensure an acceptable level of safety and quality, the hospital must identify the equipment required to meet its patients' needs … In addition, the hospital must make adequate provisions to ensure the availability and reliability of equipment needed for its operations and services. Equipment includes both facility equipment … and medical equipment, which are devices intended to be used for diagnostic, therapeutic or monitoring care provided to a patient by the hospital (e.g., IV infusion equipment, ventilators, laboratory equipment, surgical devices, etc.).

In addition, under the HIPAA, the protection of the privacy and security of "individually identifiable health information" is the responsibility of the "covered entity" (health plans, health care clearinghouses, and health care providers) that must put in place technical and nontechnical safeguards that secure electronic protected health information (ePHI) [3]. This includes the obligation to report breaches affecting more than 500 individuals within 60 days and smaller breaches annually [4].

Furthermore, both MDMs and HDOs may have additional responsibilities to report certain types of incidents (e.g., in case of patient harm (actual or potential for death or serious injury)) and may have to initiate specific corrective action as defined by national regulation [5]. Similar requirements exist in other regions and countries, sometimes with a very broad definition of the types of incidents that require reporting (e.g., issues related to performance, diagnostic accuracy, decision support software resulting in harm, connected hardware/software, human interfaces, use error, inadequate labeling and instructions, security problems) [6].

Today's business and technology reality is complex and can lead to different scenarios and sharing of incident response responsibilities, typically defined under contracts such as a HIPAA Business Associate Agreement (BAA) [7] or Responsibility Agreement under IEC 80001 [8].

The following is an overview of the main incident response scenarios that an MDM may need to support and that may require dedicated processes. Reality is often more complex, and, in any given case, gray zones may exist between the described scenarios:

1. An HDO experiences a medical device security incident and asks MDM for support. HDO is leading the response and makes all operational and legal decisions, MDM provides technical support.
2. An HDO has contracted with MDM's service organization to maintain devices on the HDO's behalf. In this scenario, MDM has more responsibilities, but the HDO is still responsible for major operational and clinical decisions and all legal decisions.

3. An MDM service and support infrastructure is impacted by an incident that could lead to the HDO's exposure or compromise. In most cases, this will be similar to MDM's own internal IT incident response procedure but will need to include additional aspects such as management of field service equipment (e.g., laptops) as well as customer communication.
4. An MDM hosts customer and/or patient data on behalf of an HDO (i.e., is a business associate under HIPAA). The MDM has an incident in its infrastructure that affects the HDO data. In this case, the MDM's incident response procedure should cover:
   - Technical recovery (in most cases, this will be similar to MDM's own IT incident response procedure).
   - Legal and communication aspects (i.e., when and how to contact the HDO whose data is affected as well as legal notification requirements, such as under HIPAA).
5. An MDM manufacturing infrastructure is compromised by a security incident resulting in production disruption or potential degradation of product quality or could lead to product infection with malware. The incident response procedure would cover downtime management and restoration of operational capabilities, management of potentially compromised products and goods, and regulatory notification in case of field recall or shortages/supply issues of critical products as it may be required, for example, under the EU's NIS2 Directive (with manufacturers of certain medical and in vitro devices as well as pharmaceuticals now being in scope) [9] or under Section 506J of the FD&C Act ("Notifying FDA of a Permanent Discontinuance or Interruption in Manufacturing of a Device") [10]. (At the time of this writing, the Healthcare Sector Coordinating Council (HSCC) is finalizing detailed playbook on the topic of Manufacturing Cyber Incident Response).

In all the above, forensic evidence about the incident should be collected, preserved, and analyzed and used as input to the MDM's postmarket surveillance collection, as a security incident is a type of cybersecurity signal that may indicate a new vulnerability or threat.

As part of their cybersecurity documentation package, an MDM should provide technical instructions to HDOs on how to respond to and recover from a cyber incident on a specific device. Operational and legal decisions are covered elsewhere (i.e., in the HDO's own procedures).

In the scenario where a fielded medical device at an HDO is involved in a security incident, responsibility for managing the incident should be contractually defined [11] and will involve both the HDO and the MDM. Although keeping abreast of incidents, new vulnerabilities, and emerging threats is a critical aspect of postmarket management and support, investigating incidents to determine their potential for harm (current or future) is more urgent than investigating newly reported vulnerabilities, although the latter will occur more frequently. Increasingly, MDMs are asked to support their HDO customers with an incident response. The actions the MDM must take, their order and timeline, the parties involved, and associated communications and recordkeeping should all be covered in an incident response plan.

Several of the above are in line with established incident response best practices and have been documented in great detail elsewhere, for example:

- ISO/IEC 27035: Information Security Incident Management [12, 13];
- NIST SP 800-61: Computer Security Incident Handling Guide [14];
- NIST SP 800-83: Guide to Malware Incident Prevention and Handling [15];
- NIST SP 800-86: Guide to Integrating Forensic Techniques into Incident Response [16];
- NIST SP 800-150: Guide to Cyber Threat Information Sharing [17];
- NIST SP 800-184: Guide for Cybersecurity Event Recovery [18];
- SANS: Creating and Managing an Incident Response Team [19];
- CMSEI: Handbook for Computer Security Incident Response Teams (CSIRTs) [20];
- ISACA: Incident Management and Response [21];
- Interpol: Guidelines for Digital Forensics First Responders [22];
- IEEE: Cyber Autopsies: The Integration of Digital Forensics into Medical Contexts [23];
- MHRA: Guidance for manufacturers on reporting adverse incidents involving Software as a Medical Device under the vigilance system [6].

Therefore, the following discussion will be limited to some general best-practice advice as well as aspects that are unique to the medical device ecosystem, as, for example, provided by the MITRE white paper "Medical Device Cybersecurity Incident Preparedness and Response" [24].

Considering the pace of technological change, the ever-evolving threat landscape, and the widening attack surface driven by the demand for increasingly interconnected devices, every MDM should accept the potential for a future cybersecurity incident as likely, if not inevitable. If this has not seemed to be the case in the past, the assumption is that several factors contributed to underreporting, ranging from our ability to actually recognize a medical device cybersecurity incident and correctly attribute it as such; the lack of device features that would give us the ability to alert, log, or report compromise attempts; and the deliberate underreporting out of reputational and legal concerns.

It is now commonly accepted that the MDM's best practice is to design a device with the assumption that it likely will be involved in cybersecurity incidents during its useful life [25]. Planning for such incidents, through device design and resilience features as well as through the appropriate support processes, is the mark of a responsible, mature, forward-thinking organization that has determined to fulfill the security expectations of regulators, stakeholders, and, most importantly, clients and their patients.

As discussed above, in a typical medical device incident response scenario (see Figure 9.1), there are two series of events taking place: the HDO's response to the incident and their triage decisions around care delivery, recovery, communication, and the MDM's response as part of its postmarket and customer support responsibilities. The former is patient-focused and care delivery-focused, while the latter is design-focused and device lifecycle-focused. The most critical and complex scenario occurs when an MDM's device experiences a cybersecurity incident in the field, for example, when anomalous traffic on its communication channel is logged and reported to a user, operator, or field technician. The MDM must support the affected HDO appropriately and must document incident, response, and response effectiveness (if applicable). Some of these postmarket responsibilities are defined by the FDA and other regulators, while many postmarket responsibilities will be determined by the contractual relationship between the HDO and the MDM (see Chapters 7 and 10). The remainder of this chapter outlines the major areas that need to be addressed by an MDM to support incident response of a fielded device.

9.2 Incident Response Plan

Incident response begins with planning. A complete set of response procedures addressing the various incident scenarios discussed above should be fully complete, approved, trained, and practiced, and the MDM should be prepared to execute them, before the device is released to market. Incident response plans may vary, for example, an incident owned and managed by an HDO where the MDM provides only a supporting role calls for different activities than a scenario where the MDM would be the process owner. The MDM's incident response plan (or set of plans) should be scalable and flexible enough to address the various possible scenarios.

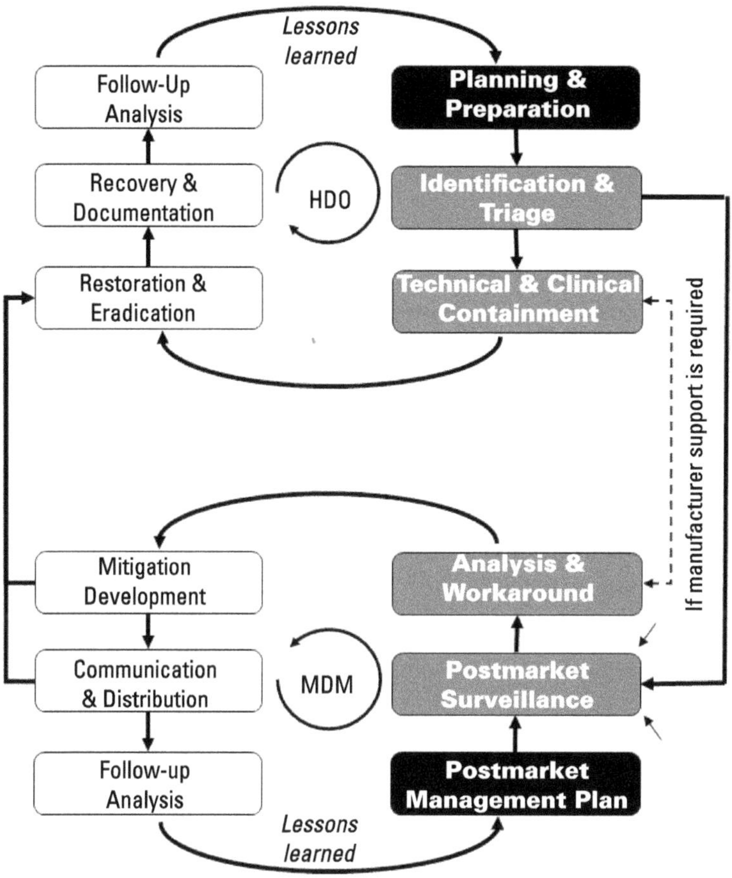

Figure 9.1 HDO versus MDM IR responsibilities and external touchpoints.

Incident response planning should harmonize with the MDM's existing risk management system (RMS) and quality management system (QMS). Although they will differ in some respects, procedures for responding to medical device cybersecurity incidents follow a similar basic structure to response planning for IT/infrastructure incidents. Derived from ISO/IEC 27002, the essential elements of a medical device incident response plan include:

- Procedures governing preparation and response for different incident types and severities. For example, a log report of anomalous comms activity with no apparent impact on the device's clinical function calls for a different type of response and warrants a different level of urgency than a report of a device undergoing manual emergency shutdown due to malfunction or aberrant behavior.

- Procedures for monitoring devices and information sources to detect, analyze, and report incidents.
- Procedures for tracking incident response activities, including activities carried out by the MDM as well as by HDOs and deputized agents in the field (which could include self-reported response activities by end-users).
- Procedures for collecting, handling, and preserving as well as dealing with forensic evidence.
- Procedures for incident assessment, decision-making, and escalation rubrics.
- Procedures for response, including:
  - Remediation and mitigation determination;
  - Mitigation design;
  - Mitigation deployment and controlled recovery, including deployment tracking and assessment;
  - Response escalation;
  - Communication and reporting, internal and external;
  - Incident closure, including post-closure debrief and follow-up [26].

Depending on the flexibility and maturity of the MDM's organization, these procedures should additionally ensure the competence of assigned personnel, including retraining existing personnel to develop the security capabilities of existing roles, as well as the availability of points of contact and reliable, efficient communication channels to all incident-response stakeholders. This requires regular maintenance of contact lists both internal and external (legal and enforcement authorities, regulators, media, resources, and information sharing).

Training procedures for the incident response plan should include mock exercises at least annually. Exercise results should be analyzed and the results used to inform updates to training procedures and to the plan itself. For further discussion of roles, responsibilities, and training, see Chapter 2.

9.3 Information Sources

MDMs should prepare to receive information pertaining to cybersecurity incidents from multiple channels. The most obvious is the cybersecurity event logging and reporting designed directly into the medical device system during the secure development lifecycle. Depending on the event and the device system, however, even this information may reach the MDM in multiple ways. A device system that routinely reports to the MDM's service monitoring infrastructure could inform the MDM "directly" about potential incidents. In other scenarios, the MDM may be informed directly by their customers (typically the HDO), by device users, field technicians, clinical engineers (HDO or contracted), a security incident response team (HDO or contracted), analysis of returned devices, an ISAO (see Chapter 8), or even law enforcement.

As information may come in through a variety of channels and may be provided by various sources, the MDM may have to rely on reviewing and correlating reports, which may or may not be clearly reported as cybersecurity incidents.

It will be necessary to train support personnel to recognize keywords (such as "security," "hacking," "attack," "infected") and other indicators in user requests for troubleshooting and support, or complaints about aberrant device behavior, that could point to a cybersecurity incident, and escalate as appropriate. With appropriate training, some of these sources—such as HDOs and field technicians—can be channeled through the MDM's existing support system. Others—ISAOs, disclosures by software and hardware suppliers, and in-house security testing—can be captured and assessed as part of the routine activities of the MDM's dedicated product cybersecurity team.

9.4 Investigation

Investigations are primarily an MDM activity initiated by an incident response trigger. Note that even though it is becoming more common for HDOs to add contract language allowing them to perform penetration testing and other security-related activities, including incident investigations, MDMs should always perform their own investigations in addition to supporting their clients' investigations. Until the exact root cause of the incident is identified, a mitigation cannot be crafted, and HDOs very rarely have the device-specific insight or the resources needed to ascertain this.

In general, an MDM's investigation will consist of the following:

1. Accumulating information about the event.
2. Attempting to reproduce the event in a controlled environment.
3. Determining the root cause of the event.
4. Assessing whether other products or versions may be affected by the same root cause.
5. Documenting and reporting the information gathered during steps 1 to 4 concurrent with the investigation.
6. Creating a final End of Action report summarizing activities and documenting the findings.

When an incident response is triggered, data-gathering activities are initiated to amass as much information about the incident as possible, including an assessment of the quality of the data and the reputation of the sources of the gathered information. Forensic quality and chain of custody need to be maintained and preserved.

Concurrent with this information-gathering effort, a technical team is enrolled to review gathered data and apply this information in a root-cause analysis (RCA) performed upon representative devices supplied by the MDM. An RCA is not the same as a vulnerability investigation: the RCA reveals why the vulnerability is present in the fielded system and how to mitigate it. This process can be hampered by insufficient or low-quality incident information, such as which models of medical devices are affected, what version of the software is being exploited, and what environmental conditions or configuration are utilized by the threat.

As the RCA progresses, temporary compensating controls (sometimes called workarounds) should be explored for dissemination to the reporting parties, HDOs, ISAOs, and relevant media.

Once the vulnerability's root cause is identified, mitigating controls should be implemented in a controlled environment, where the mitigated devices undergo complete revalidation testing to ensure that the mitigation generates no unintentional negative side effects. The technical team should also use root cause information to explore other similar medical devices created by the manufacturer where this same issue may also be present but has not yet been detected.

It should be noted that all root cause investigations should be performed on nonfielded medical devices, including any medical device that has been returned from the field. No medical devices that have been or may have been attacked should be directly involved in the investigation. Instead, if they become available to the MDM, these devices should be securely quarantined and managed in conformance with a chain of evidence policy [16, 22, 23].

Communication during all parts of the incident should not be limited to just the team conducting the investigation. At minimum, it is important to let the reporter and other stakeholders know that an investigation is underway. As the investigation progresses, the incident response plan will indicate whether additional parties—such as MDM upper management, ISAOs, regulators, or national security agencies—need to be alerted, provide a rubric for making that determination, and should ideally include prewritten templates for communicating the investigation's findings. These templates should have been reviewed, at minimum, by the MDM's public relations team and perhaps by legal as well, as part of the plan's approval process.

The question of what to communicate, when, to whom, and how should never be left up to the individual investigators to decide in the heat of the moment. Communication should be largely predefined, leaving room for customization based on incident particulars, and both template and incident-specific customization should be based on criteria defined in the incident response plan. This principle applies equally to every aspect of the incident response.

9.5 Triage

In general, the FDA, HDOs, and patients expect that manufacturers will award the highest response priority to incidents that may render the device unsafe or compromise its essential clinical performance, or, similarly, which could pivot from the device itself into connected systems and thereby impact patient safety, user safety, or the essential clinical performance of connected systems [27].

Once the root cause has been positively identified, the risk should be triaged based on the scope, potential severity, and exploitability. The primary concern should be given to preserving the essential clinical functions of the device. Confidentiality of protected information categories, such as personal information (PI), PII, and PHI should take secondary priority, with exposure of MDM confidential information such as intellectual property taking tertiary priority alongside other risks to the MDM business model (e.g., damage to MDM reputation).

All of this information needs to be presented to the MDM's senior management, along with a proposed plan for addressing the risks to devices in the field. Actions that might be recommended include:

(A) Do nothing; it was determined that the risk is sufficiently low. (Even when this is the recommended proposal, it is still the incident response team's responsibility to present the facts, findings, and recommendation to senior management.)

(B) Once a new version of the software has been validated, provide an update for the devices in the field and support update process as needed.

(C) Provide communication to customers, ISAOs, regulators, and others as needed.

(D) Perform a voluntary recall to update and/or replace affected devices.

(E) Engage in regulatory-mandated recall and field-corrective action in case of patient harm or potential for patient harm.

While a suggested course of action is crafted and suggested by the technical investigation team, it is senior leadership who must decide on the course of action to be taken.

9.6 Roles and Responsibilities

The organization of internal roles and responsibilities, including those that will be involved in a managed incident response, is discussed in detail in Chapter 2.

9.7 External Communications

Managing communications with stakeholders outside the MDM is critical to incident response. Because the general types of incidents that may occur can be anticipated, in part because the threat landscape is continuously being researched and reported on as it evolves and in part because the possible set of relevant incidents is limited by the device's design, communications details can and should be preplanned to a great extent. For each incident type, the incident response plan should provide:

1. A high-level summary describing general communications strategy for the incident type, including escalation criteria and paths for different types of incidents.
2. Descriptions of what information should and should not be communicated, to whom, and by what means, for each stage of the incident response (investigation, triage and mitigation planning, mitigation rollout, mitigation deployment and efficacy tracking, and closure).
3. A template, including sample language, for each communication.

Communications strategy, specifics, and templates require review and preapproval by at least three groups internal to the MDM: cybersecurity support engineering to ensure technical accuracy and honesty; public relations, marketing, or brand managers to ensure that messaging is clear, comprehensible to nontechnical audiences, and consistent with the MDM's overall public relation strategies; and legal or regulatory to ensure both that communications will reach all parties required by local regulations and that communication contents will limit the MDM's potential exposure as much as possible. Senior management should approve the communications plan only after it has achieved the recommendation of these three groups. Previously, many MDMs leaned towards communicating less rather than more. However, from a business impact perspective, it is now more commonly accepted that an open communication strategy is a sign of an MDM's cybersecurity maturity and responsible practices. Conversely, vague, minimal, or secretive responses are interpreted as attempts to cover up the facts and evade responsibility, triggering a loss of confidence in the MDM, which is then reflected negatively in the company's reputation, valuation, and market opportunities. Evasive communications may also be considered incomplete from a regulatory perspective, prompting additional scrutiny and even corrective action by regulators.

Depending on the impact and scope of the incident, specifics of the response may vary. Generally, during the incident response, external communication generally takes place between the MDM and six other stakeholder categories. These include those who are or who may at some point be directly affected by the incident—HDOs and patients—as well as those who are or who may become indirectly affected—MDM shareholders/owners; regulators and government safety and law-enforcement agencies; industry counterparts with systems, devices, and infrastructure that could be affected by means of the same exploit (or the same underlying vulnerability), through ISAO participation; and the public at large, who are communicated with via the press and any means the MDM uses for direct publication, such as on its corporate website, social media, or circulated news bulletin. Each of these groups has a particular interest in the incident and in the MDM's response, which differs greatly from all of the others, as well as differences in the incident-specific information that is appropriate, advisable, and perhaps legal for the MDM to provide them. Consequently, not only what is said, but how and to whom, must be evaluated well before any incident actually occurs for the MDM to maintain and enhance its business position during and following incident response.

Yes, it is possible for a well-prepared MDM to actually enhance its business position due to an exemplary response to a cybersecurity incident. Remember that we are no longer operating in a legal and business environment in which MDMs are punished or denigrated for having a device on the market that falls prey to a vulnerability exploit. If the MDM developed the device responsibly (i.e., securely), and if it responds appropriately to the incident, it will immediately become an example for consumers and professionals alike as an object lesson in how to do it right.

Effective communication with those who may be directly affected by the incident is also critical for minimizing unsafe unilateral actions that HDOs, end-users, or other field personnel might consider taking in the absence of timely information concerning the incident and the MDM's response plan. For example, an HDO might falsely believe that it needed to protect other critical infrastructure by transferring a patient from a compromised device that it believes could infect other systems, which could cause harm to that patient, when meanwhile the MDM could have a less risky mitigation option that it had simply failed to communicate clearly or in a timely manner to the HDO. The MDM should tightly coordinate and communicate all incident-related activities and work with the HDO to develop alternatives and compensating controls until the root cause and mitigating response are ultimately determined.

To that end, an MDM should consider well in advance how it intends to communicate with certain groups. Communications to regulators and to the

MDM's ISAO will generally follow established norms and expectations. Communications to HDOs, patients, and other directly affected persons will largely be determined by the specifics of the incident itself, the mitigation plan, and the messaging goals established by public relations. However, advance thought should be given to communications to MDM owners and shareholders to ensure that they are accurately and appropriately apprised without downplaying the incident, oversharing technical details, or needlessly exposing any party to legal liabilities.

Communications to the press also require consideration well before any incident occurs. Communication content should be prepared for such events over the course of several months and with rigorous review cycles being performed across a team of multiple disciplines, including technical, customer-facing communications, legal, regulatory, sales, and support.

It is recommended that the MDM's marketing, public relations, or media relations personnel cultivate working relationships with reporters who cover cybersecurity and medical devices in key publications, well in advance of any incidents, with the goal of establishing channels through which incident-related information can be released at the appropriate time. This will better enable the MDM to influence the public narrative and greatly reduce the chances of the incident becoming sensationalized or reported inaccurately.

9.7.1 Communicating Cybersecurity Risks to Patients

Cybersecurity risks, including those discovered as a result of an actual incident, must be communicated to patients and associated end-users, such as caregivers, at two general points in time:

1. During the clinical consult, typically at the same time that other potential risks to the patient, such as side effects, and risks related to the surgical implant, are communicated.
2. At any time when the MDM becomes aware of a new vulnerability that could potentially impact devices already prescribed to or used by patients.

For item 1, the MDM's obligations are generally fulfilled through labeling, plus any additional supporting materials that the purchasing HDO may require (see Chapter 7). This initial communication is not generally considered part of incident response or incident response planning.

Item 2, in contrast, is an essential element of incident response. The FDA has published a paper [28], "Best Practices for Communicating Cybersecurity Vulnerabilities to Patients," which includes findings and recommendations from the Patient Engagement Advisory Committee and input from the industry and other stakeholders. The paper included sample communications for use in incident response and recommends and explains the following best practices:

- Make the content easy for people to read and understand, including how to:
  - Keep it timely.
  - Keep it relevant.
  - Keep it simple.
  - Keep it readable for diverse audiences.
- Discuss risks and benefits.
- Acknowledge and explain the unknown.
- Make it easy for patients to find and use, including:
  - Make communications easy to find in online searches.
  - Make communications easy to view on mobile devices.

Additional examples and recommendations that MDMs may find helpful are available from Australia's Therapeutic Goods Administration (TGA) [29]. For insight into the HDO's role and perspective, the reader may benefit from reading "What Should Cardiac Patients Know About Device Cybersecurity Prior to Implantation?" available from the *American Medical Association Journal of Ethics* [30].

9.8 Mitigation Plan

As stated before, different types of incident response plans should be prepared to address different incident scenarios. For those that relate to and impact the MDM's products, it is good practice to have a prepared mitigation plan as part of a larger postmarket management plan (see Chapter 8). Incidents stemming from risks to particular components may, in some cases, be able to be mitigated in the field by temporarily disabling vulnerable features (e.g. by unplugging a communications cable, disabling a mobile device or wireless channel) until a long-term solution can be delivered. If the incident is isolated to a single device or device network, it may be possible to replace the compromised device(s) with clean or pre-updated units, as appropriate for the incident type, while the compromised device is investigated under quarantine, perhaps avoiding the need to initiate a full recall.

The mitigation plan should consist of two elements:

1. A set of preapproved generic templates describing the general procedure by which various kinds of incidents shall be mitigated, which will be used on an incident-by-incident basis.
2. The specific actions that will be performed, as well as the tools, procedures, and order that will be used to perform them, to mitigate the specific incident. These elements should be drafted by security engineering in conjunction with engineering proper and reviewed and approved by the chief security officer (or equivalent), as well as other key stakeholders as described in Chapter 2, before it is enacted.

No matter the nature and type of incident, time is essential for successful response and having a mitigation plan as developed and as detailed as is practical in advance of product release is strongly recommended. Minimizing the quantity and depth of decisions that must be made under the pressure of a live incident or exploit-in-progress is the best way to avoid mistakes or omissions in the mitigation.

Both (1) and (2) should include risk-based considerations to assess their potential impact on the device's safety and essential clinical functions. The MDM's existing risk management rubrics should work for this purpose, enabling the MDM to weigh the risks of proposed mitigations against one another as well as against the current state of the device.

Note that, because of the incident, the device under investigation cannot be assumed to still have the same baseline risk level established during premarket approvals. Depending on the potential scope of the incident (does the exploit affect a single device, a single network, an entire product line?), cybersecurity artifacts in the DHF will need to be updated based on updated pre-mitigation and post-mitigation scoring as part of incident closure.

Note that, for a large-scale and complex incident, the investigation and response may involve multiple organizations, for example, several HDOs or even MDMs (that use the same supply chain component), regulatory and law enforcement could be participating (for example, as it happened during the WannaCry ransomware attack) and that the entirety of response may even cross industries and national borders.

The mitigation plan should also define the emergency actions that may be taken or recommended by support personnel who become aware of an imminent threat or exploit-in-progress. Pre-authorized emergency actions should be limited in scope and risk-assessed just like the rest of the mitigation plan. The goal of the emergency actions section of the mitigation plan is twofold. It exists to authorize swift action to address the incident as early as possible and so, hopefully, to prevent its severity from escalating or its impact from spreading. It exists to limit the activities of those who might take such actions to controlled activities that can be evaluated in advance of incident discovery.

Finally, the mitigation plan or applicable elements should be communicated to those it will impact: both in training, for the internal roles described in Chapter 2, and during the incident itself, to those involved. In terms of outward-facing communication, it is important that affected users and other field personnel be informed as early as possible that the MDM has a plan, first of all, along with any emergency action being taken; and then any plan specifics determined to be appropriate or useful to external parties once those specifics have been approved internally.

Amid the pressure of incident response, it can be easy to overlook or undervalue the important role of communication to nontechnical stakeholders, if only for the purpose of providing reassurance in the moment and staying well ahead of the public news cycle, which could become sensationalized. The MDM should always bear in mind its goal is not merely to address the incident on a technical level, but to enhance its market position through every step of response.

9.9 Mitigation Rollout and Tracking

Deployment of software and firmware (or even hardware) mitigations is described in detail in Chapter 8. In general, and similar to any other software or firmware update, the phases of rollout are:

1. Preparing the update apparatus, including delivery mechanisms for the update package, such as a management server, as well as update documentation, including updated SBOM. This is most likely managed by the MDM.
2. Typically, responsibility for deploying the update lies with the operator (HDO), but may be contracted out to a third party, possibly including the MDM's service organization. This includes change management functions, such as preparing and confirming the prepared status of target devices or device components to receive the update (e.g., confirming that taking the device's essential clinical functions offline for the time required to deliver and install the update will not put users at risk).
3. Confirming that the update is installed and runs successfully or, if it failed, that the device reverted to its previous state. Depending on the update delivery method, this may involve both the HDO (or its agents) and the MDM.

4a. If the update failed: Reinitiating the update procedure, or canceling the update and authorizing the device for return to normal use, if deemed safe to do so.

4b. If the update succeeded: Confirming its success and authorizing the device for return to normal use, if deemed safe to do so.

5. Track the status of the update rollout across all target devices in the HDO and MDM's asset management systems.

The activity level required from users or on-site technical personnel will vary based on the device and the mitigation. For incidents that do not feel urgent or high priority to users, but that require user activity to mitigate, the MDM may find it helpful to establish incentives for complying with the update process. This is generally less expensive and more efficient and may be better received by users and the public versus sending its own agents to perform the update activities at each device location (which, in certain cases, may be impossible anyway). A possible exception may be an on-site mitigation activity that is too technically complex for the average user to perform or that could expose sensitive information to an inappropriate party.

From the MDM's viewpoint, tracking rollout progress includes status reporting on the update, both from the field to the MDM (either by the device itself, as part of the update procedure, or by users or field personnel), within the MDM's asset management program, and by the MDM to external stakeholders, including regulators (especially if the update is part of corrective action or recall), the media, and so on. Continuously collecting and reporting rollout status are necessary and may be required daily, depending on incident severity. Additionally, in the case of a mandatory recall, regulatory reporting may be required.

9.10 Coordinating with the ISAO During Incident Response

It must be assumed, until proven otherwise, that an incident that affects one device at one customer site also indicates a risk to other customers. Therefore, responsible and timely disclosure of incident-related information is of the essence to enable other participants to be proactive in preventing a similar occurrence. In theory, every cybersecurity incident should be reported to regulators. However, the FDA's Postmarket Guidance allows MDMs to report many incidents to an ISAO instead, except those that led or could lead to patient harm [31]. The MDM's analysis should show whether the incident's root cause is a one-off or a general problem due to systemic vulnerability, which should be communicated to the ISAO along with mitigation plans and actions.

Both MDMs and HDOs should be members of and can find support by using ISAOs to help with this process. The FDA encourages MDMs to actively participate in ISAOs and gives the MDM greater autonomy in the communication process (among other incentives) for doing so [31]. Communicating a security incident and incident response plan to all affected HDOs can be logistically difficult; an ISAO facilitates this communication.

ISAOs have been further discussed in previous chapters. The purpose and organization of an ISAO are described by CISA [32], and a list of approved ISAOs is provided via the HSCC [33] and the ISAO Standards Organization [34].

9.11 Incident Closure

The incident response plan should include criteria for determining when the incident has been brought to a close, a procedure for officially declaring each party's response activities closed, and a procedure for initiating follow-up or debrief. Depending on the type of incident and its impact, the MDM may need to comply with regulatory processes such device recall or field replacement action.

Incident closure criteria should include a rubric that defines whether the MDM's response to the incident was appropriate for the incident type, was warranted by its investigation, and was communicated appropriately to stakeholders. It should further include a risk matrix, similar to the design vulnerability assessment, which evaluates the pre-mitigation and post-mitigation risk status of the device for any incidents for which mitigating action was deemed necessary.

Documentation regarding incident handling and closure should be generated, distributed as required to stakeholders, and retained. The incident response plan should include a template for a final incident report, which describes succinctly, but with sufficient technical detail: (1) what the incident consisted of; (2) how the MDM became aware of the incident; (3) what steps the MDM took to investigate the incident; (4) what the investigation determined; (5) what mitigation the MDM devised to address the incident, if applicable; (6) a timetable charting successful mitigation deployment over time (e.g., within x period of time, y percentage of affected units had successfully applied the mitigation, growing to z percentage of affected units after $2x$ time… and so on, up to the date of the report); (7) any follow-up steps, if applicable, including regulatory notification, corrective action, recall, public communication, and post-incident follow-up; and (8) the MDM's justification for declaring the incident closed (based on its established closure criteria). Each section should include references to detailed documentation for each response activity. The report should be prepared by the incident response team, approved, and then issued to all appropriate stakeholders, including senior management, the ISAO, and, if needed, regulators.

The general trend of recent and forthcoming cybersecurity standards and regulations calls for MDMs to adopt a posture of continual improvement with regard to cybersecurity management, just as they already do with quality management. Incident response follow-up is a major area for such activities. Depending on the incident type and investigative findings, the MDM may deem it necessary to trigger a CAPA-like process so that findings can be fed back into the MDM's design process or organizational practices, quality system processes, manufacturing controls, and the incident response plan itself.

As the threat landscape evolves, so must the MDM's cybersecurity activities. Responsible post-incident follow-up provides an excellent opportunity for the MDM to exercise, and to demonstrate, its growing maturity.

References

[1] "Incident," NIST Glossary, https://csrc.nist.gov/glossary/term/incident.

[2] Centers for Medicare & Medicaid Services, "Pub. 100-07 State Operations Provider Certification," February 21, 2014, https://www.cms.gov/Regulations-and-Guidance/Guidance/Transmittals/Downloads/R103SOMA.pdf.

[3] Office for Civil Rights, "Summary of the HIPAA Security Rule," October 19, 2022, https://www.hhs.gov/hipaa/for-professionals/security/laws-regulations/index.html.

[4] Office for Civil Rights, "Breach Notification Rule," July 26, 2013, https://www.hhs.gov/hipaa/for-professionals/breach-notification/index.html.

[5] FDA, "Mandatory Reporting Requirements: Manufacturers, Importers and Device User Facilities," May 22, 2020, https://www.fda.gov/medical-devices/postmarket-requirements-devices/mandatory-reporting-requirements-manufacturers-importers-and-device-user-facilities.

[6] UK Medicines & Healthcare Products Regulatory Agency, "Guidance for Manufacturers on Reporting Adverse Incidents Involving Software as a Medical Device under the Vigilance System," May 15, 2023, https://www.gov.uk/government/publications/reporting-adverse-incidents-involving-software-as-a-medical-device-under-the-vigilance-system/guidance-for-manufacturers-on-reporting-adverse-incidents-involving-software-as-a-medical-device-under-the-vigilance-system.

[7] Office for Civil Rights, "Business Associate Contracts," January 25, 2013, https://www.hhs.gov/hipaa/for-professionals/covered-entities/sample-business-associate-agreement-provisions/index.html.

[8] ISO Health Informatics Committee, "Application of Risk Management for IT-Networks Incorporating Medical Devices — Part 2-6: Application Guidance — Guidance for Responsibility Agreements," December 2014, https://www.iso.org/standard/63108.html.

[9] National Cyber Security Centre, "NIS 2: A Quick Reference Guide," https://www.ncsc.gov.ie/nis2/.

[10] https://www.fda.gov/regulatory-information/search-fda-guidance-documents/select-updates-506j-guidance-506j-device-list-and-additional-notifications.

[11] HSCC Cyber Security Working Group, "Model Contract-Language for Medtech Cybersecurity (MC2)," March 2022, https://healthsectorcouncil.org/model-contract-language-for-medtech-cybersecurity-mc2/.

[12] International Standards Organization, "ISO/IEC 27035-1:2016 Information Technology — Security Techniques — Information Security Incident Management — Part 1: Principles of Incident Management," November 2016, https://www.iso.org/standard/60803.html.

[13] ISO/IEC 27035-2:2016, "Information Technology — Security Techniques — Information Security Incident Management — Part 2: Guidelines to Plan and Prepare for Incident Response," https://www.iso.org/standard/62071.html.

[14] Cichonski, P., et al., "SP 800-61 Rev. 2 Computer Security Incident Handling Guide," NIST Computer Security Resource Center, August 2012, https://csrc.nist.gov/publications/detail/sp/800-61/rev-2/final.

[15] Souppaya, M., and K. Scarfone, "SP 800-83 Rev. 1 Guide to Malware Incident Prevention and Handling for Desktops and Laptops," NIST Computer Security Resource Center, July 2013, https://csrc.nist.gov/publications/detail/sp/800-83/rev-1/final.

[16] Kent, K., et al., "SP 800-86 Guide to Integrating Forensic Techniques into Incident Response," NIST Computer Security Resource Center, August 2006, https://csrc.nist.gov/publications/detail/sp/800-86/final.

[17] Johnson, C., et al., "SP 800-150 Guide to Cyber Threat Information Sharing," NIST Computer Security Resource Center, October 2016, https://csrc.nist.gov/publications/detail/sp/800-150/final.

[18] Bartock, M., et al., "SP 800-184 Guide for Cybersecurity Event Recovery," NIST Computer Security Resource Center, December 2016, https://csrc.nist.gov/publications/detail/sp/800-184/final.

[19] Proffitt, T., "Creating and Managing an Incident Response Team for a Large Company," SANS, July 18, 2007, https://www.sans.org/reading-room/whitepapers/incident/paper/1821.

[20] West-Brown, M., et al., "Handbook for Computer Security Incident Response Teams (CSIRTs)," Carnegie Mellon Software Engineering Institute, Pittsburg, PA, April 2003, https://resources.sei.cmu.edu/asset_files/Handbook/2003_002_001_14102.pdf.

[21] ISACA, "Incident Management and Response," 2012, https://www.isaca.org/bookstore/bookstore-wht_papers-digital/whpimr.

[22] Interpol, "Guidelines for Digital Forensics First Responders," March 2021, https://www.interpol.int/content/download/16243/file/Guidelines%20to%20Digital%20Forensics%20First%20Responders_V7.pdf.

[23] Grispos, G., and K. Bastola, "Cyber Autopsies: The Integration of Digital Forensics into Medical Contexts," *IEEE*, 2020, https://www.computer.org/csdl/proceedings-article/cbms/2020/942900a510/1mLMlUZYaYw.

[24] Connolly, J., et al., "Medical Device Cybersecurity Regional Incident Preparedness and Response Playbook," *MITRE Technical Papers*, October 2018, https://www.mitre.

org/publications/technical-papers/medical-device-cybersecurity-regional-incident-preparedness-and.

[25] FDA, "Postmarket Management of Cybersecurity in Medical Devices," FDA Guidances, December 28, 2016, p. 13.

[26] International Standards Organization, "ISO/IEC 27002:2013 Information Technology — Security Techniques — Code of Practice for Information Security Controls," October 2013, https://www.iso.org/standard/54533.html.

[27] FDA, "Postmarket Management of Cybersecurity in Medical Devices," FDA Guidances, December 28, 2016, p. 15.

[28] FDA, "Best Practices for Communicating Cybersecurity Vulnerabilities to Patients," October 1, 2021, https://www.fda.gov/about-fda/cdrh-patient-science-and-engagement-program/best-practices-communicating-cybersecurity-vulnerabilities-patients.

[29] TGA, "Medical Device Cyber Security Information for Users," November 2022, https://www.tga.gov.au/sites/default/files/medical-device-cyber-security-information-users.pdf.

[30] Zeitler, E. P., and D. B. Kramer, "What Should Cardiac Patients Know About Device Cybersecurity Prior to Implantation?" *AMA Journal of Ethics*, September 2021, https://journalofethics.ama-assn.org/article/what-should-cardiac-patients-know-about-device-cybersecurity-prior-implantation/2021-09.

[31] FDA, "Postmarket Management of Cybersecurity in Medical Devices," FDA Guidances, December 28, 2016, pp. 25–26.

[32] https://www.cisa.gov/information-sharing-and-analysis-organizations-isaos.

[33] https://healthsectorcouncil.org/hic-miso/.

[34] https://www.isao.org/information-sharing-groups/.

10

The Device Security Lifecycle for Healthcare Delivery Organizations

As we progress towards growing digitalization and digitization of our care delivery system, we need to understand and manage the increasing dependency on the availability and reliability of our health IT infrastructure, including connected medical devices. We live in a healthcare world where, unequivocally, cybersecurity is directly tied to patient safety [1, 2], and where the maintenance of devices' security posture by the healthcare delivery organization (HDO) is of critical importance. Although the HDO's perspective on lifecycle management (and cybersecurity in general) is not a main topic for this book, it is important for the MDM to understand how the device will be managed by their customers and device operators and how a device's secure design, security features, supporting security processes, and the provision of cybersecurity documentation will be utilized by HDOs.

In the most general sense, and as illustrated in Figure 10.1, the main lifecycle phases from the perspective of the HDO are:

- Pre-procurement;
- Procurement;
- Deployment;
- Operations;
- Decommissioning.

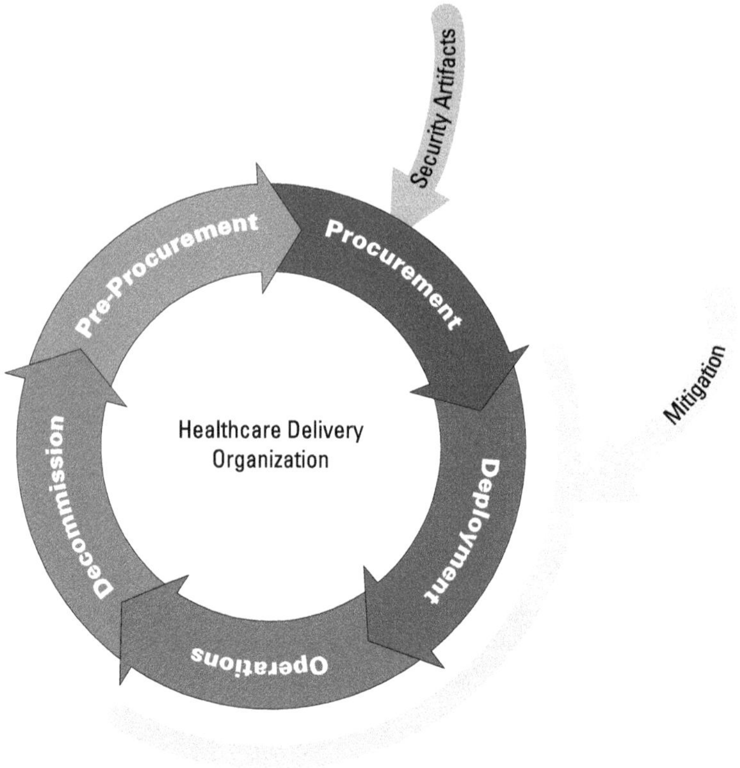

Figure 10.1 HDO lifecyle.

These are leading practices adopted by many HDO, with minor variations, to address and proactively manage medical device cybersecurity within their operations. As MDMs interface with these through several critical deliverables, they need to have a good understanding of the HDO lifecycle so as to ensure proper alignment of activities and expectations and to avoid misunderstandings and unnecessary friction.

The HDO lifecycle is centered around the actual device and its target integration ecosystem, supported by documentation and artifacts provided by the manufacturer, and in the form of risk mitigations during the HDO's deployment and operations phases.

Obviously, there can and will be differences in implementation and execution depending on the type of device and use case, type of organization, and whether the device will be based in the hospital or elsewhere outside of it (e.g., in the patient's home).

Another differentiating factor is who will perform these activities. Some or all may be provided by HDO staff, or it may be contracted out to the device manufacturer, an independent third-party service provider (commonly referred to as independent service organization (ISO)), or a combination thereof. This may lead to differences in the actual implementation processes but, more importantly, may create additional communication and process interfaces between the respective activities and between the varying stakeholders, which requires careful consideration so as not to create gaps in responsibility or disconnects in communication. It must be clearly defined which party owns the cybersecurity responsibility and which parts thereof. For example, clinical engineering (or an outsourcing partner) may be responsible for the security of the device itself, but the IT department may be responsible for maintaining servers, firewalls, and the network. This scenario would require close coordination around security responsibilities, but also activities such as integration management, network architecture, data flows, the application of security controls, implementation of risk mitigation, and incident response.

This type of generic device lifecycle has been in use for a long time and its applicability is not limited to networked medical devices, but adding cybersecurity considerations to it requires profound changes to each of the phases, as well as to the roles and responsibilities of the individual contributors.

10.1 Pre-Procurement Phase

Pre-procurement itself can be broken down into several main tasks by itself. It is a critical phase in cybersecurity planning, as this is the time to:

1. Assess security properties of existing devices and define priorities for replacement.
2. Define security expectations and requirements as a buying criterion.
3. Establish agreement among internal stakeholders and identify roles that specifically support cybersecurity (e.g., requirements development, contracting, or approval).

Pre-procurement is the phase where expectations are aligned and where directions are set. It is an opportunity to align priorities among a broad range of stakeholders, including clinicians, clinical engineering, IT and IT security, purchasing, legal, and compliance, and, to an extent, even executive management.

Under the umbrella of pre-procurement, we typically find the following activities. Some are more of a general nature and may not apply to a specific device or type, but rather define the criteria for the general, security-aware procurement activities. However, others may be specific to a device, device type, or specific purchasing project.

- *Cybersecurity governance:* Based on the overall organization's mission and objectives, governance establishes a framework of cybersecurity rules and practices, inclusive of applicable laws and regulations as well as business criteria, by which leadership defines goals and objectives, establishes responsibilities (duties, privileges, and roles), and ensures accountability, proper supervision, and control, as well as information-flow and monitoring of implementation and compliance.
- *Cybersecurity policies and procedures:* Based on the established governance, define policies, processes, procedures, work instructions, and templates as they pertain to the individual medical device cybersecurity lifecycle phases as well as the key risk management and incident response activities.
- *Risk criteria:* Define criteria for cybersecurity risk acceptability as they will be applied as part of the risk assessment.
- *Replacement planning:* Needs and requirements analysis in conjunction with assessment of current inventory, providing guidance on replacement priorities in alignment with clinical and patient care needs. As an overarching process, replacement planning provides the opportunity to align cybersecurity expectations among stakeholders and to balance cybersecurity priorities with clinical needs and budget realities.

  Replacement planning is also an important long-term planning tool to allow an HDO to align cybersecurity with organizational goals and budget strategy. This process should result in visibility of future security needs as well as engagement of stakeholders in the planning and budgeting process.
- *Experience analysis:* As an ongoing activity, it is important to continually analyze relevant cybersecurity measures, including reported issues with individual devices and device types, performance and compliance of vendors, monitoring of the cyber threat landscape, and analysis of information provided through public resources (e.g., information-sharing and analysis organizations, security research).
- *Security requirements definition:* Every HDO should have a clear definition of their security requirements that can be shared with internal and external stakeholders. Although these requirements will establish the objectives required to improve the overall security posture of devices and the device ecosystem, they also have to be rooted in reality and be based on what is practically and economically feasible. Simply establishing a criterion of "a vendor shall deliver devices that are free of vulnerabilities" is neither practical nor possible; therefore, such a requirement is not

useful. Better examples have been provided by organizations such as the Mayo Clinic [3], Indiana University Health [4], and the U.S. Department of Defense (DoD) [5].

- *Training and staffing needs:* Another overarching objective is to ensure that medical device cybersecurity activities are sufficiently supported by appropriately trained staff. This includes training users (clinical staff) in cyber-appropriate behavior [6] as well as enabling device maintenance staff to securely deploy and manage devices. The training can be general (e.g., security best practices) or device-specific or task-specific [7].
- *Publication and communication:* It is beneficial to all participants, internal and external, if the general cybersecurity expectations are communicated and everybody's expectations can be aligned. The previously provided references are good examples of how healthcare organizations can not only establish security criteria, but also efficiently communicate them to their vendor community.
- *Requests for vendor input:* An HDO may decide to query their vendor community for input on their degree of cybersecurity compliance (e.g., by requesting vendor disclosure of their security practices and processes). Many vendors have dedicated security sections on their web pages and/or have published white papers about their approach to cybersecurity. Reviewing available information helps with vendor pre-selection as well as sets the right expectations with the vendor partner.
- *Vendor and partner preselection:* Specific activities may be undertaken to support vendor and partner preselection to build a "trusted partner" list or in support of a specific procurement project. This could, for example, happen in form of a Request for Information (RFI) or similar formal or informal procurement tools. This activity can be considered analogous to the MDM's management of an approved supplier list (see Chapter 5).

10.2 Procurement Phase

Procurement activities apply to a specific device or device type in the context of an upgrade, replacement, or expansion project. It is in this phase that an HDO has the opportunity to clearly define security expectations, assess the most suitable vendor and product, and formally seal an agreement, inclusive of cybersecurity requirements, in form of a contract. It is beneficial to disclose security expectations and requirements, as well as the intent to include these in the final contract, during the bid process to set the right expectations and avoid surprises or resistance when negotiating contract terms later.

Typically, procurement includes the following activities:

- *Requirements definition:* An HDO's announcement of its intent to purchase (e.g., in form for a request for proposal (RFP)) should include security requirements, including technical, documentation, testing, communications, and procedural requirements.

  Recently, good templates for cybersecurity terms and requirements have become available. Use of these templates ensures the HDO that their requirements are based on industry-recognized leading practices, although some may want to add additional criteria that meet their situation and needs. For the MDM, the benefit is that the requirements that they are receiving have become largely harmonized (although some exceptions will always exist), which significantly reduces their effort and improves the quality of their responses.

  Specifically, the following templates should be considered:
  - The Healthcare and Public Health Sector Coordinating Councils' (HSCC) "Model Contract-Language for Medtech Cybersecurity (MC2)," with an updated version expected to be available by the time this book goes to print [8]. This document was derived from and is based on the previous work done by the Unified Medical Device Cybersecurity Alliance (Kaiser Permanente, Cleveland Clinic, Mayo Clinic, Froedtert Hospital).
  - The European Network and Information Security Agency's (ENISA) "Procurement Guidelines for Cybersecurity in Hospitals" provides a set of leading and easy-to-implement procurement practices for hospitals' acquisition of services, products, and infrastructure, including medical devices [9].
- *Device selection:* Based on vendor-provided documentation and artifacts (discussed in detail in Chapter 7), market research, and past experience, the device selection process compares the available offering against the established cybersecurity criteria. The final selection will typically be a balance between budgets, clinical needs, business plans, and cybersecurity properties. It should be the intent of an HDO to procure the most secure device possible; however, this should not lead to compromise of the device's clinical features. Typically, the final device should provide security properties that meet the established risk acceptance criteria. Any exception from that should be managed via deviation process and documented mutual agreement among stakeholders.

The device selection process may include a pre-selection process as well as a final selection step that may include review of provided security documentation, confirmation of those documents and security properties against established risk criteria, and potentially independent validation and testing of sample units.

- *Contracting:* As the final step in the procurement process, contracting provides the formal agreement on the device's security capabilities as well as, on the procedural side, the vendor's maintenance and communication processes. Ideally, the contract should cover both the device security requirements and compliance with local regulatory requirements (e.g., in the United States, the FDA Premarket Cybersecurity Guidance), as well as the vendor's ongoing support obligations in line with regulators' postmarket expectations.

Specific examples of cybersecurity contract languages can provide a baseline that may need to be adopted or enhanced to meet specific or unique requirements. Furthermore, contracts should spell out the requirements for security documentation and artifacts that will enable the HDO to manage the device through its entire lifecycle: from deployment to operations (including maintenance and management) to the end of its useful life and subsequent decommissioning.

10.3 Deployment Phase

Once the new device has been received, the following typical activities take place:

- *Pre-deployment review and testing:* To ensure conformance with contractual requirements and completeness of delivery. Specific tasks may include:
  - *Device software update:* If a new software version has become available since the device was manufactured, it should be upgraded to the latest available version.
  - *Security documentation review:* Ensure the completeness, currency, and compliance of the provided documentation.
  - *Security testing:* Before the device is placed into the production environment, a series of tests should be performed. This includes a vulnerability scan, security evaluation, or penetration testing, as well as deployment on a test network to assess the presence of potential de-

vice security compromise (there have been reported cases of devices delivered with malware already on them).

- *Remote scanning capabilities:* Test the device against the organization's commonly used tools like vulnerability scanner, risk and compliance managers, asset discovery, computerized maintenance management system (CMMS), and/or configuration management database (CMDB); establish allow or deny rules for the respective tools. If test scans raise concerns about the device's ability to withstand network interrogation, it should be excluded from future scans. However, this should then lead to the establishment of well-documented and well-followed compensating security activities (e.g., periodic manual review) in lieu of the functions provided by scanning tools.

- *Installation:* Deployment of the device in its target (production) environment, including security-specific considerations (e.g., implementation of physical security controls, network and firewall configuration). Also, at installation time, user and admin accounts (clinicians, local or remote technicians) are typically set up in compliance with the HDO's access and role-based access policy (RBAC) in a way that reflects the device's capabilities, use case, and risk.
- *Integration:* In addition to integration with the medical device network and clinical IT systems, the device needs to be integrated with and tested against security-specific tools such as asset management (CMMS, CMDB), vulnerability scanner, cybersecurity management systems (SIEM, SOC), secure remote access, credential management system, and automated risk management systems.

  In addition, especially for devices that integrate or embed off-the-shelf (OTS) components with commercial operating systems, integration with the hospital's security infrastructure may be desired or required. For example, if the device can run a vendor-approved anti-malware tool, that tool should integrated with the HDO's security management (e.g., a security operations center (SOC)) and security event detection infrastructure (e.g., a security information and event management system (SIEM)). It is critical to confirm that the device can be integrated without undue functional and reliability risks, and to establish compensatory procedures and mitigations if it cannot.
- *Implementation:* The final phase of implementation includes security-specific training for users (as needed) and maintenance staff and then ultimately the clinical go-live, at which point the device enters the operations phase.

10.4 Operations Phase

During the operations phase, certain security-relevant tasks are performed:

- *Vulnerability and threat monitoring:* The device operator should monitor commonly available sources for information about newly discovered vulnerabilities or threats. Such information can typically be obtained from the device manufacturer, government agencies (e.g., FDA, DHS, FBI, ENISA), industry peers, ISAOs, or security researchers.

  The manufacturer's obligations with regard to vulnerability and threat monitoring as well as communication should be formally agreed upon in the contract. Furthermore, to ensure an efficient response, security-specific communication processes and contacts should be identified and kept up-to-date. Supporting industry standards and best practices, such as SBOM and Vulnerability Exploitability eXchange (VEX), are currently being developed.
- *Event and incident monitoring:* From a cybersecurity perspective, one of the most critical tasks is that of ongoing detection and monitoring of security events (an observable occurrence) and incidents (the compromise of an asset with potentially adverse consequences). Such cybersecurity signals may be provided by features of the device itself, through the review of log data, or the observation of the device's environment (e.g., at the network level).

  Over the past years, passive network monitoring solutions (PNM), sometimes also referred to as anomaly detection tools, have become available. These tools observe network traffic, establish a baseline of "normal behavior," and alert to deviations that could indicate a security event. In addition, these tools help with inventory discovery and management (by identifying devices based on their network fingerprint) as well as risk management and reduction activities [10].
- *Incident response:* Should a new cyber threat or ongoing incident be discovered that poses a risk to or is already impacting a device, a specific, predetermined incident response process should be initiated. This process should include decision triage including clinical aspects (e.g., procedures for patients), roles and responsibilities, defined technical response steps (identification, containment, eradication, and recovery), internal and external communication, preservation of artifacts, and forensic information. Due to the criticality of our medical device ecosystem, the development, training, and testing of an organization's incident response plan are often-overlooked but critical tasks.

The response to a cyber incident involving medical devices requires an immediate yet careful execution as it can directly affect patient safety as well as quality and timeliness of care [11]. The impacts are complex and may not be limited to the individual department or HDO as a whole. They may even extend regionally; for example, emergency room patients may need to be diverted [12]. Specific guidance on how to plan for and respond to a cyber-attack was recently published by MITRE [13], providing a regional response framework across HDOs, state or local health departments and emergency services, state or regional law enforcement and cybersecurity partnerships, industry and trade associations, and sector-specific ISAOs or ISACs.

Specifically, MITRE's "Regional Incident Preparedness and Response Playbook" provides a customizable regional preparedness and response tool for medical device cyber resiliency, including:

- Incorporation of a baseline for medical device cybersecurity information;
- Roles and responsibilities for internal and external responders;
- A standardized approach to response efforts;
- A basis for enhanced coordination activities among stakeholders;
- Recommendations for decision making and response escalation;
- Resources that can be leveraged.

Medical device cybersecurity incident response for HDOs is discussed in more detail elsewhere [14]. Cooperation and coordination between HDO and MDM in responding to an incident is discussed more detail in Chapter 9.

- *Ongoing management:* Typically, the device manufacturer provides a preventive maintenance (PM) plan, which should include tasks related to maintaining the device's cybersecurity posture. This may include regular review of log files and password changes, as well as checks for the availability of software updates.
- *Deployment of security mitigations:* At times, the operator of the device will be made aware of the necessity of security mitigations. In many cases, this information will be provided by the manufacturer, but could also come from other sources, including observations by the operator themselves. The required mitigation may be a configuration change, software update (patch), or an external measure such as network configuration. In either case, the change must be:
    - (a) Verified as being authentic and approved (e.g., for a patch, through digital signature).

- (b) Tested before live deployment in a clinical environment.
- (c) Deployed in alignment with ongoing clinical operations and care delivery priorities (requiring formal change management).
- (d) Evaluated for its impact on dependencies across other systems.

Depending on criticality, the security mitigation may be deployed as part of the normal maintenance cycle, or, in case of a critical patch, it may need to be deployed with higher priority and outside of the normal schedule.

10.5 Decommissioning Phase

HDO cybersecurity activities pertaining to the device's useful clinical life should not be confused with cybersecurity activities triggered by EOL/EOS considerations on the MDM side. A device's clinical life usually ends when:

- It can no longer safely fulfill its intended function and provide the desired result.
- It is no longer economical, practical, or possible to be maintained.
- It is superseded by a next-generation or improved version.

In the latter case especially, it is common that such devices are resold, donated, or returned to the manufacturer; each of these pathways may have unique cybersecurity concerns for patients, HDOs, or MDMs. In any case, sensitive data will need to be purged from the device. Local regulations and laws will dictate which types of information must be included in the sanitization and the technical process and degree of data destruction required. Typically, this includes any patient information, but should also include HDO-identifying or user-identifying information as well as device and user credentials. Procedures should be developed and followed that define the expected decommissioning process [15], plus any device-specific addendums as needed, how decommissioning is to be documented, and the data types that should be removed from any device leaving an organization, for example:

- PHI and PII;
- Network configuration data including network credentials, host names, IP addresses, firewall configurations, Wi-Fi passwords;
- User accounts and passwords;
- Clinical trial and research data;

- Location history data (especially for mobile devices or devices that are used in the patient's home).

  MDMs should include required decommissioning features in the device design, and should provide information on:

  - (a) The type, amount, and duration of data held by the device;
  - (b) Protocols defining how the critical data should be removed or destroyed in accordance with regulatory requirements.

In absence of such information, the device owner and operator should take measures to destroy the device or the device's data carriers.

Additional considerations for the specific scenario of device ownership change include possible requirements around license transfer to the new owner. For this purpose, many organizations establish a formal "transfer of custody" process. The device operator should have formal processes in place to manage device decommissioning as well as perform the necessary certification and documentation steps.

Similar data sanitization processes apply when a device is temporarily removed from HDO premises (e.g., sent for repair) or is moved between locations (e.g., from one patient's home to another's home).

10.6 Special Scenarios

In the context of HDO cybersecurity lifecycle management, there are two scenarios that require special consideration:

- *Leased, loaned, or demo devices:* In principle, the lifecycle steps for any temporary device are the same as described above. However, there are two special considerations:
  - The lifecycle phases happen with a much higher frequency than with traditional owned devices, which may create some unique challenges.
  - The device owner may not be the same as the operator, and therefore, special considerations with regards to security responsibility and data sanitization may need to be applied.
- *Legacy devices:* Typically, the term "legacy" applies to devices that as a whole or in part (e.g., a key software component like the operating system) are no longer maintained by the manufacturer [16]. Although this problem also exists outside of cybersecurity (e.g., availability of spare parts), it takes on unique challenges in cybersecurity as this means that key security risk reduction strategies, such as software patching, are no

longer available. This means that complementary security mitigation measures need to be applied externally to the device, using strategies such as network segmentation, firewalls, or passive network monitoring (PNM) systems. It should be noted that this approach may, to an extent, protect against network-based security threats, but will not prevent a security compromise through non-network vectors, such as a data carrier. For more on the topic of securing legacy devices, see the Health Sector Security Council's publication titled *Managing Legacy Technology Security* [17].

10.7 Summary

Although the main responsibility for proactively securing the device and designing the appropriate security features and controls into the device lies with the MDM, the HDO has a unique opportunity to improve the security posture of its device ecosystem through the adoption of leading lifecycle management practices. This will allow an HDO to reduce their risks by (1) applying mature security processes and technologies, and (2) developing a constructive security relationship with their vendors. Failing to do so will force an HDO into a reactive cycle of security fire drills that will neither improve their long-term security posture, nor be cost-effective or resource-effective. A reactive organization is bound to fail its patients.

In this chapter, we discussed the main lifecycle management phases from the HDO perspective: pre-procurement, procurement, deployment, operations, and decommissioning. The medical device manufacturer typically interfaces with the procurement phase, which ends with the delivery of the device and its supporting cybersecurity artifacts, and the deployment and operations phases through the delivery of risk mitigation measures. It is important for the MDM to understand how their lifecycle considerations, processes, and phases interface with the HDO's lifecycle phases and functions.

References

[1] Southwick, R., "Cybersecurity and Patient Safety: Why It Needs More Attention," *Chief Healthcare Executive*, April 6, 2022, https://www.chiefhealthcareexecutive.com/view/cybersecurity-and-patient-safety-why-it-needs-more-attention.

[2] U.S. Cybersecurity and Infrastructure Security Agency (CISA), "Provide Medical Care Is in Critical Condition: Analysis and Stakeholder Decision Support to Minimize Further Harm," September 2021, https://www.cisa.gov/sites/default/files/publications/CISA_Insight_Provide_Medical_Care_Sep2021.pdf.

[3] The Mayo Clinic, "Medical and Research Device Risk Assessment Vendor Packet Instructions," October 23, 2019, https://www.mayoclinic.org/documents/medical-device-vendor-instructions/doc-20389647.

[4] Indiana University Health, "Information Security – Medical Device Security and Responsible Vulnerability Disclosure," https://cdn.iuhealth.org/resources/Medical-Device-Security-and-Responsible-Vulnerability-Disclosure.pdf.

[5] Hardman, A., and W. Martin, "Risk Management Framework for DoD Medical Devices," HIMSS 18, Las Vegas, NV, Session 136, March 7, 2018, https://365.himss.org/sites/himss365/files/365/handouts/550231590/handout-136.pdf.

[6] Grimes, S. L., and A. Wirth, "The Case for Medical Device Cybersecurity Hygiene Practices for Frontline Personnel," *Biomed. Instrum. Technol*, July 2021, pp. 96–99, https://doi.org/10.2345/0890-8205-55.3.96.

[7] Wirth, A., and C. Falkner, "Cyberinsights: Cybersecurity as a Team Sport," *Biomedical Instrumentation & Technology*, January/February 2020, Vol. 54, No. 1, pp. 64–67, https://doi.org/10.2345/0899-8205-54.1.64.

[8] Health Sector Coordinating Council, "Model Contract-Language for Medtech Cybersecurity (MC2)," March 2022, https://healthsectorcouncil.org/model-contract-language-for-medtech-cybersecurity-mc2/.

[9] European Network and Information Security Agency (ENISA), "Procurement Guidelines for Cybersecurity in Hospitals," February 2020, https://www.enisa.europa.eu/publications/good-practices-for-the-security-of-healthcare-services.

[10] Upendra, P., "Selecting a Passive Network Monitoring Solution for Medical Device Cybersecurity Management," *Biomedical Instrumentation & Technology*, December 2021, https://array.aami.org/doi/10.2345/0899-8205-55.4.121.

[11] Dameff, C., M. A. Pfeffer, and C. A. Longhurst, "Cybersecurity Implications for Hospital Quality," *Health Serv Res.,* October 2019, https://www.ncbi.nlm.nih.gov/pmc/articles/PMC6736916/.

[12] Dotinga, R., "San Diego EDs Deluged with Patients After Cyberattack," *MedPage Today*, October 30, 2021, https://www.medpagetoday.com/meetingcoverage/acep/95357.

[13] MITRE, "Medical Device Cybersecurity Regional Incident Preparedness and Response Playbook Version 2.0," November 14, 2022, https://www.mitre.org/news-insights/publication/medical-device-cybersecurity-regional-incident-preparedness-and-response.

[14] Wirth, A., and S. Grimes, (eds.), *Medical Device Cybersecurity: A Guide for HTM Professionals*, Association for the Advancement of Medical Instrumentation (AAMI), May 2018, https://store.aami.org/s/store#/store/browse/detail/a152E000006j66qQAA.

[15] Waters, C., "Information Security Considerations for Decommissioning Medical Devices," *ECRI*, January 17, 2023, https://blog.ecri.org/decommission-medical-devices-protect-phi.

[16] International Medical Device Regulators Forum (IMDRF), "Principles and Practices for the Cybersecurity of Legacy Medical Devices," May 2022, https://www.imdrf.org/consultations/principles-and-practices-cybersecurity-legacy-medical-devices.

[17] HSCC, *Managing Legacy Technology Security*, March 2, 2023. https://healthsectorcouncil.org/legacy-tech-security/.

11

Implementing and Incorporating Security Solutions and Approaches in Medical Device Systems

Considering the wide range of medical device types, variations in architectures, processor platforms, operating systems, use cases, and available power and computing resources, it is difficult to discuss specific security technologies that may or may not apply to specific devices and technical implementations. Obviously, a resource-constrained, battery-powered implantable device has very different security capabilities and needs than a room-sized piece of imaging equipment that is a system of systems in itself.

Furthermore, we often find a potential conflict between the ease of use of clinical features and security. Therefore, it is of critical importance that the correct security technology is chosen and implemented and is appropriate for the device and use case to facilitate usability in a secured environment without compromising clinical features. In some cases, security requirements may need to be subordinate relative to clinical requirements. The traditional example is a "break-glass" emergency access feature that may be deemed a poor security choice ("zero-factor authentication") but that is necessitated by clinical reality and may one day save a patient's life.

Security technologies available through commercial security products can be used as a complement to secure software architecture and implementation best practices. They should not be used to replace secure-by-design best practices, but rather to provide complementary protection or to address specific security needs, for example, to reduce the inherent risks that come with incorporating a commercial operating system as part of the medical device system.

This section will review security technologies in general terms and discuss the respective technology's advantages and disadvantages in the context of some general implementation scenarios. The terminology describing various security technologies is by no means standardized. Different vendors may market their products and features with different terms and descriptions. Table 11.1, outlining the main categories of security tools as they apply to a typical IT environment, should not be considered an absolute definition.

The remainder of this chapter will discuss a select subset of these tools and their potential applications or relationships to the security of medical device systems and/or the security of their operating environments.

11.1 Compliance and Infrastructure Management

Generally, enterprise IT environments operate a number of tools that help them to stay compliant with established IT security standards and regulations. Such

Table 11.1
Overview of Common Security Technologies Found in the IT Environment

Compliance and Infrastructure Management	
• Compliance & vulnerability management• Risk management• Configuration management database (CMDB)	
Endpoint Endpoint protection: Anti-malware HIDS/HIPS Product-specific (Mobile, IoT) Endpoint Detection and Response (EDR) Cryptography	**Network** Security gateways Intrusion Detection and Prevention (IDS/IPS) Network access control (NAC) Virtual private network (VPN) Deception (honeypots) Passive Network Monitoring (PNM)
Perimeter Firewalls IDS/IPS Web filter/gateway Web isolation SSL inspection	**Cloud** Cloud access and security broker (CASB) Zero Trust Platform Cloud-specific security solutions (workload protection, DLP, authentication)
Enterprise Security	
• Authentication• User Behavior Analytics (UBA)• Simulation & Awareness• Email Security • Access Control• Encryption & Obfuscation• Data Loss Prevention (DLP)• Web Security	
Orchestration and Response	
• Security Information and Event Management (SIEM)• Security Operations Center (SOC)• Incident Response and Remediation	

tools typically scan networks and devices and assess them against a defined set of requirements derived from, for example, HIPAA, PCI/DSS, ISO/IEC 27000, and NIST SP 800-53 and then identify and report areas of noncompliance and, to a certain extent, help to manage and deploy mitigation such as configuration changes or deployment of updates and patches.

The goals of using such tools are obvious and simple: create visibility across one's environment, identify gaps, and ultimately reduce risk. However, they are built around the assumption of a traditional enterprise environment and with managed systems running a limited number of supported enterprise OSs. This assumption is likely to introduce problems when the endpoint is not a PC or server but instead a medical device running on the same network.

Consequently, medical devices need to be capable of operating in an environment that is managed via such tools. Ideally, they should be able to constructively respond to scans and queries and participate in the information exchange and supply data about their configuration and operational state. Even if that is not possible, they should at least be able to tolerate these network operational scans and activities without reacting negatively. Furthermore, because many of these tools use agents, customers may expect that the MDM certifies certain commonly deployed management agents so that they can be placed on the device.

Commercially available tools have been in the market for decades, although their features and capabilities continue to grow. The following categories of tools have become the bread-and-butter of enterprise security management:

- *Compliance and vulnerability assessment*: Scans an IT environment for known vulnerabilities as well as assesses compliance against IT standards. Note that some medical devices may not respond well to such scans. They should be tested and, if necessary, excluded from the scan list before being put on the production network. MDMs should be aware whether their customers use this type of tools and should make sure their devices are sufficiently robust to withstand this type of interrogation.
- *Security risk management tools*: Provides automated assessment of an enterprise's security posture relative to established standards and corporate governance as well as automates risk management tasks.
- *Configuration management database (CMDB)*: An IT asset management tool that automates processes around common IT management tasks (deployment, updates, management) and provides IT asset visibility (again, medical devices may need to be excluded from being managed, especially if such a tool uses agents that are located on the target endpoints).

11.2 Endpoint Security

In traditional IT, the term "endpoint" designates either end of a data flow on a network. In that sense, a server is an endpoint (as the origin or depository of data), as is a desktop or laptop computer or mobile device where the user receives or enters data. This implies that an endpoint is typically a composition of commercial software components such as OS, databases, web browser, or email program, which then leads to a common feature set typically found in commercial endpoint security products.

Obviously, a medical device is not a desktop computer. Some medical devices are architecturally close enough that these commercial endpoint security products can be applied, either as-is out of the box or with a degree of configuration to tune it in line with device requirements and constraints.

However, there also have been examples of this approach failing and endpoint security products behaving as if they were running on a standard desktop (e.g., starting a full disk scan in the middle of a procedure), to the detriment of the medical device's clinical functionality [1].

Commercial endpoint security products usually provide a combination of detection and prevention features, as well as offer integration with external security management systems such as SIEM, SOC, or automated endpoint detection and response (EDR).

11.2.1 Anti-Malware

The best-known and most-used endpoint security technology is traditional anti-malware (historically called antivirus). Today's anti-malware has come a long way from the early days of merely signature-based virus detection and incorporates many more features. This is in response to the growth of malware sophistication and capability, as well as the explosion of the volume of malware and malware variants that started in the 2010s. To use a simple measure as an example, threat research suggests that through the use of automation and tools for virus and variant generation and obfuscation, the production of new malware and variants has grown from about 1 million a year in 2008 to 100 million or more per year today [2].

Today's commercial anti-malware products usually contain a set of features that may include:

- Signature-based scanning (the traditional antivirus);
- Reputation analysis (files, web sites, domains);
- Behavior analysis (detecting events that are typically associated with malware);
- Critical resource protection (e.g., system registry);

- Network and port controls (e.g., blocking USB autorun);
- Artificial intelligence/machine learning (AI/ML)-based detection;
- Features addressing specific security problem areas through, for example: run-time isolation, exploit prevention, account management, software-based firewalls.

This list highlights one of the main challenges of using commercial anti-malware on medical devices. Although today's commercially available products are resource-optimized, their impact on performance and behavior of specialized systems such as medical devices may be too unpredictable. However, as previously mentioned, some medical devices may be sufficiently computer-like that use of commercial software is acceptable.

Some other concerns of using anti-malware in the medical device use case are:

- The need for a secure network connection to provide timely signature updates and to utilize enhanced features like file reputation analysis (which requires comparing file signatures to a reputation database, typically cloud-based).
- Risk of false-positives, which, although a rare nuisance on a desktop computer, could have significant impact on a medical device.
- In response to new attack techniques, anti-malware features (and consequently product versions) typically develop faster than medical devices are updated. MDMs may find it challenging to maintain compatibility and ensure that the latest security features are available for their devices.
- New anti-malware product versions need to be validated by the MDM and approved for use on a given device. Keeping device and anti-malware lifecycles in sync is challenging and may delay anti-malware feature availability to devices, creating windows of vulnerability that can last for months or even years.
- An even worse scenario occurs when future new features may not be compatible with the device and therefore prevent the device from using this latest upgrade, thus leaving it vulnerable to new attack techniques.
- Anti-malware products typically require operating system integrity and currency, and may not work effectively when using older, less frequently patched, or customized OSs.
- Both endpoint and anti-malware agents are designed to inhibit the operation of the device when corruption is detected; however, in all but

the most extreme cases, the HDO's desire is for the medical device to continue its essential performance.

Yet, in some cases, anti-malware may be a viable choice of security technology due to the following advantages:

- It is a known technology to the HDO and therefore faces a lower acceptance threshold (or may even be asked-for in bids and contracts).
- HDOs typically standardize on a specific anti-malware solution and therefore it may be preferential as it integrates well with their existing security management environment.
- Easy to install and easy to deploy (but a degree of device-specific configuration may be required).
- As these are commercially available products, licensing can be delegated to the HDO.

Whether the anti-malware product is MDM-supplied or HDO-supplied, in most cases, the MDM would need to validate that the anti-malware product does not adversely affect the device. Based on that, the MDM would then typically supply installation and maintenance instructions.

It is a reasonable approach to remove or disable (through configuration) certain anti-malware features to reduce system resource allocation, improve timing and predictability, and generally reduce any potential compromise of device performance or features. For example, assuming that most medical devices do not provide (and should not need) email or external web browsing, it is perfectly reasonable to turn off dedicated email and browser protection features in the anti-malware product. Other configuration steps may include restricting resource-heavy tasks like signature update or full disk scan to certain times or to less-critical operating states of the device.

However, caution should be taken that the anti-malware product is not configured to such an extent that its ability to secure the host device is compromised. For example, turning off certain advanced scanning, detection, or prevention features may reduce the security capabilities of the software as these advanced features are there for a reason, or, making the assumption that certain file types or directories are not at risk and excluding them from scanning may be fine today, but will this still be correct after 10 or more years of system operation?

To summarize, reasons that MDMs may select commercial anti-malware products for their devices are: (a) the device is sufficiently resourced and computer-like that this is a valid design decision, (b) it is a known entity to HDOs, often in line with their enterprise security standards and procurement requirements, (c) it can delegate license purchase and management to the HDO, and (d) it allows the MDM to "check the security box." However, (d) may be a less desirable (and less defensible) rationale and MDMs should be encouraged to look at more appropriate security technologies, as discussed in the next section.

11.2.2 Host Intrusion Detection and Prevention Systems

If we take the (admittedly somewhat simplified) view that anti-malware uses a deny-list (based on a continually updated virus signature list), then HIDS/HIPS can be introduced as allow-listing, that is, rather than trying to detect and prevent certain unwanted behavior (like anti-malware does), these products control and allow only known good behavior and forbid everything else. This, too, is a simplified description—both types of products have evolved over the years and in today's commercial products have some overlap in security features.

The advantage of anti-malware is that it is a perfect fit for the common desktop and server environments where there are certain commonalities in architecture (e.g., the same or limited number of operating systems in use), yet high variance in the individual device, with every user running different programs and performing different tasks. The ability to be easily deployed and managed across such variance is designed into today's products, but this comes with significant overhead and product complexity.

In the medical device environment, we have the opposite situation. Every device of a specific model is identically configured when it leaves the factory and normally there is the desire to keep each device exactly that way (this is why these devices are sometimes referred to as dedicated purpose systems). This is the perfect scenario for using HIDS/HIPS. Although it does require up-front engineering effort to configure the rules engine for each device model and version, once that is done, it can be easily reproduced and will perform very well on each device in the field.

Typically, implementing a HIDS/HIPS requires configuration of policies that define allowed system behavior. This includes executable and process permissions, communication behavior, and port access. Once the allow-list policies have been defined, they only require change when the system is updated. Practically, that means that any HIDS/HIPS policy changes can be applied together with the system update (or patch) and no changes are required or desired in between.

One limitation should be mentioned, and that is if a device needs to be configured locally at the time of deployment, for example, at the HDO or the patient's home (e.g., for network configuration or installation of additional software), these changes may be difficult to anticipate when developing the HIDS/HIPS security policies in advance.

Specifically, the generally recognized advantages of HIDS/HIPS (relative to anti-malware) are:

- Significantly lower resource consumption and smaller memory footprint, providing predictable behavior and minimal resource impact.
- Tight system control and prevention of changes by malicious actors or unintentional changes due to operator mistakes.
- No requirement of signature updates (policy changes can be supplied at the time of the software update), therefore supporting disconnected use cases.
- No dependence on operating system currency, typically even supporting legacy operating systems going back one or sometimes two generations.
- As it is not a reactive security technology, it is typically even effective against unknown and zero-day exploits.
- In addition to protecting OS and other commercial products, it can equally protect custom applications and files.
- Enables fine-tuning of system resource allocation as well as provides granular access control management.
- Reduced manufacturer lifecycle management pressure (fewer patches); allows for better control of release schedules.
- In many cases, can be applied as a compensating control beyond a device's end-of-support horizon.
- Improved reliability through deliberately applied constraints, privilege management features, and tamper protection.

In addition to application- (or executable-, or process-) based allow-listing and behavior control, commercial products provide: network, port, device, and file controls; software firewall; sandboxing and memory controls; intrusion and exploit detection/prevention features; user account and configuration controls; and security incident logging and audit support.

Today's commercial HIDS/HIPS products contain features to make implementation easier for example by providing out-of-the-box policies for typical OS environments or by providing a "learn" mode to observe normal system behavior (including custom processes) and then turn these into predefined policies. Such a policy baseline takes care of the majority of the configuration work and can be fine-tuned further as part of the engineering process.

From the manufacturer perspective, HIDS/HIPS enables not only more secure devices that are protected from malicious and nonintentional (well-meaning insider) change, but also reduces lifecycle management pressures (fewer reactive patches are required), allowing the manufacturer to gain better control of the release schedule and focus on the commercial aspects associated with updates and patches.

11.2.3 Cryptography

One of the most common and powerful security tools is cryptography. However, it is also one of the most complex, as demonstrated by many misguided or incorrect implementations (hence why Chapter 12 is dedicated to a deeper discussion of this topic).

Cryptography addresses aspects of information protection such as confidentiality (by converting plaintext to cipher text), integrity (by signing data with cryptographic hashes), and authenticity (by establishing unique identities). The common areas of cryptography implementation include:

- Algorithms that perform actual cryptographic functions such as encryption or signing.
- The key(s) used by the algorithm to perform the cryptographic calculation.
- The key infrastructure that defines the hierarchy, anchors the individual keys to a trusted root, and supports the generation, provisioning, revoking, and updating of keys.
- The management processes that supports everything from key generation, provisioning, and lifecycle as well as algorithm maintenance and update in case of deprecation.

Often cryptographic functions are transparent to the user and are provided through seamless integration with commercial technologies, such as full disk encryption or web browser to web server communication via HTTPS. However, when designing cryptography into medical devices, these commercial technologies may not be suitable as they may not meet the unique requirements of the use case, such as:

- Resource limitations given through device physical size and available battery power.
- Timing and real-time requirements of data flows.
- Disconnected use case as devices may not be network-connected at all times.
- Unique error and exception handling requirements as with different types of devices different continuity considerations may apply.

- Use of unique technology platforms or interfaces for which commercial crypto solutions may not be readily available.
- Protection of secret key material (e.g., during provisioning), especially in cases where a third-party manufacturer is used.
- Point-to-point connections (without the internet) that require secure authentication and protocols.

Implementing cryptography requires complex trade-off decisions to account for device capabilities, clinical use cases, and deployment infrastructure. Unfortunately, there is also ample opportunity for implementation errors with any of the above and unfortunately, examples of unsuitable or incorrect implementations are plenty. Again, this is why cryptography is discussed in much more detail in Chapter 12.

11.3 Perimeter and Network Security

Although in most cases network security is part of the operators' responsibility (HDO or contracted service provider), MDMs need to understand best practices to (a) be able to design devices that can be operated on a secure network, and (b) if applicable, provide a secure system of systems that reduce the risk of connecting insecure networks as part of their functionality.

By no means, however, should an assumption about the device operating on a secured network be used as a substitute for securing the device itself—neither for the total lifecycle security designed into the device from inception nor as a long-term mitigation for a vulnerability identified during postmarket. The FDA has stated that manufacturers should work from the assumption that HDO networks and patient home networks are "hostile" [3]. Although network security should be part of an organization's overall defense-in-depth approach, it cannot and should not be used as the primary measure to secure the device. Another way of looking at it is that network security's primary purpose is to secure the network and protect the devices from network-based threats, but not to secure the devices themselves. The network can only protect the devices insofar as the devices also protect the network.

Typical elements and components of network security are network architecture, firewalls, IDS and IPS, honeypots, and data loss prevention (DLP). More recently, medical device-specific passive network monitoring (PNM) or anomaly detection solutions have established themselves in the market, providing security monitoring of network traffic plus a number of other security and risk-management features.

11.3.1 Network Architecture

The commercial network security tools discussed later in this section are complementary to a secure network architecture. A technical discussion of this would exceed the objectives of this book and has been covered elsewhere. It should be stressed, however, that proper network segmentation is an important aspect of any security architecture and provides an additional (but not perfect) layer of security. Good network segmentation does improve protection against network-based attack vectors but does not protect against air-gap attacks, for example, via a USB data carrier. In either case, segmentation would help with the containment of an attack or outbreak (i.e., it provides a degree of damage control by limiting the scope of an attack).

Decisions on how to segment a network vary and may be done based on applying multiple criteria:

- *Functional integration:* Segmentation should support and not hinder the data flow between systems. Complex integration across network boundaries, although not totally avoidable, should be minimized.
- *Organizational structure:* May align, to an extent, with functional integration needs and if so, this can help to align security responsibilities with organizational responsibilities.
- *Geographic distribution:* Segmentation based on location (building, floor, on/off campus).
- *Risk and value:* High-risk and high-value systems should be cordoned off from main networks, providing higher levels of protection.
- *Redundancy:* With segmented networks providing containment of a potential attack or outbreak, devices supporting the same clinical function should be segmented off to ensure that partial functionality is maintained (e.g., if a hospital has six cath labs, it may be advisable to break them up into two network segments of three each to avoid total shutdown of interventional imaging services in case of a cyberattack, such as a malware worm that could spread to all devices within a segment).

Final architectural decisions should be a balance between the criteria suggested above and practical considerations. There may be conflicting priorities between, for example, functional integration and the need for redundancy. A comprehensive and well-guided risk assessment approach is advisable and is, for example, as described in the U.S. Department of Veterans Affairs (VA) "Medical Device Isolation Architecture Guide" [4]. Although a bit dated from a network technology perspective, this resource does outline a viable decision process for risk-based network segmentation.

The more segmented the network, the more secure it is, but unfortunately it is also more difficult to implement and maintain. This is certainly true for traditional virtual local area networks (VLANs) built based on traditional bridges, routers, switches, and firewalls. With the recent advent of software-defined networks (SDN) and supporting commercial products, the task has become easier and made the implementation and management of highly or micro-segmented networks much more feasible.

11.3.2 Firewalls

Firewalls protect and control data flows and access to the enterprise network (as a boundary to the outside) as well as between internal network segments. In a sense, firewalls restrict access from one network (or segment) to another, or from an enterprise network to the internet, with the goal of improving the security posture of the network.

Firewalls translate and enforce an organization's network security policies via firewall policies and are traditionally implemented as hardware appliances but can also be virtual or software-based (e.g., as part of a commercial anti-malware product installed at an endpoint). Due to the critical security function of firewalls, they should always be hardened to prevent unauthorized changes that could weaken their security function. Firewall change management requires a deliberate planning and implementation process to avoid extreme policies (whether too restrictive or not restrictive enough) and to ensure that the firewall policies can be properly maintained over time.

Depending on the use case and purpose, firewalls vary widely in architecture and functionality and the different types not only vary in their detection and prevention capability, but also in their ability to log events and provide alerts. The more complex and capable the firewall, the higher its impact on bandwidth, or the higher the hardware effort required to maintain a given bandwidth.

Some of the more common firewall types are:

- *Packet-filtering firewall:* A fairly simple model that uses basic rules and access control lists (ACLs) to make simple accept/reject decisions on a packet-by-packet level. Although efficient, they require careful design and maintenance of the firewall rules and have limitations with regard to their security capabilities (e.g., they cannot detect an application-specific vulnerability in a data stream that is sent/received as part of the communication between one legitimate application and another). Due to their design and rule-based operation, they are very susceptible to vulnerabilities stemming from configuration error or maintenance drift.

- *Stateful firewalls:* Typically work on the OSI Transport Layer to provide more advanced and state-dependent filtering (i.e., they are capable of more complex and granular decisions; a common example is that of a user visiting a website). A stateful firewall would allow the traffic from the web site back to the user's browser since it was initiated from there. However, it would reject any random traffic from the outside that was not previously logged within the firewall's state table. Again, there are some security limitations; for example, a stateful firewall would not protect from a compromised website visited, knowingly or unknowingly, by an internal user.
- *Proxy firewalls:* Act as a middleman (hence, proxy) and perform deep inspections of the messages leaving and entering a network. It also isolates internal endpoints from the external network as it replaces the internal address with the address of the proxy, such that internal devices are never directly exposed to the external network and instead remain anonymous. These two functions represent major security advantages over packet-filtering and stateful firewalls. Actual implementations vary and may provide different levels of inspection. A firewall may be capable of (or configured to) inspect application-level traffic and, for example, make decisions about files attached to an email. In cases where traffic may come from an unknown application, (e.g., an FTP request), a proxy firewall may still fall back on simpler rules (e.g., based on allowed IP addresses). Although very good for security, proxy firewalls may result in performance (bandwidth) or functional limitations.
- *Next-generation (NG) firewall:* All the previously discussed types have one common limitation, and that is that they depend on known risks and threats or predefined rules (as defined by their policies) to make decisions. They may be able to detect known bad IP addresses, suspicious files, or malware, but will likely fail if confronted with unknown and/or new threats. Next generation firewalls introduce a new concept by executing programs or opening files in a controlled environment and observing their behavior (commonly referred to as "exploding") prior to permitting or denying them. For example, an email-attached file is captured by the firewall and executed in a virtual environment to see if it behaves maliciously and only if it is benign is it passed on to the email recipient. Obviously, even this NG firewall does not solve all security problems, but it addresses the problem of the unknown malicious file. Still, as always in cybersecurity, it is an arms race and cyber adversaries have responded to NG firewalls by delivering malicious content in individual pieces rather than all at once (requiring that the firewall's isolation

environment be smart enough to assemble the individual components). Even then, smart malware may be virtual-aware and detect that it is being executed in a virtual environment and may not "detonate" its payload.

Typically, firewalls are deployed as boundaries between network segments or between the internal and external networks (internet). However, there are also cases where firewalls are used for just one specific device: any system or device that is protected with a commercial endpoint security solution (anti-malware or HIDS/HIPS as discussed above) has a built-in software firewall that should be turned on unless performance or predictability concerns would prevent that. Also, devices that have inherent vulnerabilities that cannot be addressed (e.g., running a post-end-of-support operating system or other legacy component) can be protected by a dedicated, external hardware firewall. This would reduce the likelihood of an external attack succeeding; however, the legacy device would remain vulnerable to non-network attack vectors, for example, USB data carriers (air-gap attack).

From an implementation perspective, firewalls can be added across several security control points and come in a number of form factors, such as a physical device or virtual appliance. Also, many commercial OTS software components, such as OS, anti-malware, or HIDS/HIPS, may come with a built-in software firewall that can be implemented on the device level. Lastly, there are low-cost firewall appliances that can be dedicated to a specific device to add an additional layer of protection (although, from an economic perspective, this is probably a more feasible approach for large devices such as imaging equipment than for battery-operated devices such as infusion pumps). It is critical that the differences and respective limitations between the various types and form factors are understood and be taken into consideration. Simply stating "we have a firewall" is not sufficient.

11.3.3 Intrusion Detection and Intrusion Prevention Systems

Network intrusion detection systems (IDS) and intrusion prevention systems (IPS) detect malicious network activity based on comparing observed behavior against what is considered normal behavior or identify specific patterns that are associates with a typical attack. IDSs are simply focused on detection, logging, and alerting, whereas IPSs would take action and actually interfere with or prevent suspicious activity.

The choice between IDS and IPS depends on the level and type of risk a network operator is willing to accept. An IPS may stop good traffic in a false-positive situation, which may or may not be acceptable. However, an IDS would insert a human decision into that process so that the preventive action could consciously be decided upon, but this would delay action and allow the adversary more time on the network.

Common attack activities that IDS/IPS systems can detect and/or prevent are: external port scans and similar reconnaissance activity, denial-of-service attacks, network policy violations, and activity typically associated with a network exploit.

IPS/IDS solutions commonly use traffic rules and/or a set of signatures that would identify abnormal and malicious behavior. The quality of detection highly depends on having up-to-date rules and signatures loaded into the system. Modern and more advanced systems introduce an approach that passively monitors network traffic and from that discern device data and characteristics, thus improving asset visibility, enabling risk management, and detecting anomalies relative to normal network traffic patterns that could indicate a security compromise.

Traditionally, IDS/IPS differ from firewalls as the latter relies on a set of static rules (the firewall policies) to manage network connections (i.e., it is a reactive but preventative approach to intrusions). An IDS/IPS only intervenes once an intrusion has been detected. With today's advanced adversaries, both approaches have unique advantages in certain attack scenarios, which helps to explain why the lines between these two security technologies is blurring and modern systems contain elements of each.

11.3.4 SSL Inspection

One challenge every network security technology faces is that of detecting malicious content in encrypted traffic, be it for security purposes (e.g., IDS/IPS) or privacy protection (e.g., DLP). Although encryption of data is an important security tool, it is increasingly being used by threat actors to disguise the content of malicious and/or captured data, for example, malware being sent into an organization or information being exfiltrated [5]. As a simple example, if adversaries do not want to be flagged when exfiltrating source code or patient data, they may be able to send it via an encrypted SSL or TLS channel. Unfortunately, that is a technique increasingly used in advanced attacks.

One way to address at least part of the problem is to introduce a network-based SSL inspection system that essentially terminates the encrypted traffic at the network boundary, decrypts and inspects it on the appliance, and then reencrypts it before sending it on to the destination. Such commercial tools exist and can be readily deployed; however, two concerns need to be addressed: (1) degradation of bandwidth and introduction of latency, and (2) maintaining security by not reducing the level of encryption (i.e., encrypt outgoing traffic at the same strength as incoming) [6].

11.3.5 Security Deception (Honeypots)

Honeypots are decoy computers on a company's network that attempt to entice malicious users to attack them and thus enabling them to collect and analyze malicious activity. They are deliberately set up to be attractive to an attacker, either by displaying security vulnerabilities that could be exploited or by pretending to have valuable information stored in them. Commercial offerings, sometimes paired with a management service, are available to help set up honeypots as part of the larger security operation, including the required capture and analysis tools. Some products can even be set up to look more realistic by mirroring existing system types, such as medical devices. Although honeypots are not a defensive security tool by themselves, they are a useful tool in the arsenal of an enterprise's larger security strategy and certainly are an important forensic tool.

When setting up a honeypot, one needs to be conscious of the legal boundaries between enticement and entrapment. Legally speaking, honeypots can be used to track a criminal that is already committing a crime (e.g., broke into my network) but cannot be used to motivate somebody to commit a crime that they were otherwise not planning on (e.g., by publicly advertising the honeypot's apparent offerings).

11.3.6 Passive Network Monitoring

Traditional asset management solutions fail to address the unique medical device security use case. Configuration management database (CMDB) tools are well suited to track and manage IT assets but may fail to capture medical devices and their required properties. As previously mentioned, they rely on network scans and possibly agents, both of which will not work for many medical devices. Computerized maintenance management systems (CMMS), as they are typically used on the clinical engineering side, are more designed to support traditional inventory and maintenance activities and do not support aspects of cybersecurity.

That gap has been filled with passive network monitoring (PNM) solutions (historically also referred to as anomaly detection tools). Commercially available tools have been around for many years and have reached maturity. The initial idea of these tools was simple: if I cannot scan and interrogate a medical device for its security properties or events, then I can resort to passively monitoring the device's network traffic, establish a baseline of normal behavior, and detect abnormal behavior. Over the years, however, these tools have significantly developed beyond this simple security task and now provide a rich set of features related to medical device security risk management, including:

- Pre-procurement risk assessment.
- Asset inventory, including security relevant parameters.

- Traffic mapping and usage/utilization analysis.
- Network-based security event detection.
- Ongoing security risk assessment.
- Vulnerability management, prioritization, and mitigation.
- Guided risk mitigation following device-specific best practices.
- Forensic data capture.
- Configuration management and control (e.g., policy management, NAC policy creation).
- Security documentation libraries (such as MDS[2], recalls, vulnerability disclosures).
- Integration with existing enterprise management systems such as CMDB, CMMS, compliance and risk management, security event detection.

Due to the passive nature of these tools, they integrate seamlessly across an HDO's medical device inventory and support devices of differing ages and security maturity, making it an attractive solution for hospitals to address their medical device security shortcomings.

11.4 Cloud Security

In a sense, the cloud is just another computer and, therefore, tools applied here are similar to the enterprise tools discussed earlier in this chapter. However, due to the difference in technology platforms and architectures as well as differing security responsibilities across the XaaS stack, the enterprise equivalent of such cloud security tools stand alone as a unique offering.

Any cloud deployment needs risk management, server security, access and authentication, and encryption. Ideally, these tools, although implemented in the cloud, should not offer a different management platform than the on-premises tools; for example, I do not want to have to configure different security and compliance rules for my cloud servers and my local servers. We may not be quite there yet, but directionally this is where the industry is moving.

One unique tool set that has developed over the past year is the cloud access and security broker (CASB), a group of products that combine several desired cloud security functions under one umbrella. It acts as a security policy enforcement point and combines management of enterprise security policies, access and authentication, inventory and risk management, encryption, logging and alerting, anti-malware, and similar functions.

11.5 Enterprise and Orchestration Tools

Enterprise security tools support a wide range of common, enterprise-wide security and risk reducing tasks, such as access control and authentication, automated security training, simulation and awareness, encryption, and email and web security. The range of commercial offerings is large, here just a few examples to provide some highlights.

11.5.1 Data Loss Prevention

Data loss prevention (DLP) solutions are more focused on information protection than pure security. Commercial DLP solutions provide the capability to identify data, at rest as well as in motion, that meet certain criteria (e.g., patient data, account numbers, intellectual property). Once the data is identified, protective actions can be taken, such as the data being removed/blocked, encrypted, and alerts created.

DLP products provide both protection against malicious activity (e.g., data exfiltration) as well as careless user behavior (e.g., copying data to a USB stick or emailing sensitive data to an outside entity). Products vary in complexity and features, allowing an enterprise to implement their preferred information protection rules using a DLP policy across traditional endpoints, cloud storage, and transport mechanisms such as email or actions such as copying or printing data.

In the context of medical devices, DLP products are helpful as they may provide insight on data transfers that are in violation of corporate policy. For example, an MDM may copy data from a device for maintenance purposes but may inadvertently copy PHI, identifiers, or network credentials. MDMs may use DLP on their corporate networks to protect their intellectual property (design documents, source code, research, and clinical trial data).

11.5.2 Security Management Tools

Security management, orchestration, and response are provided through SIEM systems, SOC, and incident response and remediation.

The above may not be directly applicable to medical devices, outside of exceptional cases, but medical device software engineers and service personnel need to be aware that medical devices will be operated in an IT environment that probably uses at least some of these tools. For example, a medical device may be exposed to an asset discovery or vulnerability scan and the device's operation should not be affected by it. Should there be concerns that a device may react negatively to scans or interrogation by these IT tools, these should be explicitly communicated, and appropriate configuration and management instructions should be provided.

11.6 Securing Communication Mediums

11.6.1 Bluetooth Low Energy

Over the last several decades of medical device development, there has always been some communication medium currently in vogue. The RS-232 Serial port with a DB25 connector was replaced by a DB9 connector, which was succeeded by USB, which mutated to USB OTG. Following that, we had a series of wireless media, such as Zigbee, IRDA, and proprietary protocols.

Then, beginning in 2010 with the ratification of the Bluetooth Core Specification v4.0, Bluetooth Low Energy (BLE) was added into the Bluetooth universe along with the original form of Bluetooth BR/EDR. (BLE and Bluetooth BR/EDR are very different in just about every way, but both are defined and maintained by the Bluetooth SIG.)

We are not going to dive into the technical details—there are many websites [7] that discuss the various technical aspects of BLE—but the definitive text is the Bluetooth SIG's Bluetooth Core Specification [8]. However, it can be difficult to read and locate specific topics in this very large document. A much more readable source of technical information is Mohammad Afaneh's free eBook *Introduction to Bluetooth Low Energy* [9] or NIST's *Guide to Bluetooth Security* [10].

11.6.1.1 Legacy BLE Specifications

Lesson 1: Don't use BLE v4.0 or v4.1 for anything. The key exchange process is badly flawed, allowing a paired/bonded BLE connection to be taken over by an attacker. Lots of online documentation exists describing this vulnerability, as well as easily-downloadable exploit software. Moreover, Data Length Extension (DLE) is not supported until BLE v4.2; DLE is desirable because it increases the BLE payload size from 27 bytes to 251 bytes. This is useful for cryptographic operations, which typically require large numbers, as well as faster firmware updates and log downloads, and increasing overall BLE throughput from approximately 6 kbps to 270 kbps. (You can get even more throughput by using BLE v 5.0 or greater; the 2-Mbps link speed can push it up to 1,400 kbps, assuming embedded BLE radios are used for both sides of the connection [11].)

The use of BLE versions older than 4.2 indicates an older BLE radio, which probably costs more than a new BLE radio, is closer to EOL/EOS, and will not support many of the other BLE features the device may need to leverage. Therefore, any new medical device should be designed based on BLE v4.2 or above and has been designed to select Secure Connections Only Mode (meaning it is forced to Security Mode 1 Level 4, as recommended by NIST) [12]. This is as secure as you can make standard BLE.

Unfortunately, this is not good enough. There have been many published attacks [13] against the BLE stack, and while using legacy connections is much worse than v4.3's secure connection (which utilizes ECDH), both have had successful attacks against them. One of the more recent attacks on BLE, dubbed SweynTooth [14], yielded 12 new vulnerabilities as shown in Table 11.2.

These vendors' components are deployed in a large collection of devices, including medical devices from major MDMs, not all of whom are identified in the disclosure. This is also a good example of the challenges of vulnerability communication. What is the best way for an HDO to receive appropriate vulnerability information relating to specific devices as it relates to an underlying technology vulnerability? Just because the device runs a particular software does not provide enough information about the actual vulnerability to the device, as it largely depends on the actual implementation in each device.

11.6.1.2 Securing BLE

Not discussed in the SweynTooth disclosure mentioned above, but nevertheless a vulnerability in the design of some devices, is that when a BLE radio's CPU is utilized to perform the medical device's essential performance concurrent with BLE communications, an inherent conflict can occur. In all BLE SOC radios, when a BLE event (reception or transmission of a BLE packet) occurs, the MCU in the BLE radio interrupts the execution of the application-layer software to process the BLE event in the BLE stack. This means that for milliseconds the main application stops functioning, along with whatever medical device functionality that it was supposed to perform. So a simple attack is to pair to the medical device where the minimum connection interval is set to 4,000 ms (as an example) and then request a change to the connection parameters to a minimum connection parameter of 7.5 ms (lowest possible setting defined in

Table 11.2
SweynTooth Vulnerabilities

Result	Vulnerability	BLE Vendor	CVE
Crash	Link Layer Length Overflow	NXP Cyprus	CVE-2019-17519 CVE-2019-16336
	Truncated L2CAP	Dialog	CVE-2019-17517
	Silent Length Overflow	Dialog	CVE-2019-17518
	Public Key Crash	TI	CVE-2019-17520
	Invalid L2CAP Fragment	Microchip	CVE-2019-19195
	Key Size Overflow	Telink	CVE-2019-19196
Deadlock	LLID Deadlock	NXP Cyprus	CVE-2019-17060 CVE-2019-17061
	Sequential ATT Deadlock	STMicroeletronics	CVE-2019-19192
	Invalid Connection Request	TI	CVE-2019-19193
Defeats Secure Connections	Zero LTK Installation	Telink	CVE-2019-19194

BLE). This leaves the MCU with very little time to perform the intended application, potentially too little. As mitigation, the application should be tested under stress with the connection period at 7.5 ms, or the BLE stack should be configured to never allow the minimum connection period to drop below a predetermined working period.

It should be noted that as BLE is typically defined, the central BLE radio (typically a smartphone) is the ultimate authority on the connection period; however, this just means that the attacking BLE connection will terminate, as the central BLE will be set to 7.5 ms while the medical device BLE is still at 4,000 ms (or some lower threshold significantly above 7.5 ms).

This clearly demonstrates that trust cannot be conveyed by means of standard OTA security as provided by BLE. Furthermore, all the OTA weakness examples above assumed that this topology was an embedded BLE radio communicating to another embedded BLE radio. If the topology was instead an embedded BLE radio communicating to a smartphone, then it does not matter what OTA security exists, because the data stream between the BLE radio in the smartphone and the app is in plaintext and has no protection applied to it (neither confidentiality, nor integrity, nor authenticity). Any app or malware on the smartphone (Android and iOS) can sniff and inject data from/to that data stream.

This indicates that for a medical device, a second layer of security controls must be put in place. The steps to be followed must be performed in order (refer to Chapter 12 for further discussion of each technique):

1. Authentication between the app and the medical device ("Are you the correct app?").
2. Authorization between the app and the medical device ("What role are you connecting as?" and "Are you an authorized device/user?").
3. Key exchange between the app and the medical device.
4. Enable encryption between the app and the medical device.
5. Depending upon the type of encryption being utilized, it may or may not be necessary to support integrity checking of the BLE payloads with an additional key.

There are many variations on this approach. For example, in step 3, the app could connect to a server to reference a unique key for key exchange. Likewise, the app may never have any knowledge of the key, and the encrypted payload would go to a server for decrypting and then selective plaintext would be returned to the app via a secure TLS connection.

Another approach is that, when the key is exchanged, it would then simply serve as a seed key to use in a highly ephemeral key structure, where the key would change every packet or every minute. These types of structures are one of the most secure approaches, commonly known as Perfect Forward Secrecy (PFS) or the double-ratchet algorithm [15], which can serve as a basis for lightweight embedded solutions where cryptographic accelerators in hardware can be leveraged.

One last point about BLE is that while there have been cheap BLE sniffers available for years, these are low-level tools that take a lot of effort and experience to yield any results. Certainly, $10 SDR BLE dongles can be cost-effective in addressing simple BLE issues. Good high-end tools cost approximately $10,000 to $40,000, but the increased cost of a good BLE sniffer quickly pays for itself when trying to root out more complex issues.

11.6.2 Other Media

Other frequently seen, traditional communication mediums in medical devices include serial (RS-232), USB, Inductive, and MedRadio Communications as well as Ethernet, 802.11x Wi-Fi, and cellular baseband. These popular mediums all share the common trait of being highly insecure legacy mediums, coming from a time when most MDMs (as well as regulators and HDOs) did not consider security. Thus, when you find these types of media in use, the medical devices incorporating them generally have few or no security features at all. Some of these inherited security concerns can be mitigated via a security overlay (e.g., end-to-end encryption) but some concerns remain, such as the protocol's ability to withstand a DoS attack.

No communication medium is secure by nature. Only the capacity for additional layers of security to be enabled inside of the medium matters. So, starting with the basics, the system must be able to perform a set of protocol steps to support secure data exchange:

1. Authentication;
2. Authorization;
3. Key exchange;
4. Enable encryption;
5. Depending upon the type of encryption being utilized, it may or may not be necessary to support integrity checking of the payloads with another key (via a hash function).

The five steps above should be utilized on any connections carrying PHI, PII, IP, or command-and-control capability. They can be accomplished by using standard secure protocols (such as TLS or HTTPS in their most recent and therefore most secure version) or by implementing end-to-end encryption between the applications running on the respective devices (e.g., between the device, the smartphone app, and the cloud service). End-to-end encryption makes any security implementation independent of the transport layer, its configuration, or number of transport hops. For example, a home care device may connect to the patient's smartphone vie BLE, the smartphone connects to the local cell tower or the home Wi-Fi router, which connects to a data center via public internet or telecoms infrastructure. Ensuring a secure configuration of all points and secure connection across all steps cannot be guaranteed or controlled, and therefore an end-to-end approach is advantageous.

11.6.2.1 USB Connections

USB connections are omnipresent in our digital universe; however, like the other media presented in this section, it can be easily and cheaply sniffed and injected into (e.g., as man-in-the-middle attacks performed by a USB capture device). So, as with any communications medium, the five steps above need to be performed to protect USB communications.

USB also has a unique weakness. When it initially enumerates to a commercial OS (Windows, MacOS, Linux), it can identify as a whole host of various devices, including a human interface device (HID) like a keyboard or mouse, a mass storage device such as a flash drive, or a network interface. Furthermore, a USB device can masquerade as all of these things at the same time, thus enabling the following attacks against a commercial OS:

- It can pretend to be a keyboard and "type in" prewritten commands.
- It can inject malware into your system via its mass storage connection, simply by taking advantage of the OS's autorun feature (many medical device security incidents of this type have been reported).
- It can act as a network interface and open a connection to a compromised or malicious website hosting a drive-by infection package.

This is why it is very difficult to harden any commercial OS platform when USB is not disabled. CERT has provided guidelines addressing this [16]. Many commercial security products (anti-malware or HIDS/HIPS, discussed earlier in this chapter) include security features specifically designed to mitigate the USB attack vector on systems with commercial OSs.

11.6.2.2 Inductive and MedRadio

Inductive and MedRadio are frequently seen in implanted medical devices as they communicate well through human tissue (unlike BLE, which needs more power to propagate through tissue due to its high operating frequency of 2.4

GHz). In implantable devices, the most important design requirement is complying with battery life constraints, due to the physically small space in such devices and the difficulty of replacing batteries or devices (as that would require a surgical procedure in most cases). One common design approach used to maximize battery life is to put certain functions into sleep mode when they are not needed, for example, the radio.

With either of these media, one possible attack scenario is to just keep "waking up" the implant repeatedly. This consumes much more battery charge than anticipated usage (which might be once every few months in a doctor's office), leading to early battery depletion and the need for surgical remediation to replace the implant, and this comes with all the usual risks that accompany surgery, as well as associated costs and stress. In the case of MedRadio, there is a 2.4-GHz wake-up signal that can be addressed to a specific type of device, or just a general "wake up anyone who can hear me" signal. In either case, there is no built-in security as the specific wake-up call is in plaintext and can easily be sniffed with an SDR and replayed. Or the attack can simply utilize the general wake-up call, which has an 8-bit Company ID field with a default value of 0x01. All that is needed to attack this company ID is to just run through all possible 255 values to wake the radio up. For an attack such as this, there is no practical attack range, meaning that a whole house, HDO, or neighborhood could be subjected to a high-power RF wake-up call every few seconds.

One possible approach to mitigate these types of risks is to limit the amount of time that the radio stays awake, and possibly even create a back-off period where the radio cannot be woken for a set period after its previous wake event. (Note that MedRadio's main data channels do not support any native OTA encryption, so encryption must be performed at the application layer.)

In the case of inductive communications, there are common myths held by many MDMs (further discussed in Chapter 13), such as the belief that a proprietary protocol is secure simply because it is proprietary, when it is not [17]. There are also misconceptions about induction's typically short maximum communication distance acting as a security mitigation, because it would require the attacker to be in close proximity to their target. In reality, communication distance is a function of coil diameter and transmit power, so instead of a wand-style physical format with a range best measured in centimeters, a briefcase-sized coil with a powerful transmitter could be used to attack over a greater distance—from across the building, down the street.

With any of these media, MDMs must follow the five steps previously laid out, with good strong cryptographic primitives (encryption, authentication, and integrity protection) to secure the device's communications.

11.7 Hardware and Physical Interface Security

This is another one of the many areas where an embedded device differs from the security used in IT/MIS environments. Unauthorized persons can be prevented from walking freely around a data center (and doors and locks are a perfectly plausible security control), yet an embedded device can be easily procured (as it has been demonstrated repeatedly) by security researchers or potential adversaries, who could then analyze the device's vulnerabilities at their leisure.

The nature of attacks at the physical level is not as widely discussed, so frequently MDMs downplay the possibility of them occurring or ignore their potential, writing them out of scope.

11.7.1 Hardware Hacking 101

Emphasis on cybersecurity purely from a network perspective results in blindness to a significant portion of the device's attack surface. Physical attacks are routinely used and could have devastating effect. So, let's walk through a playbook of how a physical attack could proceed.

First, look at the device. What obvious conclusions can you reach by examining the device? Are there wired communications with obvious connectors, such as RJ11 Ethernet, USB, or DB9 serial ports? If so, these are your first area to look for vulnerabilities to attack. Plug them in, see what you get, perform communications at all the common baud rates, perform a port scan, and see if they left the telnet service running. Nonstandard connectors may be present—they just take a little more work after the enclosure has been opened.

Look online at the MDM website. What do they say about maintenance and servicing? Is there some clue about utilizing these ports? Go look at older similar devices from this MDM. Look for more details about usage, communications, and service credentials (there have been reports of training manuals available online that contained service usernames and passwords). Try these credentials on your unit. It is common for MDMs to continually reuse credentials that were defined 40 years ago, like "admin"/"123456" (some attempt to justify this practice by claiming that field service reps would not be able to remember new ones).

Is there some manufacturer-authored downloadable executable, such as a patient app or a physician programmer? If so, install it on a given attack platform (such as a PC) and reverse-engineer the executable using decompilers. Then run the app and sniff the communications. Is there any form of authentication or authorization? Is anything being transmitted in plaintext (e.g., legacy DICOM or HL7 or plain ASCII)? Are there any obvious key exchanges taking place? If you have multiple units that can be attacked, try this on several units and analyze the data for any differences in the communications that might indicate unique authentication at work. Did the executable ask for any unique data items like the unit's serial number? This can point to the algorithm used for unique authentication.

Go after the executable. Most likely it is written in a managed language; if so, use a decompiler to return it back to human-readable code. What you are looking for are hardcoded credentials or packet structures such as a CONST data structure of commands. Are there "holes" in the commands? Try those values. What about command values above and below the value in the CONST data structure? Anything else that looks interesting, like hidden functionality accessed by clicking on a specific blank part of the screen? (Amazingly, we have seen this done more than once.)

Look at the device's technical documentation (including labeling and manuals). Is there an FCC ID number on it, indicating a wireless RF communications medium? If so, search the FCC's website for the background technical documentation, such as band or frequency of operation, type of modulation, and pictures of the PCBA. Identify major active components from those pictures. Start to build up your knowledge about the electronics being utilized and download datasheets from the component manufacturers, especially errata sheets. While you are there, explore any development tools they may have available or white papers on security or firmware updating. Perform general searches of the internet for known attacks against any of the major components you identified in the pictures.

If the device has a standard wireless interface, start attacking that by connecting to it, fuzzing the interface, and performing port scans.

All of that, and we have not even opened the enclosure yet. You may not need to, you may already be "in." Bear in mind that a successful attack may not necessarily mean that data is accessed or compromised. As discussed before, the attacker can also do harm by depleting batteries, triggering a reboot, overloading data traffic, or "bricking" the device.

We have squeezed this untouched device for about as much as we can get from it. Let's continue our quest by delving deeper into the device. Remove the enclosure. Sometimes this is a simple as using a screwdriver or Allen wrench, but sometimes the MDMs thought that they were being tricky and utilized "secure" screws. This hardware is anything but secure. For $13, you can purchase a kit to defeat almost all of them (there are some very small screws utilized in smartphones that requires a different kit to access), or maybe the MDM wanted the device to really be sealed—waterproof and dust-proof—so they ultrasonically welded the enclosure together, or even glued it together. In that case, you will want to reach for your handy Dremel rotary tool or a zero kerf pull saw or cheaper equivalent. These can be used to slice open the case at the obvious parting line of the enclosure (do not just go hacking the enclosure up, stick carefully to the parting line). The pull saw is best; it is slower and has less chance of damaging the PCBA(s) inside the enclosure. Or did the MDM put a lock on the enclosure? You can drill it out (makes a mess of metal flakes inside the enclosure), but you may realize that the MDM did not use unique keys for each unit. It is just a matter of identifying the correct common key that was utilized, and these are always available to the public, such as the A126 key or the CH751 key. A $4 key is not much of a security control.

Once inside the enclosure, carefully remove the PCBA(s) and start documenting everything. Note the laser branding on the large components and PCBA-mounted connectors and test points. The silkscreen labels on the PCBA can be a big help as well.

The connectors could be a JTAG, SWD port, or serial port. Investigate to see if they were left active. If you can use a JTAG or SWD port to extract the binary contents of the program memory from the MCU, then we have hit the hacker's jackpot.

Did that serial connection appear to display some data during power-up of the device (that is why you do not want to damage the PCBA while opening the enclosure)? Find the correct baud rate and read what is being output on that port. You are trying to identify the boot loader. Also, check the semiconductor manufacturer's website to see if there is a boot loader that can be downloaded.

A common bootloader is UBOOT; if this is the boot loader being utilized, try the MD command [18] to drop the contents to the display and capture it. Parse the captured information back into a binary format and decompile this binary file back to readable C code [19].

(Note that this is not the only thing you can do at UBOOT console. Colin O'Flynn has an excellent write-up on exploiting UBOOT [20, 21].)

Returning to the PCBA, examine the components to see if a nonvolatile memory is being utilized. This might be mapped to program space or contain other interesting values.

Continuing to go deeper, the next step is to attempt to decap the MCU with the goal of exposing ROM, Flash, or JTAG fuse (so JTAG can be reenabled) [22–24]. Granted, this is not as critical of a vulnerability as it used to be. With the increased use of small geometries on the die, any die below about 90 nm means the equipment to perform micro probing becomes very expensive, maybe not practical for a low-budget adversary but certainly possible for a nation-state actor targeting IP. However, as MCUs shrink to 40 nm and below, flash memory can no longer be on the die, thus creating a different vulnerability.

An MCU side-channel attack is an attack that exploits unintended information exposure to infer additional sensitive information exposure, such as cryptographic keys or other shared secrets. The specific unintended information exposure comes from unexpected areas, such as the amount of current being drawn by an MCU [25, 26], the RF emanations coming from an MCU, or timing differences. All of these occur while some crucial activity is being performed by the MCU. Typical activities to monitor are the comparison of multibyte values, moving of encryption keys in memory, the encryption process, any values being manipulated in a predictable bit order. With these side-channel attacks, the goal is to expose the secret values, literally bit by bit. To avoid timing attacks, decisions for critical values should have operations on those values take constant time, no matter the value of the content.

For power monitoring, an oscilloscope can sometimes be utilized, but special-purpose hardware exists to make the attack easier and quicker to perform [27]. These types of attacks have yielded some impressive results [28]. As indicated earlier, any predictable pattern can be exploited in a side-channel attack, so randomization can be an effective mitigation [29].

11.7.2 Hardware Security Considerations

As recently as 10 years ago, finding an MCU that had any hardware-based security tools or protections was a rare event; fortunately, this is no longer the case. Besides being faster than pure firmware alternatives, and consuming less RAM and program space, there is no better anchor for your set of security mitigations than a hardware root of trust. This is true even considering the occasional hardware vulnerability being disclosed; it is still the more reliable and most secure approach.

Some of the more useful tools provided by modern MCUs that are now at the security professional's disposal include:

- *Secure boot:* Confirms the integrity of some or all of the executable program space before transferring execution to it. This can be the base of a chain of integrity checks, where the boot code is checked and then, when it is executed, the OS is checked, and then it, in turn, checks the application code.
- *Cryptographic engine accelerator:* A set of cryptographic functions typically including symmetric and/or asymmetric encryption algorithms of various bit widths, as well as keyed and unkeyed cryptographic hash engines.
- *True random number generator.* A very useful tool for creating values for authentications and unique keys.
- *Memory management unit.* Offers the ability to map the addressable memory space for certain specific functionality, linking executable, read-only data.
- *Side-channel attack protection.* Adds uncertainty into execution timing and power draw to help to obfuscate internal operations.

- *Physically unclonable function (PUF):* Each MCU has its own unique value that never changes, ranging in size from 64 to 128 bits in width. These can be useful for uniquely addressing a specific device or acting as a seed in other identification schemes.
- *Secured debugging ports:* Prevents JTAG/SWD connections to the MCU to extract the contents of internal program memory for reverse engineering. These are available in a fairly wide range of options, from protecting read, to protecting write, disable (reenabling it erases flash), and reenable based on hashtag value of the internal program space vector table.
- *Die geometry:* A nice side effect of semiconductor manufacturers seeking smaller die geometries is that as you pass the 90-nm size, the die gets very difficult (and expensive) to microprobe. However, as we reach 40-nm geometries, we will no longer be able to have flash memory on the same die, which creates vulnerabilities for external flash memories.
- *Address randomization:* A few MCUs have a feature where external flash program space is accessed in a random fashion, with the key generated uniquely for each MCU. This means that the contents of the external flash cannot be mass-programmed but need to be programmed via an interface into the MCU. This does make it difficult to easily extract the executable for the external memory.
- *Virtualized-secure CPU:* Some MCUs (such as ARM cores and TrustZoneM) are now supporting an isolation technique where more trusted code can be run completely isolated from less trusted code. This can be used to contain cryptographic keys, as well as to perform dynamic integrity checks of the program space code, to prevent injection attacks.
- *Cryptographic secrets protection:* Utilizing secure and tamper-protected memory as for example found in the trusted platform modules (TPM) of modern processors and MCUs.
- *Trusted execution environment (TEE):* Operating independently of the main OS to provide secure and execution of cryptographic functions.

All of these features should be considered and leveraged in your device to the fullest practical extent.

11.8 Other Security Considerations

When commercial technologies intermix with dedicated-purpose devices such as medical device, a slew of security problems need to be considered and addressed. The following sections will discuss several of the more critical of these issues.

11.8.1 Hardening Common OSs

Hardening a commercial OS refers to the process of enhancing its security posture by reducing potential vulnerabilities and minimizing the attack surface through the use of secure configurations. The goal of hardening is to make the device more resistant to unauthorized access, exploitation, and data breaches.

Typical steps needed to harden a commercial device include:

- *Disable unnecessary services and ports:* Turning off unnecessary services and closing unused ports reduces the potential attack surface and limits the exposure to possible exploits.
- *Keep up with firmware and software updates:* Keeping the device's firmware and software current is crucial for addressing known security vulnerabilities and improving overall system security.
- *Use access controls:* Implementing proper access controls ensures that only authorized individuals can access sensitive data or perform critical operations.
- *Disable default accounts and settings:* Default accounts and settings are commonly known and targeted by attackers, so they should be changed or disabled during the hardening process.
- *Use strong authentication:* Enforcing strong and unique passwords for user accounts, utilizing MFA, or implementing biometric authentication can significantly enhance security.
- *Encrypt:* Data should be encrypted both at rest and in transit to protect sensitive information from unauthorized access or interception.
- *Use firewalls:* Configuring software firewalls to filter incoming and outgoing network traffic helps to block malicious requests and potential threats.
- *Use secure boot processes:* Implementing secure boot processes ensures that only authenticated and authorized software can run on the device, preventing malicious code execution during startup.

This is just a general (and very incomplete) list. Fortunately, there are free standard collections of the rules need to harden specific platforms available from various authorities, including the OS vendor itself, government agencies, organizations like OWASP or SANS, or the larger software security community (e.g., GitHub or Reddit). When hardening a platform for use in a specific medical device system, an MDM should attempt to implement the entire collection of rules and test to understand their impact on the system. Some tuning of the rules may be needed to be able to properly execute your application.

Sets of hardening rules are grouped together in what is called a Security Technical Implementation Guide (STIG). The most popular collection of hardening rules is maintained by the U.S. Department of Defense (DoD). In their collection are hundreds of individual STIGs, and each STIG could contain numerous rules that cover specific security configurations and settings for various technologies, applications, and platforms. The exact number of DoD STIG rules at any given time may depend on the technology landscape and the DoD's evolving security requirements, as new technologies are introduced or existing ones are revised to address emerging threats.

11.8.2 Utilizing Smart Phones and Other Off-the-Shelf/"Bring-Your-Own" Devices in Medical Device Systems

Off-the-shelf (OTS) commercial devices such as tablets and smartphones are a fairly recent innovation in medical device systems, replacing the previously utilized "purpose-built" interface devices or gateways. Typically, these purpose-built devices were physically large and had limited user interface options, short battery life, and no wireless connection to the internet.

This all changed with the inclusion of BLE wireless interfaces in smartphones and later in tablets. This allowed manufacturers to save the cost of developing a purpose-built device and gave the consumer a superior device in all regards, including the user interface, battery life, and a wireless connection to the internet, together with the BLE connection to a medical device. Connected medical devices can upload patient data to the cloud, facilitate real-time remote monitoring by healthcare providers, and even permit the patient's physician to modify medical device settings remotely, allowing the patient to avoid an in-person clinic visit or the doctor to avoid a house call. Generally speaking, these devices have been a huge boon for the manufacturer, care providers, and the patient.

Both tablets and smartphones can be used in medical device systems in one of two ways: Bring-your-own-device (BYOD) or kitted.

- BYOD is just what it sounds like: The user supplies an OTS device as the platform to execute a manufacturer-supplied app. The range of acceptable OTS devices is specified by the manufacturer, typically including Android and iOS devices where some range of acceptable version of the platforms and the operating systems is specified.

- Kitted is when the manufacturer supplies an OTS device along with the medical device, fully preconfigured, most likely locked down to prevent other, nonmedical uses, and ready to execute the manufacturer's app, which the manufacturer has previously installed during manufacturing.

Both approaches have pros and cons. With a kitted device, the manufacturer has complete control of its app's run-time environment, so much so that the physical device is considered a medical device accessory by the FDA and needs to be maintained just like the medical device itself. This results in the need to remotely collect operating logs and have the ability to upload patches and updates to the kitted device. Kitted devices need to be hardened to prevent malware from affecting the device and are usually placed into kiosk mode (if available/sufficient) or otherwise locked down such that the only app that can be run is the manufacturer's medical device app. This adds to the postmarket burden on the manufacturer, because either (a) large quantities of the kitted device need to be purchased and stored to support years of production, or (b) an ongoing program to revalidate the manufacturer's app on new platforms and new OS versions need to be implemented. This can result in substantial changes to the original app to accommodate changes in the platform, OS, and so on.

With a BYOD device, the FDA considers the BYOD to be a platform, which the manufacturer is not responsible for maintaining as it would be with a kitted device. Instead, the device app is considered SaMD and needs to undergo the usual testing to ensure best practice mitigations have been put in place for the app. Validation testing is usually constrained to a handful of OTS models and operating system versions, but since this is an ever-changing situation, the manufacturer will be performing ongoing testing on new platforms and new version of the operating systems as they are released (similar to Scenario B for a kitted device). In some cases, manufacturers have coded into their app the ability to identify the platform and OS version and warn end users if the app has not been tested on their particular device/OS or has been tested and found incompatible.

Frequently, we see the use of both kitted and BYOD devices in a medical device system. A clinical tablet may be kitted for a surgeon to test and configure an implant, or for a technician to perform troubleshooting, and a patient app will be available for BYOD smartphones.

Tablets and smartphones are playing a significant role in medical device systems and healthcare settings, providing various benefits to patients, healthcare professionals, and organizations. They have found a permanent home, to the point where many users would not even consider returning to the old days of purpose-built user interfaces and give up the following advantages:

- *Patient monitoring and health tracking:* Tablets and smartphones can run applications that allow patients to monitor their health conditions and track vital signs. For example, patients with chronic diseases can use mobile apps to record and share their blood pressure, blood glucose levels, and heart rate with their healthcare providers for remote monitoring.
- *Telemedicine and telehealth:* Tablets and smartphones facilitate telemedicine services, enabling remote consultations between patients and healthcare providers. Through videoconferencing, accumulated medical device readings, and secure messaging platforms, doctors can diagnose, treat, and advise patients without requiring them to physically visit a doctor (or vice versa).
- *Clinical decision support.* Healthcare professionals can utilize mobile apps that offer clinical decision support, assisting with diagnosis, treatment plans, and drug interactions, thus improving patient care and safety.
- *Medical Education and Training.* Due to the flexible user interface found on OTS devices, patients can be easily and safely trained by placing the OTS device into training mode, where they can get accustomed to the app interface and medical device controls without (yet) affecting any of the real settings on the medical device.
- *Remote monitoring:* OTS devices can act as control interfaces for various medical devices, such as insulin pumps, pacemakers, or home ventilators, allowing patients and healthcare providers to manage and monitor these devices remotely.
- *Mobile health research and data collection.* Researchers can use OTS device to collect health data from study participants, allowing for large-scale data collection and analysis for medical research.
- *Software patching and updates.* The OTS device can act as a conduit for update images targeting the medical device while it is in the field.

It is essential to emphasize that when incorporating tablets and smartphones into medical device systems, security and privacy considerations must be paramount. Sensitive health data, communications, and data at rest must be encrypted and have ensured integrity to comply with regulations and safeguard patient information.

11.8.3 Software as a Medical Device

The International Medical Device Regulators Forum (IMDRF) defines software as a medical device (SaMD) as [30]: "software intended to be used for one or more medical purposes that perform these purposes without being part of a hardware medical device." This encompasses a wide range of types of software, including, for example, clinical decision support, machine learning (ML), and AI, applications to view diagnostic results and images on a smart device, and certain home health apps.

SaMDs have some unique characteristics that require specific attention and consideration from a cybersecurity perspective, including:

- Impacts from differences in hardware platforms on which the user installs the SaMD.
- Management and maintenance responsibility (including of the hardware and operating system platform) typically lie with the user.
- Distribution and deployment are difficult to control, which may cause challenges to identify and manage specific instances of said device, such as for user notification or update purposes.
- Colocation, resource sharing, and potential data-sharing between SaMD and nonmedical apps running on the same device or the same network segment.

In order to comply with regulators and customers' cybersecurity expectations, standards such as IEC 62304 "Medical Device Software—Software Life Cycle Processes" can be utilized to guide a risk-based approach across the SaMD lifecycle. Taking a quality system as well as a systematic systems engineering and lifecycle management approach to SaMD development will ensure reliable and repeatable delivery of devices with the desired and maintainable security posture. Specifically, IMDRF points to the importance of considering the sociotechnical environment, the technology and system environment, as well as information security with respect to safety considerations and recommends to:

- Ensure adequate processes during planning, design, development, testing, and release; document how cyber risks are managed and mitigated; and establish risk control traceability.
- Implement formal postmarket surveillance processes to continually monitor to detect and respond to changes in device risk posture or environment. This includes, but is not limited to, device log files, customer feedback, complaints, market studies, focus groups, and service records and includes gathering of specific feedback relating to new cyber threats, vulnerabilities, or incidents as they may be provided by suppliers, customers, government agencies, or security researchers.

Specific software engineering practices and technologies that can be applied include, for example:

- Use of license or other key technologies (such as unique cryptographic identities) to prevent copying and operation of unauthorized versions of the SaMD.
- Partitioning critical functionality in a back-end server behind authenticated access.
- "Phone home" feature to maintain database of SaMDs in operation as well as to collect usage, event, and log data.
- (Secure) OTA features for updates, including user notifications or automatic update during downtime.
- Containerization of SaMD application to create separate, isolated runtime environments.
- Procedural enforcement, including EULA acceptance, customer disclosure feature.

Technically, the statement that "SaMD is just another medical device" is correct but considering some of the unique use and implementation challenges of SaMD, as discussed above, a set of specific considerations applies to meet the desired level of security at release and maintain that level during operation.

11.8.4 Multiple Function Devices

FDA Policy and Considerations for Multiple Function Devices stated that [31]: "a product with multiple functions that contains both a software non-device function and a device function, FDA may assess the impact that the software non-device function has on the device function when assessing the safety and effectiveness of the device function (section 520(o)(2) of the FD&C Act). FDA believes that a similar approach should be used for the assessment of all multiple function device products." In other words, when submitting for regulatory approval of a device that includes one (regulated) device function, as well as a function that normally would not require premarket review, "FDA may assess the impact of 'other functions' when assessing the safety and effectiveness of the device functions-under-review of a multiple function device product."

Specifically, in the risk assessment process, manufacturers should mitigate the other functions' impacts on performance, safety, and effectiveness and document identified risks, including safety and cybersecurity risks. The interconnection and interdependencies between the device and other functions should be included in the hazard analysis, and appropriate controls should be created to reduce the adverse impacts. Further, FDA recommended separation of the device function in architecture, design, and implementation. Accomplishing this may necessitate logical and/or architectural separation, code, and data partitioning.

While considering cybersecurity risks, such separation of the device from the other function(s) may be necessary to mitigate cybersecurity risks. Specifically, the manufacturer should consider:

- Possible impacts on device performance;
- Any limitations of the device function;
- Appropriate customer hardware and software specification(s);
- Recommended actions by the end user;
- Identification, evaluation, and mitigation of any additional risks, including cybersecurity risks.

FDA will assess if there is impact on the safety or effectiveness because of the other function and, if there is an impact, whether it would it increase risks or have an adverse effect. Such considerations should be included in the hazard analysis. Considerations should include whether the other function and the device function:

- Shares computational resources, data, or other relationships.
- Provides input data that is used for a critical calculation and the device relies on results from the other function.
- Shares code, memory, storage that are necessary for proper execution.
- Creates interference due to processing time or memory requirements.
- Shares programming pointers.
- Could have conflicting privileges or interrupts.

The manufacturer should provide documents, including those that assess the impact of the other function(s) on the device function, including:

- Indications for use;
- Device description;
- Labeling;
- Architecture and design;
- Device hazard analysis;

- Requirements and specifications;
- Performance testing.

FDA has clearly stated that [32]: "Manufacturers are responsible for identifying cybersecurity risks in their devices and the systems in which they expect those devices to operate and implementing the appropriate controls to mitigate those risks. These risks may include those introduced by device reliance on hospital networks, cloud infrastructure, or 'other functions'."

11.8.5 "Unregulated" Devices

We are seeing a large increase in the desire by patients as well as manufacturers to incorporate nonmedical devices into the larger care delivery process, and therefore into the medical device ecosystem. While this may be desirable from a patient's point of view, as it centralizes more information in one place, it brings with it a number of cybersecurity concerns.

Although most of this book specifically deals with safety, efficacy, and/or business risk as a result of a cybersecurity risk, and consequently focuses on safety regulations such as provided by FDA, EU MDR/IVDR, TGA, and others, we have to assume that for one reason or another and through one regulation or law or another, every device, or its data, is regulated in some way.

Regulatory complexity is one of the challenges that makes it difficult to get cybersecurity right. For example, in the United States, HIPAA regulates healthcare providers' responsibilities with regard to the protection of PHI, whereas the FDA regulates device manufacturers' responsibilities for cyber safety. Yet the manufacturer has no direct regulatory obligation to design a device that complies with HIPAA (although it does make good business sense to do so), and the HDO has no regulatory incentives for cyber safety (although general care delivery considerations as well as business considerations warrant it). The situation gets even more confusing if the device does not fall under existing medical device regulations or its data is not regulated by HIPAA (e.g., it is not considered PHI, or its operator is not a HIPAA-covered entity). With today's complex consumer-driven health devices, data flows across private and public networks and is handled and stored by a variety of service providers, including the manufacturer. This can lead to significant regulatory complexity.

Considering this, the term "unregulated device" is misleading and will, in most cases, require more thorough regulatory and legal analysis. Here is a partial list of laws and regulations that may need to be considered for manufacturers of nonmedical (perceived as unregulated) devices:

- If the device manufacturer provides data services (e.g., fitness tracking, analysis, aggregation/trending), then certain state or federal privacy laws may apply. Many U.S. states have citizen privacy laws [33] that apply to any (not just health) data, so do many countries and regions, for example GDPR for EU residents.

- Under the California IoT law (SB 327), manufacturers of connected IoT devices must ensure "reasonable security features," as appropriate for the device and the data it collects, to protect the device and any information contained therein from unauthorized access, destruction, use, modification, or disclosure [34].
- Section 5 of the Federal Trade Commission (FTC) Act stated that "unfair or deceptive acts or practices in or affecting commerce… are… declared unlawful," which is generally interpreted as to include failure of an organization to adequately safeguard computer systems and infrastructure. Further, misleading consumers into believing that their private data is secure could be interpreted as an "unfair or deceptive act" [35].
- The Security and Exchange Commission (SEC) has recently adopted a final rule requiring the disclosure of cybersecurity incidents and also insight into the company's cybersecurity risk management, strategy, and governance. To summarize, the rule requires:
    - Form 8-K disclosure of material cybersecurity incidents within four (4) business days of the company's determination that the cybersecurity incident is material.
    - Annual disclosure in Form 10-K regarding the company's cybersecurity risk management and strategy, including the company's processes for managing threats and whether risks from threats have materially affected the company.
    - Annual disclosure in Form 10-K regarding the company's cybersecurity governance, including oversight by the board and management [36].

    In light of some of the recent high-impact security incidents, the SEC has already demonstrated its willingness to use its oversight powers to investigate and enforce cybersecurity issues [37].
- In the EU, the Radio Equipment Directive 2014/53/EU (RED) has recently added cybersecurity to Article 3.3 (d, e, and f) [38]. The Directive covers radio equipment essential requirements for safety and health, including specific technical features for the protection of privacy, personal data, and against fraud as well as compliance regarding the combination of radio equipment and software.
- The EU Cyber Resilience Act (CRA) is set to enter into force in the second half of 2024 and manufacturers will have to place compliant prod-

ucts on the Union market by 2027. It excludes MDR/IVDR regulated medical devices but will affect many other health and fitness devices and services. Its objective is to secure products by ensuring that they contain fewer vulnerabilities and that manufacturers support security throughout a product's lifecycle. It also provides for conditions allowing users to take cybersecurity into account when selecting and using products with digital elements. The specific security objectives are: (a) improve security throughout the lifecycle, (b) ensure a coherent cybersecurity framework for hardware and software, (c) enhance the transparency of security properties, and (d) enable businesses and consumers to use products securely [39].

In addition to existing and evolving laws and regulations, there is progress among standards organizations to address the need for better security outside of the regulated device space. As one example, the IEEE/UL recently developed IEEE 2933: "Standard for Clinical Internet of Things (IoT) Data and Device Interoperability with TIPPSS—Trust, Identity, Privacy, Protection, Safety, Security." This standard addresses the need for Clinical Internet of Things (CIoT) data and devices, including wearable and home devices, and their interoperability with healthcare systems and other devices in the connected healthcare systems. The standard is guided by, and identifies best practices, around the six TIPPSS principles:

- *Trust:* Devices can assume that connected devices are reliable and trustworthy partners.
- *Identity:* Devices have a consistent method of identifying each other.
- *Privacy:* Ensure that device, personal, and sensitive data are kept private.
- *Protection:* Protect devices and users from harm—physical, digital, financial, and reputational.
- *Safety:* Provide safety for devices, infrastructure, and people.
- *Security:* Maintain security of data, devices, and people.

Other useful guidances for IoT device security practices are:

- IoT Security Foundation publications provided by the IoT Security Foundation, including a "IoT Security Compliance Framework," "IoT Security Assurance Framework," as well as specific topics such as "Software Bill of Materials for IoT" and "Vulnerability Disclosure Best Practice Guide" [40].

- National Institute for Standards and Technology (NIST) "Cybersecurity for IoT Program" [41].
- Cybersecurity & Infrastructure Security Agency (CISA) "Shifting the Balance of Cybersecurity Risk: Security-by-Design and Default Principles" [42].

The above provides a representative set of examples on how the topic of cybersecurity for critical infrastructure industries and devices has become a much broader topic than just affecting the traditionally regulated industries. In addition, what used to be well-defined operating and data location environments are changing as devices are moving to the patients' home and data is moving into the cloud. Responses to this technological change is evident through the wide range of new laws, regulations, standards, and practices that have been developing. This provides a path forward to a better secured future in which consumers can trust their devices and data, at the same time, this also has created a confusing and complex array of laws and regulations through which device manufacturers have to navigate. Because of the high level of exposure, even in what are today "unregulated" industries or products, regulations are already developing and are expected to tighten, especially for devices that deal with personal information and/or pose a high economic or national security risk.

11.9 Conclusion: How Are MDMs Making Use of These Technologies?

Although the situation is improving, many of the medical devices being sold into the market today still have insufficient security mechanisms. In some cases, this is due to the manufacturer's concern over the risks posed by such security tools to the essential operation and reliability of their medical device system, such as erroneously detecting malware due to updated signature databases, opening ports to support these security tools, and, as with any additional software, the risks from supply chain attacks. MDMs have often relied on the HDO's security measures, and it was generally but incorrectly thought that network segmentation was enough to secure a medical device. While it is certainly desirable to isolate a networked medical device and firewall it, this does not necessarily mean that the device is sufficiently secure to prevent:

- Other computers and devices on the network from inadvertently accessing the device and causing it to misbehave;

- Malware introduced by a data carrier (e.g., USB thumb drive) or by an infected laptop;
- A hacker on the hospital network taking advantage of devices' weak security posture and delayed patching or EOL operating systems.

Firewalls and network segmentation alone are not adequate security measures. Even these measures may be unavailable, or inadequately implemented, for all medical device deployment and use scenarios (e.g., devices operating in a home healthcare environment).

Table 11.3 summarizes the applicability of selected key security and which entity typically would implement it (MDM or HDO).

Today, MDMs and regulators are embracing a secure-by-design approach to medical devices. However, the long lead time to introducing updates into already fielded devices and to incorporating new security technologies into devices that are far along in design and testing means that many more devices are

Table 11.3
Overview of Security Technology Applicability

	Technology	Use Case	Trade Off	Note
MDM	Anti-malware	Protect commercial and application software	Resource impact; need for updates; increased attack surface through connectivity; false-positives	Suitable for computer-like systems; known entity to customers
	HIDS/HIPS	Protect commercial and application software	Up-front engineering effort to develop policies; limited field flexibility	Suitable to protect commercial OS platforms, even on resource-limited or EOL systems
	Cryptography	Protecting data confidentiality, integrity, and authenticity	Resource requirements; developer training; protection of keys/certificates	Requires supporting PKI infrastructure to manage keys
HDO	Network segmentation	Separation of critical systems	Effort to manage and maintain; does not prevent physical/local attacks such as USB compromise	May not deter sophisticated hacker; does provide outbreak containment
	Firewalls, gateways	Separation of critical systems	Does not protect from physical/local attacks such as USB; requires maintenance	May not deter sophisticated hacker; does provide incident containment
	PNM	Network-based device security	Relatively new technology but maturing; risk of false-positives	A good alternative to manually managed IoT networks

likely to reach the market with insufficient security features. Additionally, it will take some time to properly secure the fleet of medical devices already in the field (with many devices still in use, the only way to secure them is to replace them with secure-by-design devices). As more cybersecurity solutions become available that focus on addressing the unique requirements of medical devices, MDMs may be able to offload some of these concerns to organizations whose core competency is medical device and data security. This will free up medical device vendors' engineering resources to continue to innovate and focus on the device's clinical functions.

Security, unfortunately, has its cost—whether due to additional engineering or testing resources required, procedural and documentation overhead, or licenses for third-party security tools. However, the earlier in a product's lifecycle security is implemented, the lower the total cost of security will be over the entire clinical life of the product, even if it may increase the purchasing price for the device itself. Until all medical devices are secure by design and can withstand potentially hostile network and physical environments, the best that we can do from the network side is provide perimeter defenses and reactive security. Many issues will only be uncovered after they have potentially impacted the function of the device and patients will be at greater risk of direct harm or delays in care. While the current state of medical device security generally remains inadequate, MDMs, global regulators, and companies that provide proactive security solutions are taking these security concerns seriously and progress is being made. Things are moving in the right direction, but the industry still has plenty of work to do.

References

[1] FDA, "MAUDE Adverse Event Report: MERGE HEALTHCARE MERGE HEMO PROGRAMMABLE DIAGNOSTIC COMPUTER" and "MAUDE Adverse Event Report: MERGE HEALTHCARE MERGE EYESTATION SYSTEM, IMAGE MANAGEMENT, OPTHALMIC," July 31, 2023, https://www.accessdata.fda.gov/scripts/cdrh/cfdocs/cfmaude/detail.cfm?mdrfoi__id=5487204, https://www.accessdata.fda.gov/scripts/cdrh/cfdocs/cfMAUDE/detail.cfm?mdrfoi__id=5415351&pc=NFJ.

[2] AV-TEST, "Malware," November 2023, https://www.av-test.org/en/statistics/malware/.

[3] FDA, "Cybersecurity in Medical Devices: Quality System Considerations and Content of Premarket Submissions Guidance for Industry and Food and Drug Administration Staff," September 27, 2023, https://www.fda.gov/media/119933/download.

[4] Haislip, H., et al., "Medical Device Isolation Architecture Guide," U.S. Department of Veterans Affairs, August 2009, http://s3.amazonaws.com/rdcms-himss/files/production/public/FileDownloads/2013-Medical-Device-Isolation-Architecture-Guide-2009.pdf.

[5] HelpNet Security, "85% of Attacks Now Use Encrypted Channels," December 19, 2022, https://www.helpnetsecurity.com/2022/12/19/attacks-encrypted-channels/.

[6] Durumeric, Z., et al., "The Security Impact of HTTPS Interception," *NDSS*, February 27, 2017, https://www.ndss-symposium.org/ndss2017/ndss-2017-programme/security-impact-https-interception/.

[7] https://novelbits.io/academy/.

[8] Bluetooth SIG, "Bluetooth Core Specification 5.2," December 31, 2019, https://www.bluetooth.com/specifications/bluetooth-core-specification/.

[9] Afaneh, M., *Intro to Bluetooth Low Energy*, November 22, 2018, https://www.novelbits.io/introduction-to-bluetooth-low-energy-book/.

[10] Padgette, J., et al., "Guide to Bluetooth Security," *NIST Publications*, January 19, 2022, https://www.nist.gov/publications/guide-bluetooth-security-2.

[11] Afaneh, M., "Bluetooth 5 Speed: How to Achieve Maximum Throughput for Your BLE Application," September 6, 2017, https://www.novelbits.io/bluetooth-5-speed-maximum-throughput/.

[12] Padgette, J., et al., "Guide to Bluetooth Security," *NIST*, May 2017, https://nvlpubs.nist.gov/nistpubs/SpecialPublications/NIST.SP.800-121r2.pdf.

[13] Zuo, C., et al., "Automatic Fingerprinting of Vulnerable BLE IoT Devices with Static UUIDs from Mobile Apps," CCS '19, London, UK, November 11–15, 2019, https://web.cse.ohio-state.edu/~lin.3021/file/CCS19a.pdf.

[14] Garbelini, M., S. Chattopadhyay, and C. Wang, "Unleashing Mayhem over Bluetooth Low Energy," https://asset-group.github.io/disclosures/sweyntooth/.

[15] Greenberg, A., "Hacker Lexicon: What Is Perfect Forward Secrecy?" *Wired*, November 28, 2016. https://www.wired.com/2016/11/what-is-perfect-forward-secrecy/.

[16] Walters, P., "The Risks of Using Portable Devices," US-CERT, 2012, https://www.us-cert.gov/sites/default/files/publications/RisksOfPortableDevices.pdf.

[17] Halperin, D., et al., "Pacemakers and Implantable Cardiac Defibrillators: Software Radio Attacks and Zero-Power Defenses," *IEEE Symposium on Security and Privacy*, 2008, https://www.secure-medicine.org/hubfs/public/publications/icd-study.pdf.

[18] Das U-Boot, "The U-Boot Documentation," https://www.denx.de/wiki/U-Bootdoc/BasicCommandSet.

[19] gmbnomis, "uboot-mdb-dump," July 17, 2016, https://github.com/gmbnomis/uboot-mdb-dump.

[20] O'Flynn, C., "Getting Root on Philips Hue Bridge 2.0," July 12, 2016, https://colinoflynn.com/2016/07/getting-root-on-philips-hue-bridge-2-0/.

[21] Ronen, E., et al., "IoT Goes Nuclear: Creating a ZigBee Chain Reaction," https://eprint.iacr.org/2016/1047.pdf.

[22] Davidov, M., "Microcontroller Firmware Recovery Using Invasive Analysis," *Duo Security*, March 26, 2018, https://duo.com/labs/research/microcontroller-firmware-recovery-using-invasive-analysis.

[23] Riddle, S., "Decapping Chips," September 8, 2016, https://seanriddle.com/decap.html.

[24] Clark, J., "Don't Try This at Home: Decapping ICs with Boiling Acid," June 24, 2016, https://labs.f-secure.com/archive/dont-try-this-at-home-decapping-ics/.

[25] Witteman, M., "Secure Application Programming in the Presence of Side Channel Attacks," November 2018, https://www.riscure.com/uploads/2018/11/201708_Riscure_Whitepaper_Side_Channel_Patterns.pdf.

[26] Wu, M., et al., "Eliminating Timing Side-Channel Leaks Using Program Repair," *ISSTA'18*, Amsterdam, Netherlands, July 21, 2018. https://arxiv.org/pdf/1806.02444.pdf.

[27] O'Flynn, C., "A Framework for Embedded Hardware Security Analysis," DalHousie University, June 2017, http://hdl.handle.net/10222/73002.

[28] O'Flynn, C., and Z. Chen, "Side Channel Power Analysis of an AES-256 Bootloader," *CCECE 2015,* Halifax, NS, Canada, 2015, https://eprint.iacr.org/2014/899.pdf, https://eprint.iacr.org/2016/1047.pdf.

[29] Gamaarachchi, H., and H. Ganegoda, "Power Analysis Based Side Channel Attack," January 2018, https://arxiv.org/pdf/1801.00932.pdf.

[30] FDA, "Software as a Medical Device (SaMD)," December 12, 2018, https://www.fda.gov/medical-devices/digital-health-center-excellence/software-medical-device-samd.

[31] FDA, "Multiple Function Device Products: Policy and Considerations," July 2020, https://www.fda.gov/regulatory-information/search-fda-guidance-documents/multiple-function-device-products-policy-and-considerations.

[32] Giantsidis, J., "FDA Releases Guidance on Cybersecurity in Medical Devices," *Med Device Online*, May 31, 2022, https://www.meddeviceonline.com/doc/fda-releases-guidance-on-cybersecurity-in-medical-devices-0001.

[33] Desai, A., "US State Privacy Legislation Tracker," *International Association of Privacy Professionals (IAPP)*, August 4, 2023, https://iapp.org/resources/article/us-state-privacy-legislation-tracker/.

[34] California Legislative Information, "SB-327 Information Privacy: Connected Devices," September 28, 2018, https://leginfo.legislature.ca.gov/faces/billTextClient.xhtml?bill_id=201720180SB327.

[35] FTC, "A Brief Overview of the Federal Trade Commission's Investigative, Law Enforcement, and Rulemaking Authority," May 2021, https://www.ftc.gov/about-ftc/mission/enforcement-authority.

[36] U.S. SEC, "Cybersecurity Risk Management, Strategy, Governance, and Incident Disclosure," September 5, 2023, https://www.sec.gov/corpfin/secg-cybersecurity.

[37] Jones, D., "SEC Notifies SolarWinds CISO and CFO of Possible Action in Cyber Investigation," *CybersecurityDive*, June 26, 2023, https://www.cybersecuritydive.com/news/sec-solarwinds-ciso-cfo-orion/653864/.

[38] European Commission, "Radio Equipment Directive (RED)," https://single-market-economy.ec.europa.eu/sectors/electrical-and-electronic-engineering-industries-eei/radio-equipment-directive-red_en.

[39] European Commission, "Cyber Resilience Act," *Policy and Legislation*, September 15, 2022, https://digital-strategy.ec.europa.eu/en/library/cyber-resilience-act.

[40] IoT Security Foundation, *Best Practice Guidelines*, 2015-2023, https://iotsecurityfoundation.org/best-practice-guidelines/.

[41] NIST, *NIST Cybersecurity for IoT Program*, 2016-2023, https://www.nist.gov/itl/applied-cybersecurity/nist-cybersecurity-iot-program.

[42] CISA, "Shifting the Balance of Cybersecurity Risk: Security-by-Design and Default Principles," April 13, 2023, https://www.cisa.gov/news-events/alerts/2023/04/13/shifting-balance-cybersecurity-risk-security-design-and-default-principles.

12

Applying Cryptography to Medical Device Systems

12.1 Overview: Cryptography for Medical Devices

Cryptography is used to protect various types of systems and their data across many industries, use cases, and product types. Cryptography-based technologies are used to protect confidentiality and ensure the integrity and authenticity of information—not just financial information or PHI, but also less obviously sensitive information that flows between devices and applications to perform technical functions in the background. Information has become the lifeblood of many industries; without it, businesses would come to a halt, and cryptography is the preferred method to protect that information. However, medical devices have unique requirements when it comes to incorporating cryptography into devices and applications, as we will discuss in the course of this chapter.

Although cryptography is an integral technology that can be used to improve a device's security posture, any cryptography used must carefully consider the impact on the end user and on the device's utility, functionality, performance, and reliability.

Cryptography should be utilized to protect the essential performance of a medical device system as well as the confidentiality of the patient data; however, cryptography should not to utilized to prevent patients from being able to access their data. It can sometimes be challenging to balance protecting the medical device system while only exposing data to the patient.

Medical device system security mitigations need to be designed to not impede the intended operation of the medical device. Mitigations such as long passwords or physical security keys, for example, could slow or deny access to critical functionality if it is forgotten or lost or if the patient is experiencing cognitive impairment due to his or her ailment.

Another area of potential impact on the essential performance of a medical device is when cryptography-based mitigations add additional computation cycles. This could cause delays to functionality that may exceed tight timing requirements. This is most commonly found in communications where timing requirements for transmitted data can be very strict and where the time required to encrypt/decrypt data streams would add unacceptable delays. A variation of this "crypto overhead" problem occurs when additional computation results in a higher battery drain, thereby reducing the device's operational life.

These are just a few examples of how cryptography can have negative, unintended side effects. However, if implemented with a specific medical device's usage in mind, the result can be a more secure, and therefore safer, device that is not as vulnerable to hacking, can operate properly in various and potentially hostile network or physical environments, and can even perform more predictably due to increased trust of the data that drives the medical device's operation. This is just one reason why medical device regulators place so much emphasis on "designing in" security from the ground up and in consideration of the security/effectiveness tradeoff.

Cryptography can, and typically will, be complemented by physical security, platform security (e.g., protecting the operating system), and network security, as addressed elsewhere in this book. The focus here will be on three main data security concepts facilitated by cryptography: confidentiality, integrity, and authenticity. These concepts help to protect data and applications from compromise. The data being protected can include anything that the device gathers, generates, stores, or transmits while actively performing tasks for a patient, sitting idle, or being serviced, and can cover the range of identities, therapy settings, diagnostic results, configuration data, and software code. Cryptography can also be utilized to protect executable applications that can range from small firmware images running on microcontrollers, to full-featured operating systems and applications running or devices built on general-purpose computers.

It is not the intent of this section to provide an in-depth review of cryptography engineering, as that has been done elsewhere [1]. It will rather focus on providing a basic foundation of cryptographic concepts and specifically addressing implementation and related maintenance challenges an MDM will need to solve due to the uniqueness of medical device use cases.

There are a number of caveats that concern the effectiveness of encryption, hashing, and digital signing. For instance, just to name a few, cryptography can be broken if the secret key is leaked [2], if the underlying algorithms have a weakness [3], if a random number generator is not random enough [4], or if a successful side-channel attack leaks secret key information [5]. These are only a few of the potential weaknesses and common implementation errors in cryptography. Unfortunately, there are many examples of failed cryptography implementations, even in protocols (e.g., TLS 1.0) and standards (e.g., WEP).

As with all security, there is no such thing as perfectly secure, but an analysis of all of the device requirements should lead to a solution that is sufficiently secure for the given medical device capabilities and its security needs. An MDM's security goal is a system that is impractical to compromise because the effort would be too time-consuming or too resource-intensive. However, even this is a moving target because, as computing power increases (e.g., faster, cheaper CPUs or through major changes such as the advent of quantum computers), some algorithms in use today may become obsolete (deprecated) because they may become too easy to break. As a result, they may no longer be deemed secure enough for a given application and use case. Consequently, crypto algorithms need to evolve along with emerging technology and newly discovered weaknesses [6]. Cybersecurity, unfortunately, is an arms race.

12.2 Introduction to Algorithms and Keys

Before data is encrypted, it is referred to as plaintext or cleartext. Encryption is a process that uses an algorithm that renders the plaintext into a form that is practicably unreadable. Encrypting plaintext results in a "binary blob" of encrypted values, commonly referred to as "ciphertext." The process of reverting ciphertext to its original plaintext is called decryption, and this process requires the correct key.

To successfully create ciphertext, a cryptographic key is used in conjunction with an encryption algorithm. Keys are large numerical values; and, generally speaking, for any given encryption algorithm, the larger the key size (in bits), the more secure the resultant ciphertext. However, this comes with the price of higher computational resources needed to utilize them. Cryptographic keys, together with a cryptographic algorithm, are used in tandem to produce the ciphertext. In modern cryptography, the keys are the actual secret and the algorithm is publicly known, as compared to earlier forms of cryptography that had no key and relied on the algorithm being kept secret. Using a sufficiently long key with an appropriate algorithm vastly widens the set of possible combinations an attacker would need to try and puts a brute force attack out of the practical reach of today's computers (where the time needed to crack encryption by brute force is measurable in centuries).

It should be noted that the length of the key value alone is not sufficient to constitute effective encryption; the algorithm needs to be evaluated as well.

This is especially true for the two different classes of encryption known as symmetric and asymmetric, where asymmetric key values need to be much larger than symmetric key values to achieve a similar level of secrecy. We will discuss the different classes of encrypting algorithms in the next section. Meanwhile, publishing the algorithm allows its mathematics (and implementation in code) to be scrutinized for flaws by cryptographic experts. This is the only process to ensure the cryptographic effectiveness of a published algorithm. Many years of exposure to expert scrutiny are needed before the algorithm can be considered reliably secure. The necessity of utilizing "known good" cryptography is further discussed later in this chapter.

In addition to the algorithm, the protection of the key becomes of utmost importance, and good cryptography implementations go to great lengths to ensure that keys are secure when created, distributed, provisioned, shared, stored, and revoked. When correctly implemented, cryptography provides the following:

- *Confidentiality* is the property of data not being available or disclosed to unauthorized persons or processes. Confidentiality of medical device data is achieved through encryption (i.e., transforming plaintext to ciphertext). By encrypting data, we can ensure that it remains private and is only accessible to those who are intended to have access to it (i.e., authorized users, processes, and devices).
    - *Privacy:* This is a subset of confidentiality and occurs when data is accompanied by supporting information, sometimes referred to as metadata. This metadata can take many different forms and may not initially be obvious as important information being leaked, such as PHI. As an example, a wireless RF MAC address could be used to identify the manufacturer to which the address is licensed, which can lead to learning the nature of the device and the type of treatment it provides, thus exposing the patient's medical condition.
- *Integrity* is the property of data not being subject to alteration or destruction by unauthorized persons, processes, or conditions [7]. The integrity of medical device data can be determined via cryptographic hash functions, which can be either keyed or nonkeyed (keyed hash functions use an algorithm that requires a key to calculate the hash; unkeyed hash functions use an algorithm that does not). By hashing a particular set of data, and storing the resulting hash, the hash can always be recalculated and, if the recalculated hash matches the original hash, then there is high confidence that the data has not been modified.

- *Authenticity* is the property of being genuine and being able to be verified and trusted, that is, of establishing confidence in the validity of a transmission, a message, or the message originator [7]. The authenticity of medical device data can be provided utilizing digital signatures. There are multiple ways that this can be implemented.
  - *Authentication:* The means of proving an individual, device, or system's identity and providing the technical means to establish trust between parties.
  - *Nonrepudiation:* Ensure the integrity and origin of data in such a way that the integrity and origin can be verified and validated by a third party as having originated from a specific entity [7]. This includes ensuring that the sender is provided with proof of delivery and the recipient is provided with proof of the sender's identity so that neither can later deny having processed the information.

12.2.1 Secret Key Cryptography

Cryptographic algorithms are categorized based on the type of algorithm and required number of keys used for encryption and decryption and are additionally classified by the methods of application and utilization. Secret key cryptography (SKC), also called symmetric encryption, uses a single, identical key for the encryption and decryption function.

Symmetric encryption has advantages. It is fast, uses relatively short key values, and is particularly effective for encrypting data at rest. In this case, the key can stay within the system (e.g., stored in hardware-protected memory) that hosts the data, and the risk of public exposure can be kept low.

This becomes a different problem for data in transit. For a sender and receiver to connect securely using symmetric encryption, they must select a key together (or if one party selects the key, it must be securely exchanged with the other party) and keep it concealed. If they are in separate physical sites, they can pre-share the key value or trust a messenger or some closed medium to prevent the exposure of the key value while it is being shared between the two parties, see example in Figure 12.1. These pre-shared symmetric encryption key arrangements are the easiest to understand. A symmetric key can be created, much like generating a random number, and that key can then be given to any other party that wishes to encrypt and decrypt data for one another. If the distribution of these keys between the parties is insecure, then the keys can be compromised. If this key is ever leaked, by the initial key distribution mechanism or by any of the parties using it, then all data encrypted using that key is no longer secure. Once the key is compromised, any encrypted data that

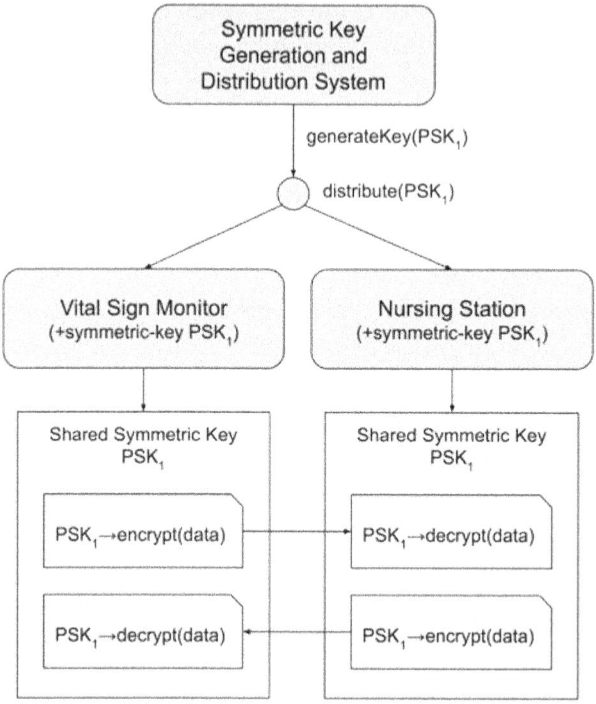

Figure 12.1 Symmetric key generation and distribution system.

was stored in the past and any data encrypted in the future should be assumed to be exposed.

The effectiveness of a particular data security solution is dependent on the selection of the underlying cryptography algorithms. A very common symmetric key encryption algorithm in use today is known as the Advanced Encryption Standard (AES) [8, 9]. There are a number of different modes that AES implementations can use; some are less secure than others. Many popular hardware platforms include AES hardware accelerators; this makes this encryption algorithm popular, especially for performance-critical applications. A newer stream cipher for symmetric key encryption known as ChaCha20 is also gaining popularity [10] as an alternative to AES. Also, ChaCha20's resistance to timing attacks and better software-only performance make it attractive; however, as of today, it is not widely supported with hardware accelerators.

When using pre-shared symmetric keys, each symmetric key should be unique for each device. This will limit the impact in the event that a particular device's symmetric key is compromised. This approach results in a more complex process of key creation, sharing, storage, and verification. There are two cases where the use of symmetric encryption is particularly prevalent:

1. Where there are only two devices that are communicating;
2. Where asymmetric cryptography is utilized to securely create and share a unique symmetric session key. This leverages the speed of symmetric cryptography with the superior key management of asymmetric cryptography. This approach is, for example, used in SSL/TLS internet communication protocols.

12.2.2 Public Key Cryptography

Public key cryptography (PKC) uses an asymmetric algorithm that uses a pair of keys, one key for encryption and another for decryption. The encryption key is public and can be freely shared, while the decryption key is private and needs to be protected, which is much easier since the private (secret) key remains with the receiving party and does not need to be shared. This type of key exchange is commonly used on the internet and supports a broader range of use cases, including ones where there are multiple devices that need to communicate and to coordinate the key exchange process. This scheme is called the public key infrastructure (PKI) and is enabled via the use of digital certificates. Anyone with a duplicate of a public key can then encode data, without the need to establish trust or care about the protection of the public key, but only the owner of the private key can read the encoded data. While the public and private keys are mathematically linked, its practically impossible to derive the private key if all you have is the public key.

As an additional layer of protection, keys should have a limited lifespan and a new key pair should be generated at regular intervals to make it harder for a cryptanalyst to derive the private key, as well as to limit the exposure and potential impact should a private key become compromised.

The fundamental benefit of public key cryptography is the relative ease of creation and sharing of keys. For example, if two parties that wish to communicate securely have their own asymmetric key pairs, the parties can share their public keys with one another and use the combination of their own secret key and the other party's public key to derive a shared, symmetric key. This key can then be used to encrypt and decrypt data between the two parties, much like the pre-shared symmetric key. In addition to encrypting the data, the asymmetric key pairs can also be used to create a digital signature to authenticate data and ensure that it came from the party that owns the private key.

Key management (creation, sharing, revocation) become simpler compare to symmetric cryptography. However, a corresponding disadvantage is that asymmetric encryption requires more system resources (computational cycles, memory, battery capacity) and/or time to perform the required calculations.

There are many asymmetric key algorithms to choose from, but commonly used ones fall into two categories. RSA, or Rivest–Shamir–Adleman (currently the most widely used) [11], and Elliptic Curve Cryptography (ECC), which is becoming more popular [12]. ECC keys are often chosen over RSA because they can achieve the same level of protection with smaller key sizes, and their calculations perform better in implementations where hardware acceleration is not available.

12.2.3 Public Key Encryption with Ephemeral Keys

A frequent practice for using the benefits of asymmetric cryptography but overcoming some of its disadvantages is public key encryption with ephemeral keys, see an example in Figure 12.2. This process is the same as PKC except that the public key that is used to derive a shared, temporary, symmetric secret key that is newly created for each communication session (or even more frequently). The trustworthiness of this new, temporary key can be established by signing it with a trusted long-term key pair, LTK, before it is shared. This form of key agreement ensures that each secure session uses a unique shared symmetric key to encrypt the data. This means that any data that is captured is only at risk of compromise if the symmetric key for that particular session has been compromised, which is unlikely due to the short-lived nature of the symmetric key. This is what is known as Perfect Forward Secrecy (PFS) [13]. PFS does not solve the concern that the LTK public key originally shared between the parties could have been intercepted by a third party attempting to execute a man-in-

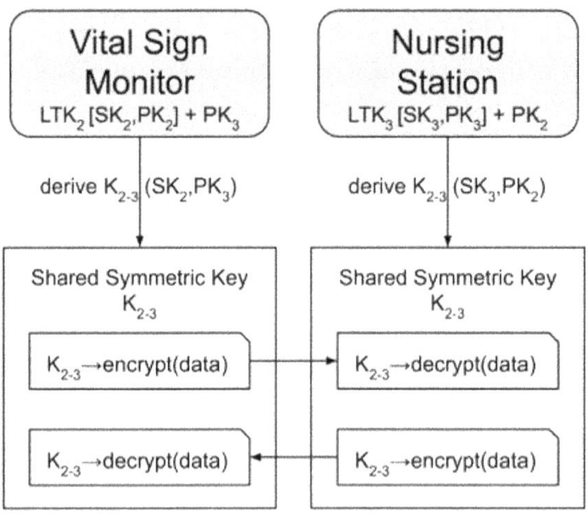

Figure 12.2 Public-key encryption with ephemeral keys.

the-middle attack. In this scenario, there is still no way to confirm that the public key belongs to a trusted party and the data could still be leaked, unless there exists a root of trust that can establish the trust of the shared public keys.

Public key encryption with ephemeral keys (EKs), plus a root of trust to establish trust of each party's long-term public keys, can help to prevent a man-in-the-middle attack. Given a root of trust, we can verify that any public key shared with us, along with an accompanying certificate, is trusted. By evaluating the certificate-chain for that public key, we can determine if it has been vouched for by a higher, trusted party and that a third party has not intercepted the public key and sent their key to perform a man-in-the-middle attack. Figure 12.3 shows the key-agreement process for deriving a shared, symmetric key using public key encryption with ephemeral keys.

A similar process can be used with symmetric keys to make them ephemeral. Assuming that an initial symmetric key has been shared via some previous method, that key can be utilized in a symmetric encryption, after which the key is processed through a KDF or HMAC hashing algorithm to produce a predictable new symmetric key for the next round of encryption. This process continues indefinitely with any given symmetric key being used to encrypt a limited amount of data. This is also a form of Perfect Forward Secrecy.

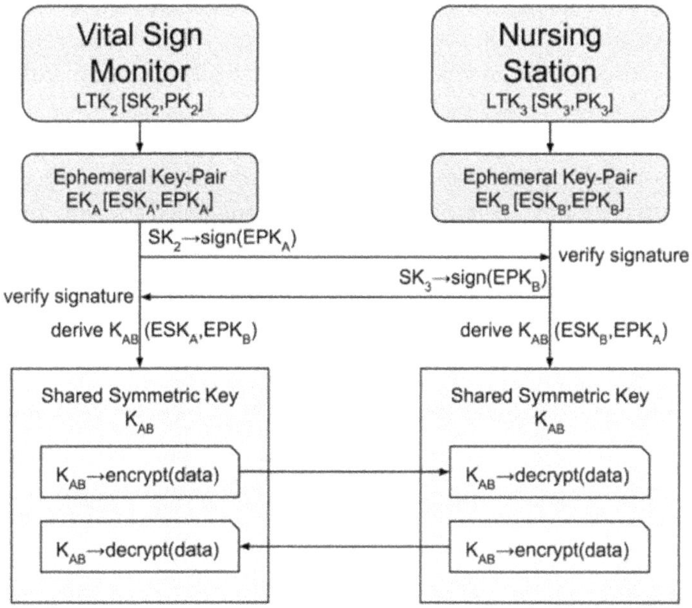

Figure 12.3 Public-key encryption with ephemeral keys based on temporary asymmetric key-pairs.

Even with this level of confidence that the public key of another party is trusted, as always, there are weaknesses that must be understood and mitigated. If a third party is able to gain access to a higher key pair in the root of trust hierarchy, then that third party can sign any key it wishes into the root of trust at that level and deceive the parties in subsequent key exchanges. In a root of trust, it is essential that the risks of key exposure are properly assessed and that there is a strategy to deal with the possible leak of signing keys that can fool parties into believing that data is secure (because it appears to have originated at a trusted source, when it did not). As in symmetric cryptography where the shared key needs to remain secret, so too with an asymmetric key pair, the private key needs to be protected so to remain secret.

12.2.4 Protecting Keys

Protecting the private key of an asymmetric key pair as well as symmetric secret material is essential to maintaining the integrity of the overall cryptographic process and implementation. Often, secret key material is kept in a key store that encrypts it so the keys cannot be readily compromised. However, the key store must be able to be unlocked, and the secrets decrypted for them to be used. While in use, the secret keys should be kept in unencrypted storage for the shortest possible amount of time, and any memory that contains the keys should be properly and promptly cleared.

It can be difficult to keep key stores secure since the information to unlock the key store must be accessible to the application that needs to use the keys. This means that if attackers gain access to the device, they could find out how to unlock the key store and access the secret keys. One way to limit the exposure of secret information is to use hardware implementations to store and lock away the keys so that they can never be read. The keys can be used exclusively within the hardware implementation to sign, verify, encrypt, and decrypt, such that there is high confidence that the secret key material will remain safe even if an attacker would gain access to the device. As always, unique keys should be utilized for each device to minimize the impact if secret or private keys are exposed.

A further useful strategy is to assign a master key to each device that is only used for the purpose of deriving keys that support the individual cryptographic functions (e.g., for encryption of messages). That way, functional keys can be regenerated on the device and the risk of exposure of the master key is minimized.

12.2.5 Hash Functions

Cryptographic hash functions are mathematical algorithms that take an input data set and compute an output value (called a digest) that represents the contents of the input data set. This digest is of a fixed length and the value of the digest will always be the same for the same input data set. The following list of cryptographic hash attributes is what makes a hash a strong cryptographic primitive.

- *Nonreversibility:* Also known as a one-way function. A hash makes it exceedingly difficult to reconstruct the original data set from the hash digest.
- *Avalanche effect:* A change in just any one bit of the input data set should result in a change to at least half of the bits of its hash digest. In other words, when the data set is changed slightly, the output of the hash digest should change significantly and unpredictably.
- *Determinism:* A given input data set must always generate the same hash digest. This is inclusive of the key when using keyed hashes.
- *Collision resistance:* It should be exceedingly difficult to find two different data sets that create the same hash digest value.

A nice side benefit of hashes is that they are relatively immune to quantum computing-based attacks.

There are two major types of cryptographic hashes: nonkeyed hashes and keyed hashes. Nonkeyed hash algorithms include:

- BLAKE;
- MD5;
- Whirlpool;
- SHA-1;
- SHA-256.

Nonkeyed hashes are used to create stand-in digests for the original data set, without exposing the data set. This is, for example, typically used in password lists. The hash digest can accompany the data set so that the integrity of the data set can be vetted at a later time.

Keyed hashes include:

- BLAKE2;
- Poly1305-AES;
- MD6;
- HMAC;

- CMAC.

A keyed hash takes as its input both the data set and a symmetric key value. The digest will always produce the same value from the same data set and key value pair. Keyed hashes are used to create digests for the original plaintext, which confers authenticity of the source of the plaintext due to the unique nature of the symmetric key value. As with a nonkeyed hash, a keyed hash can accompany the data set so that the integrity of the data set can be vetted at a later time. Keyed hashes can be particularly useful in two-party authentication schemes, such as a challenge-response method.

They can also be used for conveying authenticity:

- One approach is a hybrid of symmetric and asymmetric cryptographic functions. Initially, the original data is hashed using a nonkeyed hash (such as a SHA 256) and the resultant message digest value along with the original data is encrypted using asymmetric encryption. The receiving party can verify this signature by decrypting the cipher text using their private key and comparing the hash's message digest value to a recomputed nonkeyed hash's message digest as performed upon the received data. The hash ensures the integrity of the data and the successful asymmetric decryption ensures that the data came from the intended source and not from a forged or unauthorized source.
- A completely asymmetric approach involves the use of RSA or digital signature algorithm (DSA), this can be thought of as an asymmetric keyed hash where a private key is used to generate a digital signature, and this can then be verified by using the paired public key. This can be confusing for many as the process seems to be backward from the normal use of a key pair during encryption/decryption where the process of encryption uses a public key and decryption uses the private key. It should be noted that this approach does not need to have the hash's message digest encrypted.
- A completely symmetric approach involves the use of keyed hashes (such as HMAC or CMAC), where the original dataset is submitted to the hash along with a cryptographic key, this produces the hash's message digest, which can then later be utilized to recompute the keyed hash on the received dataset using the same shared key as was used to create the message digest. It should be noted that this approach does not need to have the hash's message digest encrypted.

12.3 Introduction to Certificates and Signatures

Digital certificates are designed to streamline the process of establishing whether a public key belongs to the alleged owner/holder. In that sense, a certificate is a form of credential. Digital certificates let their owners digitally stamp the content with their identity and protect it from tampering. The identity that originates the information is used as a significant measurement of trustworthiness.

In theory, digital certificates prevent threat actors from impersonating trusted parties by certifying that a public key belongs to a specified entity. If public keys are delivered directly from their owner to another entity, there is a risk that the message could be intercepted and the public key swapped for an alternative (a man-in-the-middle attack). The solution is to transmit and receive public keys through a reliable and dependable third party that is able to provide trustworthy assurance that the public key truly belongs to the entity with which you are connecting. Instead of sending the public key directly, you ask the trusted third party to integrate it into a digital certificate. The trusted third party that issues digital certificates is called a certificate authority (CA). Digital certificates are also known as public key certificates because they give you assurances about the ownership of a public key used in an asymmetric key scheme. A digital certificate contains the public key for an entity and is a verified statement that the public key belongs to that entity.

The specific functions of the certificate generation process include:

- Establish identity: Associate, or bind, a public key to a user, system, or (medical) device.
- Secure confidential information (e.g., encrypting the session's symmetric key for data confidentiality).

A digital certificate typically consists of three main elements:

- A public key;
- Certificate information:
  - Identifying information about the user, such as name and user ID;
  - Expiration date;
  - Name of the authority that issued the certificate (and who is therefore vouching for the identity of the user or device);
  - Serial number;
  - Any pertinent policies describing how the certificate was issued and/or how the certificate may be used;
  - Supplementary information.

- Digital signature(s) of the certificate issuer.

The most widely accepted certificate format is defined in the International Telecommunication Union Telecommunication Standardization Sector (ITU-T) Recommendation X.509.

12.3.1 X.509 Certificates

An X.509 certificate is a digital certificate that uses the international X.509 PKI standard [14] to verify that a public key belongs to the user, device, or service identified within the certificate. Certificate types are continually evolving; for the latest guidance on types of certificates and their uses, refer to the NIST guidance [15, 16].

All X.509 certificates contain the following data:

- *The X.509 version number:* This identifies which version of the X.509 standard applies to this certificate, which affects what information can be specified in it.
- *The certificate holder's public key:* The public key of the certificate holder, together with an algorithm identifier that specifies which cryptosystem the key belongs to and any associated key parameters.
- *The serial number of the certificate:* The entity (application or person) that created the certificate is responsible for assigning it a unique serial number to distinguish it from other certificates it issues. This information is used in numerous ways; for example, when a certificate is revoked, its serial number is placed on a certificate revocation list (CRL).
- *The certificate holder's unique identifier (or distinguished name (DN)):* This name is intended to be unique across the internet. A DN consists of multiple subsections and may look something like this: CN=Jane Doe, EMAIL=jdoe@xyz.com, OU=Engineering, O=XYZ, C=US (CN = Common Name, OU = Organizational Unit, O = Organization Name, C = Country).
- *The certificate's validity period:* The certificate's start date/time and expiration date/time; indicates when the certificate will expire.
- *The unique name of the certificate issuer:* The unique name of the entity that signed the certificate. This is normally a CA. Using the certificate implies trusting the entity that signed this certificate. (Note that, in some cases, such as root or top-level CA certificates, the issuer signs its own certificate.)

- *The digital signature of the issuer:* The signature using the private key of the entity that issued the certificate.
- *The signature algorithm identifier:* Identifies the algorithm used by the CA to sign the certificate.

12.3.2 CAs and Registration Authorities

CAs are the repositories for public keys and must be a trusted and recognized authority that issues certificates. When a sender needs an intended receiver's public key, the sender must get that key from the receiver's CA. A CA generates certificates and digitally signs them utilizing the CA's private key. By means of the CA's public key, anyone wanting to validate a certificate's validity confirms the issuing CA's digital signature, and therefore, the integrity of the contents of the certificate and, most importantly, the public key and the identity of the certificate holder.

In general, a Registration Authority (RA) refers to the people, processes, and tools applied to support the registration of users with the CA (enrollment) and ongoing administration of users. The RA may perform vetting—the process of verifying that a public key belongs to its claimed owner. An RA is a human entity—a person, group, department, company, or other association. A CA is often software that is used to issue the actual certificates to its computer users.

12.3.3 Digital Signatures

As described in Section 12.2, there are multiple ways to ensure authenticity using digital signatures. Public key digital signatures offer authentication and data integrity. A digital signature also offers nonrepudiation, which means that it prevents the sender from alleging that they did not send the communication.

The purpose of the digital signature on a certificate is to confirm that the certificate information has been attested to by a trusted, preauthorized person or entity. The digital signature does not attest to the authenticity of the certificate, but rather to the fact that the verified identity information is bound to the public key.

To obtain a digital certificate, you must ask a trusted CA to issue one. You provide your public key, proof that you possess the corresponding private key, and some specific information about your device. You then digitally sign the information and send the whole package—the certificate request—to the CA. The CA then performs some due diligence in verifying that the information you provided is correct and, if so, generates the certificate and returns it [17]. This process establishes what is known as a chain of trust. Following the chain of trust enables a chaining engine to verify the issuer of each certificate in a chain until it finds a trusted root (root of trust) and therefore decides to trust the device the chain attests, or else hits a break in the chain (an absent or expired certificate), or finds a circular attestation loop with no trusted root. In either of these latter cases, the device's attestation chain would be invalid, and the device should not be trusted.

12.3.4 Expiration and Revocation

Certificates are only beneficial while they are valid. In most organizations and all public key management infrastructures (PKIs), certificates have a restricted lifetime, known as a validity period, which consists of a start date/time and an expiration date/time. Forcing certificates to be reissued regularly constrains the period in which a device would be vulnerable should a certificate become compromised.

Certificates are expected to be usable for their entire validity period. Often, they are valid for the entire lifetime of the medical device to which they are assigned (not to exceed the validity period outlined by NIST) [18], or else must have the means to perform renewals or reissuances of certificates prior to expiration. When the certificate expires, it will no longer be valid, as the authenticity of its key/identification pair is no longer ensured.

Additionally, MDMs may choose to design in the ability to limit functionality of the device if a certificate has expired or is revoked. Depending on the criticality of treatment to the patient/user of the device, the device may be allowed to "fail open." From the security perspective, this is not the ideal approach, but may be required in some cases to ensure continued delivery of care.

A better approach may be to design a more nuanced fail-open condition, where some limited functionality remains available at your own risk until the expired certificate situation can be addressed. The final decision on how to handle expired certificates should be based on risk-benefit analysis and the device's clinical use profile (i.e., comparing the risk of operating without a valid certificate with the risk of not having a particular clinical function available). However, every effort should be made to prevent users from ending up in this situation.

Alerts can be designed based on the certificate's expiration date to provide sufficiently early advanced notification, together with instructions for getting the certificate renewed or reissued. These alerts should be automated, begin well in advance of expiration, and escalate as the expiration date approaches if no action has been taken. For some devices, and in situations where updating the certificate on-site is neither desirable nor practical, contingencies should be provided to ensure uninterrupted care, such as swapping out the device for one with a valid certificate.

There are also situations where it is necessary to invalidate a certificate prior to its expiration date, such as when the certificate's corresponding private key has been compromised. This is called revocation. A revoked certificate is much more suspicious than an expired certificate. Expired certificates are equally unusable, but the risk level signaled by a revoked certificate is much higher since revocation means that it is likely that the certificate has been compromised.

A revoked signature is a slightly different scenario as it indicates that the signer no longer trusts that the public key and identification information belong together, or that the certificate's public key (or corresponding private key) has been compromised. A revoked signature should carry as much weight as a revoked certificate.

In a PKI environment, the communication of revoked certificates is typically achieved via a data structure called a Certificate Revocation List which is published by the CA in a central repository that can later be referenced in the CRL Distribution Point of the Digital Certificate extensions. The CRL contains a time-stamped, validated list of all revoked, unexpired certificates in the system. To prevent the list from getting too long, revoked certificates persist on the list only until they expire and then are removed. The CA distributes the CRL to users at a regularly scheduled interval (and potentially off-cycle, whenever a certificate is revoked), to reduce the chance that users might unwittingly operate a device with a compromised certificate.

Revoked certificates and signatures call for an escalated response as appropriate for the device use case. For example, a revoked certificate that manages the communication between a pacemaker and programmer should not stop the pacemaker from pacing but may stop the programmer from making changes to the device's settings. Further, any certificate condition such as revoked or expired status should be communicated to the user, operator, and/or manufacturer as it may indicate a security incident that requires an appropriate response (see Chapter 9).

12.4 Introduction to PKIs and Root of Trust

The reliable use of cryptography requires trust. While secret key cryptography can ensure message confidentiality and hash message digests can ensure integrity, none of this works without trust being established and maintained throughout the life of a device, key, user, or set of systems. This can be accomplished by an organization-specific root of trust, which assists in preliminary identity authentication while issuing new keys or digital certificates. Roots of trust provide manufacturers complete control over the verification of the identities to whom they issue an encryption key for a medical device, a user, or its interconnecting networks.

As discussed previously, one of the core difficulties resulting from using cryptography to protect sensitive data is the complexity of managing the lifecycle of encryption keys. This includes key generation, storage, import/export, distribution, rotation, replacement, backup, revocation, suspension, destruction, and auditing. The security of any given cryptographic system depends on the security of its secret keys, since it is possible that, with a secret key, adversaries can gain access to all the other parameters of the cryptographic system. If the key is compromised, it is possible that the entire cryptographic system is compromised.

A root of trust is essentially a hierarchy of asymmetric encryption keys, usually long-term keys (LTKs) whose trust is established through PKI by signing one another's public-key material to form a trust chain [19]. These signatures, in the form of public key certificates, provide a high level of confidence that the keys higher up in the hierarchy have vouched for the keys beneath them. This indicates that the keys can be trusted, much like a military chain of command. In a chain of command, each person knows who they trust as their commanding officer and, if their commanding officer issues a command, they are aware that their commanding officer was either given that command by someone higher up in the chain or has the authority to issue that command themselves.

See Figure 12.4 for a simple example of a root of trust key hierarchy where each node in the hierarchy vouches for the one below it by signing a digital certificate that creates a trust chain.

In a root of trust, we determine the trust hierarchy, by evaluating what are known as public key certificates. Among other things, these certificates contain metadata about an asymmetric cryptographic key pair, the public key of that key pair, and the certificate source or issuer. Since we know that a cryptographic signature can only be constructed by the secret key in an asymmetric key pair, if

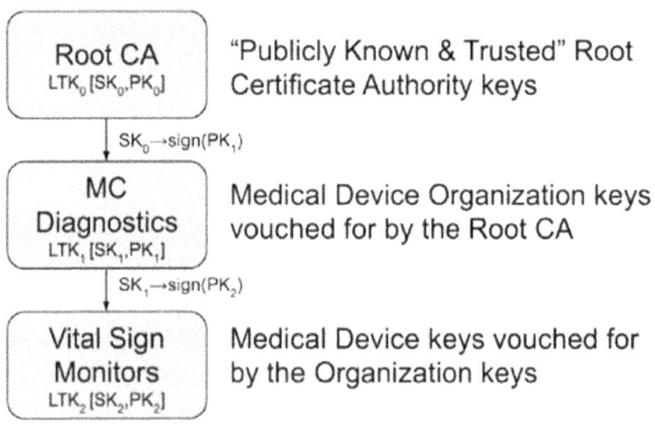

Figure 12.4 Simple root of trust key hierarchy.

the signature can be verified, then we know that the information in the certificate has been vouched for by the key that signed the certificate.

Usually, the signature in a certificate is generated by the key that is one level higher in the chain from the public key that is in the certificate. For instance, the certificate for the medical device's keys in the diagram above would be signed by the organization key, thus establishing that the device's keys should be trusted by the organization. In this example, there is one special certificate that is signed differently from the rest. The first key, or root key, has no key above it in the hierarchy. This root key signs its own certificate, known as a self-signed certificate, and establishes what is called a root certificate. This root certificate establishes the ultimate trusted key in the hierarchy. As long as a root certificate is stored in a secure location on a medical device, where it cannot be tampered with, then this root certificate can be used as a trusted starting point, or trust anchor, to determine the level of trust of any other key that has a certificate in the root of trust hierarchy.

Once trust of the asymmetric cryptographic key pairs can be established, that trust can be extended to data that has been signed or encrypted by those keys. Verifying digital signatures is a fairly straightforward concept. Given some data and a signature for that data, a cryptographic operation can verify whether the signature, which is generated by a particular public key, matches the data. If the signature can be verified, then there is a high degree of confidence that the data was signed, or vouched for, by that particular key. Therefore, if a certain key is trusted, based on its certificate-chain in the root of trust, then the data signed by that key can also be trusted.

Depending on the class of the device and its geographic market (i.e., under which regulatory authority it falls), there may be requirements for higher security standards. Medical device manufacturing can be complex and can include subassemblies from external suppliers or outsourcing of the entire manufacturing process to a third party. This makes it especially challenging to keep secret materials, like root of trust and certificates, protected from exposure. Although individual scenarios vary, there are a variety of strategies that can be used to mitigate these concerns:

- Employ an IT third-party service that can work directly with the manufacturing facility to ensure that the keys are implemented securely, although this can get expensive at high device volumes and could prevent sensible design decisions like providing multiple certificates per device.
- Generate keys directly on the device and only export the public key. Further, this key could act as a master key to generate other certificates to perform individual cryptographic functions or would allow for functional keys to be regenerated to reduce risk of exposure.

- In addition, keys embedded directly into the silicon of the medical device chips are unique and cryptographically strong and limit the impact after the compromise of a single device. The digital certificates can then act as a root of trust that future communications can rely on for the authenticity of the connection.

Lastly, defining a key hierarchy requires a number of technical but also nontechnical considerations that reflect business and organizational needs. Considerations should include, for example, key hierarchy across cryptographic functions, product lines, manufacturing locations, and potential third-party relationships, all of which may be changing through the production life of a product (e.g., through acquisitions and divestitures).

12.4.1 Hardware Security Modules

One highly recommended option to manage cryptographic keys is a hardware security module (HSM). An HSM is also known as secure application module (SAM), secure cryptographic device (SCD), hardware cryptographic device (HCD), or cryptographic module. These are all names for a secure, tamper-evident, cryptographic processor designed to manage and protect the lifecycle of cryptographic keys and to execute encryption and decryption sequences. It presents a leading practice and provides a high level of security in terms of confidentiality, integrity, and availability of cryptographic keys and any sensitive data processed with it.

FIPS 140-3, "Security Requirements for Cryptographic Modules," is a standard aimed at validating the effectiveness of cryptographic hardware [20]. There are various reasons why a HSM must meet the FIPS validation standards, including preventing unauthorized users from accessing sensitive information, protecting against any unauthorized activity in private systems, preventing changes from being made without an administrator's knowledge or permission, and detecting errors immediately, before sensitive information has been damaged or compromised.

12.4.2 PKI

A means to share cryptographic keys between devices that had no other practical means to establish trust in the keys to be shared was devised by combining the use of cryptographic primitives, X. 509 digital certificates, and a hierarchical structure of trusted servers called "Certificate Authorities." This became known as a PKI and is the basis for all trust relationships on the internet. Medical devices often need to be able to send encrypted data so that both users and other devices can trust that the data is authentic, uncorrupted, intended for them, and originated from the expected (trusted) source. This can be accomplished by delivering unique credentials to each device in the form of digital certificates. A PKI trust anchor or root certificate establishes a chain of trust that is used to validate other certificates signed by it up the hierarchical chain to the root certificate authority (CA).

A complete PKI framework consists of various elements, such as a CA, an RA, policies, directories, and management systems. A CA is an entity that signs, issues, and may even store digital certificates. An RA is an entity that verifies and attests to the identities of other entities. HSMs or cloud-based solutions with HSMs may perform the root of trust function. HSMs are employed to generate and protect the root and its subordinate CA cryptographic keys. The root CA should remain offline and only be brought out of a secure environment with multilevel controls when signing a subordinate CA. Each certificate is linked and trusted by the root of trust, which is the basis for the PKI.

A properly established CA is vital to maintaining trust throughout the certificate lifecycle. The compromise of a CA cryptographic key can result in a wide variety of issues, including but not limited to placing all previously issued certificates at risk of compromise, and/or the issuance of rogue certificates. Therefore, it is vital to properly manage all keys in order to maintain that trust.

PKI then is a framework that consists of a set of predefined policies, roles, and procedures for generating, managing, utilizing, distributing, storing, and even revoking digital certificates. A PKI establishes and operates the utilization of digital certificates and public keys for encryption. PKI assists in the processes of encrypting the transmission of sensitive data connecting the medical device and the host system.

12.4.3 Maintaining a Root of Trust

The general function of a root of trust (RoT) is to provide a trust anchor for a certificate chain, and it can reference:

- The hardware used in a medical device used to protect its associated cryptographic keys, in which a fixed-function root of trust is controlled by the firmware and can perform a specified set of functions such as encryption, certificate validation, and cryptographic key management.
- The logical protections provided by digital certificates chain of trust and their associated cryptographic keys, specifically the sensitive/secret or private keys.
- The hardware utilized for protection of sensitive/private cryptographic key(s) at the root level associated with a root CA in a PKI certificate chain of trust that signs other CAs (commonly issuing CAs) that remains offline unless it is used temporarily for signing an issuing CA.

From a hardware perspective, there are several recognized methods of establishing and maintaining trust and integrity of a device. Some of the well-known methods involve anchoring the digital keys and certificates into the silicon of the medical device during the physical assembly at the manufacturing facility. The credentials serve as a constant and unmodifiable root of trust, or device identity, that the original manufacturer, operator, or service provider can trust.

However, there is a critical difference between establishing that trust and maintaining it. Throughout the lifecycle of the device, there is a separate and subordinate lifecycle of the cryptographic keys and digital certificates associated with that device that also needs to be considered. As part of trust establishment, including the ability to maintain the trust of those that rely on that trust (relying parties) is a fundamental component of managing a PKI. If the PKI has no formal policy defined, the relying parties cannot make an informed decision whether to trust certificates issued by its CA. This is critically vital when certificates are distributed to third parties.

To assist with securing and maintaining that root of trust, best practices regarding the storage of sensitive material (cryptographic keys and secrets) are in a hardware device known as an HSM. HSMs are available for both offline and online storage. They hold sensitive information in tamper-evident, validated, and/or certified hardware (a physical computing device that safeguards and manages sensitive digital keys for strong authentication and provides cryptographic processing).

The primary complication in implementing root of trust is guaranteeing that setup is done properly at the assembly line. Conventional factory environments need to have trained cybersecurity personnel handling provisioning operations. Additionally, provisioning processes and workflow supporting technical implementations can be designed to minimize the risk of compromise during provisioning (further discussed in Chapter 6).

12.4.4 Key Lifecycle Management

For a root of trust to properly manage the lifecycle of the keys in its hierarchy, careful thought must be given to how long keys should be valid and how likely it is that keys may be compromised. Keys that are securely stored in an HSM or an air-gapped network and used to extend the root of trust, known as signing keys, have a low risk of being compromised because security protecting those keys can be tightly controlled. For such keys, the main factor influencing how long the keys should be valid is likely the strength of the cryptography itself and how long it might take an attacker to break the cryptography with brute force (the work factor) and expose the secret key.

However, the keys on an asset that is connected to a public network or is physically accessible to potentially untrusted individuals have a much higher risk of being compromised. In order to determine how long these keys should be trusted before rekeying the asset, a somewhat subjective calculation can help to inform a reasonable key validity lifetime. Other factors may also influence how practical it is for an asset to perform a rekeying process, for instance, how available a connection is to the PKI, required to approve and create certificates for the new keys.

If keys within a root of trust are compromised before their predetermined validity time frame expires, then the keys may need to be actively revoked so that assets in the field know that the keys should no longer be trusted. This revocation process can be difficult to implement and maintain over time and has the same connectivity concerns as rekeying [21].

12.5 "Don't Roll Your Own Crypto"

Some software or hardware development teams might determine that selecting an open-source cryptography library, figuring out how to use it and then optimizing it for a given use case, as well as developing their own key management system (root of trust, provisioning, protection, and lifecycle management) is easy. It is just math and some software engineering, right? And how many engineers don't like a good challenge?

In reality, however, this is not about keeping your engineers entertained but all about building a cryptographic system that delivers the desired security requirements while meeting the device's use case within the given design constraints. Getting cryptography programs right can be very difficult, even at the most basic level. Aside from simply writing code that is correct and free of bugs, like memory leaks, there are plenty of things that can go wrong cryptographically when writing software functions to perform what seem to be straightforward mathematical operations. Side-channel attacks, such as timing attacks, are often difficult to avoid as they require the software engineer to think in ways that are counterintuitive to their normal development practices. The 100% correct implementations of cryptographic algorithms, from a mathematical perspective, still yield broken cryptographic systems that leave data and/or systems vulnerable to attacks. Often these attacks are not even difficult to carry out.

There are many challenges in implementing a cryptography solution that sufficiently mitigates the risk of exploitation. Special attention must be given to attacks on the cryptography library itself, hardware implementations, weaknesses in key management, and misconfiguration.

The preferred route for implementing cryptography in software applications is to utilize common open-source libraries, such as OpenSSL, LibreSSL, libsodium, or WolfSSL, among others, in order to leverage the countless hours of design, robust implementation, testing, fixes, and scrutiny that are behind them. Even with security experts writing or contributing to such libraries, there are often vulnerabilities that are discovered that must be patched in order to keep data secure. When vulnerabilities are discovered in a common open-source library, the likelihood that those vulnerabilities will be quickly patched is relatively high given the broad adoption and the large, open-source communities surrounding them.

To illustrate our point, we have provided some notable examples of failed cryptography implementations in Chapter 13.

12.6 Introduction to Cryptographic Concepts: Conclusion

This section is intended to introduce basic concepts used in modern cryptography such as symmetric (secret key) and asymmetric (public key) cryptography, the primary challenges as they relate to cryptographic infrastructure, including root of trust and PKI, as well as establishing and maintaining trust through the application of practices and technology. In the following section, we will discuss how these principles can be applied to medical devices.

Properly securing medical devices through the use of cryptography can be challenging. There are many cryptography algorithms, software libraries, and hardware implementations to choose from, each with its own set of features and configuration options with inherent advantages and disadvantages. Moreover, building an ecosystem around the device that enables the proper management of cryptographic keys (over the lifetime of the device) to facilitate cryptography operations is nontrivial, as discussed in the previous sections.

In order to secure medical device data with cryptography, there are a few common, core concepts that are required. At the heart of modern cryptography are asymmetric keys, used for public-key cryptography, and symmetric keys, used for symmetric encryption. Asymmetric keys enable digital signing algorithms for proving that the data is intact and signed by a particular key or party. Symmetric keys enable encryption, which is used to efficiently encode data in order to keep it private [22, 23].

These cryptographic tools can be used to enable more complex capabilities like a root of trust, to prove that certain keys, and their digital signatures, came from a trusted party. In addition to the root of trust, special algorithms, such as key-exchange algorithms, can be used to securely establish symmetric encryption keys between two parties so that they can send data back and forth privately.

12.7 Designing Cryptography Solutions for Medical Device Use Cases

Cryptography is a powerful tool that addresses many security use cases, and regulators have identified it as one of the leading technologies to be considered for medical devices. However, as discussed in this section, implementing cryptography is quite complex and requires careful consideration including algorithm selection, key type and strength, implementation, use case and resource constraints, lifecycle management, and supporting infrastructure.

12.7.1 Regulators' Expectations

Not surprisingly, regulators have recognized cryptography as one of the essential security techniques that can be used to keep medical devices secure and patients safe. For example, in its "Cybersecurity in Medical Devices: Quality System Considerations and Content of Premarket Submissions" (final, September 2023) [24], the FDA stated the need to "Provide a reasonable assurance that the device and related systems are cyber secure" and that "Cryptographic algorithms and protocols are recommended to be implemented to achieve the secure by design objectives."

To ensure the security of devices as well as the security and reliability of the cryptographic implementation itself, the FDA defined the following control categories (summarized), not limited to but directly and indirectly related to cryptography:

- Authenticity, integrity, and authorization: strong authentication for connections, users, and updates;
- Cryptography: industry-standard algorithms, key generation, distribution, management, and protection; avoid exploitable configurations and/or implementations;
- Code, data, and execution integrity: authenticity of software, firmware, and configurations;
- Confidentiality: protect sensitive information, including cryptographic secrets, credentials, data at rest and in transit;
- Event detection and logging: detect and log security compromise, capture forensic evidence, user notification, failure behavior;
- Resiliency and recovery: protect critical functionality and data; retention and recovery;

- Updatability and patchability: update in a secure and timely manner; update process reliability; provide for the rapid verification, validation, and distribution of patches and updates.

12.7.2 Implementing Medical Device Cryptography

We have already discussed "Don't roll your own crypto," but there is so much more to a cryptography implementation that is secure and reliable, meets the requirements of the given use case, and withstands the test of time.

Components to be considered and correctly specified for a given implementation:

- *Algorithm:* The mathematical formula that, in combination with a cryptographic key, is used to encrypt, decrypt, sign, or hash data.
- *Algorithm implementation:* The actual software library that performs the cryptographic calculation, ideally supported by hardware accelerators.
- *Algorithm lifecycle:* Any good crypto implementation should allow for update of the cryptographic code as appropriate for the use case.
- *Cryptographic key(s):* A string of data (numbers and/or letters or their binary representation) that, together with the data to be protected, is processed through a cryptographic algorithm.
- *Key protection:* Protection of key material at rest (e.g., on a device) and in transit (e.g., during provisioning or update) for asymmetric encryption private keys and symmetric encryption secret keys.
- *Key lifecycle management:* Ability to update keys as provided by use case, desired level of security, and economic and practical limitations.
- *Certificate:* Contains a public key and additional metadata (e.g., issuer or expiration date).
- *Certificate provisioning:* The process of providing an initial identity to a device (usually during production).
- *PKI:* Everything required to establish and manage public key encryption, including:
    - CA;
    - RA;
    - Central directory (secure key storage and index);
    - Certificate management system;
    - Certificate policy;

- Root of trust: ability to confirm key: (a) as foundation within a device (hardware root of trust) on which all secure operations rely; (b) CA root of trust as the ultimate trusted root certificate within a PKI, being the foundation for the overall trust structure.

Here is a quick checklist of all the aspects that need to be considered when implementing a cryptography solution. Note that, for larger manufacturers, some of these decisions should be made across multiple product lines to ensure consistency and optimize maintenance:

- *Scalability:* Supporting low-volume to high-volume deployments at an acceptable cost.
- *Key management infrastructure:* Securely support desired deployment models.
- *Cost:* Direct (e.g., hardware, certificates, licenses) and indirect (e.g., key lifecycle management and revocation, redesign of failed implementation).
- *Resources:* Available memory, CPU, battery, bandwidth.
- *Timing considerations:* Real-time or other timing constraint; key validity duration; time stamps.
- *Security:* For example, prevention or mitigation of secret key material exposure or breach or crypto algorithm deprecation.
- *Crypto agility:* Foreseeing changes in technologies or security risks.
- *Liability:* Regulatory (e.g., noncompliance) and legal (e.g., lawsuits).
- *Regulatory and standards compliance:* Regional cybersecurity regulations and guidances; customer-specific requirements.
- *Data control and retention:* Support for audit and forensic activities.
- *Aspects of trust:* Technical trust through a unique and verifiable identifier (e.g., root of trust); public trust that could be impacted through security incidents (e.g., damage to reputation).
- *Deployment and maintenance:* Manufacturer and/or end-user supported.

A systematic and comprehensive approach will support all aspects of a secure crypto implementation, leading to the desired outcomes not only as a result of the design but also providing reasonable security for the future of the device.

Cryptographic security functions provided should achieve:

- *Access and authentication:* Device identity and trust; secure remote and local access.
- *Information protection:* Authenticity; protection of data confidentiality at rest and in transit.
- *Code protection:* Code signing; secure boot; trusted data.

Achieving this requires correct implementation during:

- *Device design:* Desired capabilities; key protection; crypto management.
- *Manufacturing:* Provisioning; hardware root of trust; safeguards; processes.
- *Operations:* Management of key lifecycle and exception and error handling.

Underlying crypto design decisions include:

- *PKI:* Design and implementation; CA root of trust.
- *Certificate:* Structure and format.
- *Crypto design:* Algorithm type and key strength; algorithm and key management.

Cryptography is a powerful technology that MDMs can leverage to ensure information confidentiality, integrity, and authenticity. However, understanding cryptography on a mere technical level is not sufficient. Medical devices require certain unique considerations.

Typical consumers of the internet and computer systems use computer hardware and software applications that rely on cryptography to protect their computers as well as their data. When cryptography breaks down in one of these systems, the worst-case scenario is the loss of sensitive data such as Social Security numbers, credit card numbers, bank information, and account credentials. This may seem horrible and can have far-reaching personal consequences, but still, such an event does not place a person's life in immediate danger. With a medical device, a failure in the cryptography upon which it relies can mean personal harm or even fatality. This personal harm can happen through direct impact on the patient (e.g., wrong dosage, incorrect diagnosis) or indirectly by impacting a healthcare provider's ability to deliver care in a timely manner.

Some of the unique requirements for cryptography in medical devices include:

- Cryptographic algorithm constraints (key, signing, hashing, encryption) (e.g., resource limitations and performance requirements).
- Failure modes (e.g., desired behavior when a certificate expires, is revoked, or a signature fails to verify).
- Certified configuration (e.g., signing the information that drives a medical device's functions).
- Secure, cross-organizational interoperability (such as an imaging system from one vendor talking to a PACS from another; from a protocol perspective, this is supported by DICOM, but how about cryptographic key management across vendor boundaries?).
- Key and certificate management challenges, including provisioning (e.g., during manufacturing of the device by a third party, or key revocation or update).
- Connectivity limitations (e.g., devices may be intermittently connected which may conflict with key verification or update).

The next few sections will discuss each of these requirements in more detail.

12.7.2.1 Cryptographic Algorithm Constraints

Medical devices have a wide range of different hardware and capabilities. On one end of the spectrum, there are low-resource devices, including embedded devices such as pacemakers or insulin pumps, that need to be small enough to be surgically implanted into a patient and use a single battery for its entire lifetime. On the other end of the spectrum, there are large capital equipment devices, with multiple, high-performance computers that are tied into a building's power system. One important thing remains constant across all of these devices: they are meant to treat or diagnose patients and do so reliably and safely. Therefore, the stability and security of the devices' data and operation are essential.

The capabilities of a device's hardware can constrain how effective its data security can be. Cryptography is based in mathematics and, while computers are great at computations and math, cryptographic algorithms can be relatively demanding in terms of processing power and can only be optimized to a limited extent without compromising the security of the algorithms. Sometimes, the secret keys and root certificates that drive a device's data security need to be secured in a way that requires hardware and software capabilities that are not necessarily available in low-resource medical device hardware.

When attempting to do cryptography on small, embedded microcontrollers, processing power can be too restrictive to make certain cryptographic operations practical. These smaller devices are often designed to operate with low power consumption so that they do not drain a battery too quickly. While this is usually enough processing power to control sensors, process sensor data, and establish unencrypted communication channels, it is often not enough to perform cryptographic operations fast enough, or with acceptably low power consumption, to be practical for the use case of the medical device. Smaller devices that do implement cryptography functions for securing data often rely on hardware acceleration to make data security efficient enough to be practical for the devices' use cases.

However, although capital equipment that is attached to building power mains may seem to have plenty of processing power to spare, there are often still challenges when introducing cryptography. Some of these larger devices process large amounts of data at high speed and with critical timing requirements. In order for the device to operate properly, it may need to process and transmit data within a very short period of time with minimal latency and high predictability. Introducing any extra calculations or processing steps, as required to perform cryptography, may cause the device to not meet its performance targets or real-time requirements or could even lead to malfunction. For this reason, even high-resource medical device hardware may need a boost from cryptography hardware implementations. Often AES encryption modes are used in these scenarios, since acceleration through AES hardware implementations is more readily supported.

Regardless of the processing power available to the device, the protection of critical cryptographic secrets or configuration may require specialized hardware capabilities or crypto elements. Otherwise, critical data like secret key material and root certificates can be compromised, causing data to be leaked and/or altered and erroneously trusted, thus opening the opportunity for an attack on the device.

12.7.2.2 Failure Modes and Exception Handling

The handling of a failed cryptography operation must be carefully considered for each type of medical device and use case. A certain cryptographic failure mode (e.g., a failed certificate verification) in one type of device may protect a patient from harm, whereas the same failure mode in a different device type could cause harm. It is important that general-purpose cryptography libraries are implemented and configured to achieve the appropriate failure modes for each type of medical device and associated clinical use case.

Two real-world examples can illustrate the unique requirements for failure modes of the cryptography implementations in medical devices. In the first case, consider a linear particle accelerator, or linac, commonly used for radiation therapy. This device can destroy tumor cells without making an incision by firing a beam of radiation into a patient's body. If a patient receives the correct radiation dose at the correct location, then it can destroy a malignant tumor and save a patient's life. If the patient receives too high a radiation dose and/or a dose in the wrong area of his or her body, it can cause irreparable harm or even death. In order to protect the patient, a linac device manufacturer might use cryptography to sign a treatment plan to ensure that it was not corrupted or maliciously changed before the machine imported the plan and turned on its radiation beam. In this circumstance, we would want the device to refuse to deliver a treatment if it was unable to verify the signature on the treatment plan. This may delay the treatment, but it would also likely save the patient from being harmed by a corrupted or maliciously altered treatment plan.

In the second case, consider a pacemaker, which is used to maintain a patient's heart rate to prevent cardiac arrest. The pacemaker is loaded with settings that adjust the pacemaker's operation for the particular patient's heart condition. Doctors can adjust these settings as needed using a specialized programmer. To prevent corrupted or maliciously altered settings from being applied to the pacemaker and harming the patient, a programmer can sign the settings before they are transmitted and the pacemaker can verify that the settings came from a trusted programmer. A common failure case in cryptographic signature verification is that a signature is good, mathematically, but the certificate for the key that generated that signature has expired. This can cause the pacemaker to reject the new settings. Since a failure to update the settings of the pacemaker could cause immediate harm to the patient, the medical professional should have some form of override mechanism to force a settings update even if the cryptographic operation, signature verification, fails. Ideally, an alert of some kind should be propagated from the pacemaker and/or programmer to a monitoring system to log the event for root cause analysis and remediation so that future settings updates can be performed with valid certificates and without the doctor being required to break protocol and force an update.

12.7.2.3 Certified Configuration

When a medical device performs its functions, its operation should be as deterministic as possible. The more predictably a medical device operates, the safer and more effective it is at achieving its desired outcome with the patient. From a cryptographic perspective, ensuring consistent and specified behavior can be achieved, in part, by providing the medical device with trusted, intact software and configuration data, and using trusted, well-defined security and operation parameters for its data services. This applies to internal data services, for securing data stored on a disk and other persistent storage media, or transport security, for safely sending data over a network.

There is the risk that configuration parameters for a medical device are established in an insecure manner. While sometimes there is corruption protection or checksum verification built into configuration mechanisms, it is far less common that configuration data is signed cryptographically in a way that establishes high confidence that the configuration data that the device relies upon is not only intact but that it came from a trusted source. Properly secured configuration data that establishes device settings such as calculation parameters, calibration values, network allow lists, security parameters for different data types, and acceptable cryptographic algorithm lists decreases the attack surface of the device and can help to establish more deterministic behavior. Using simple checksums prevents configuration files from being changed accidentally but only cryptographically signed files are protected from malicious change.

12.7.2.4 Secure Cross-Organization Interoperability

Often, the subject of secure, cross-organizational data interoperability is raised in the context of medical devices that must share data but belong to different companies (MDMs) or organizations (HDOs). For example, if a computed tomographic (CT) scanner must send imaging data to a PACS so that doctors can diagnose a patient, how does the doctor know that the images they are looking at originated from the correct CT scanner and have not been altered prior to delivery? Cryptography, in the form of digital signatures and a root of trust, can give doctors confidence that the images they are using to diagnose a patient are intact and were produced by an appropriate, trusted medical device [25].

Secure, cross-organization interoperability is more difficult than constructing a data security ecosystem for a single organization and their medical devices. Incompatibilities may exist between the different types of cryptography used in the different organizations' devices. While efforts are currently being made to address this sort of secure interoperability between medical devices, this issue has not yet been solved. Standardization of the cryptographic ecosystem and protocols surrounding medical devices may help to address these challenges.

12.7.2.5 Key and Certificate Management

Managing keys and certificates to facilitate trust and data security for medical devices can present unique challenges compared to the more common PKI use cases. Most commercial PKI platforms in the industry today lack a comprehensive solution that will work in all medical device use cases.

Often, medical devices lack the ability to leverage PKI platforms' software development kits, written in higher-level programming languages, in order to facilitate key and certificate management operations. This may result in additional burdens on the MDM's engineering teams to implement these PKI operations on their own and figure out how to make it compatible with existing PKI infrastructure that was not designed with medical device use cases in mind.

Existing PKI platforms may also lack support for key algorithms, or for efficient digital certificate formats, which will work in the unique hardware configurations that are often found in medical devices. A medical device manufacturer may determine that, due to hardware or performance constraints, they have a limited set of options for the digital certificate sizes or cryptographic key algorithms that will work for their device. PKI platforms may not support generating certificates that are small enough, or with the required key algorithm, resulting in no viable option for leveraging existing PKI solutions for the medical device.

Other considerations that need to be addressed when building or using a key management system include:

- Management of key lifecycle, including managing crypto period (the time span during which a specific key is authorized for use) and revocation operations.
- The complete key lifecycle is comprised of (not all are required in all cases): generation, registration, storage, distribution and installation, use, rotation, backup, escrow, recovery, revocation, suspension, and destruction.
- Reliable and secure deployment of keys during the production cycle of the device.
- Proper management of key lifecycles requires a combination of a technical and procedural approach (e.g., a technical infrastructure to protect the keys and procedures to manage access to that infrastructure). The effort required will vary depending on the nature of the key: it will be high for a root of trust key, whereas it can be lower for a temporary key. Methods commonly applied include:
    - Procedural: separation of duties; dual control; split knowledge.
    - Technical: physical security; access security; user-based and role-based access management; considerations for high availability and business continuity.

Detailed guidance on how to set up a PKI infrastructure and manage key lifecycles has been provided, through, for example, NIST SP 800-32 (Introduction to Public Key Technology and the Federal PKI Infrastructure), NIST SP 800-57 Parts 1-3 (Recommendation for Key Management), and NIST SP 800-130 (A Framework for Designing Cryptographic Key Management Systems).

12.7.2.6 Connectivity Limitations

Connectivity limitations that impact key management and the use of cryptographically secured communication may occur in a number of scenarios and use cases:

- Devices operating on a segmented or isolated network that limit connectivity with a PKI infrastructure.
- Intermittent device connectivity is where a device may only be network-connected after an extended stand-alone period.
- The device may be connected via a low-bandwidth connection that may impose certain data transfer rate restrictions as it relates to the use of commercial keys or certificates.
- The device may be connected via a proprietary and/or point-to-point interface.

For security reasons, medical devices often run on segmented or isolated networks and may not have the connectivity that most existing PKI platforms require. For instance, a device may not be able to make a direct connection to hosted PKI infrastructure in order to generate its certificates and acquire certificate revocation information. In this case, extra infrastructure, or PKI gateway systems, may be required to make PKI processes work for a medical device.

Although individual devices and their respective integration architecture vary widely, in general, the following requirements need to be addressed in the limited connectivity scenario:

- Type of key and certificate used, based on device compute power and bandwidth.
- Initial certificate deployment: typically during the production process, but in unique cases, other solutions may need to be developed.
- Key use during operation or communication, based on device capability and available bandwidth. Depending on use case, this could range from a simple point-to-point connection to managing connectivity in a limited sub-net.
- Key rotation, revocation, or suspension. This may not be addressable, or may not need to be addressed, in all cases.

Depending on the specific device architecture and use case, these operations could be managed at the time of connectivity to a network or proprietary device. For example, when managing an implantable or patient-worn device, the manufacturer could embed the device certificate (of the appropriate form factor) during production. During device operation, that certificate would then be used to ensure confidentiality, integrity, and authenticity of the device communicating to an external programmer. That programmer could potentially act as a gateway in cases where the device certificate would need to be updated or revoked. However, practical and hardware limitations may outweigh the added security benefits such a process would provide, and this may not be a required feature for all device types. The specific implementation needs to be assessed on a case-by-case basis based on a risk/benefit analysis.

In a similar scenario, device certificates could also be used to manage and ensure the integrity and authenticity of firmware updates deployed to the device.

In practice, there are a large number of device types, use cases, and possible connectivity limitations that are not addressed in this high-level discussion. Hopefully, the provided examples and general discussion provide enough insight on this specific topic.

12.8 Cryptography in Low-Resource Devices

The previous sections discussed how cryptography can be applied to improve the security posture of medical devices. We already touched on the need to recognize that some devices may have limitations that require a different or modified approach in order to enable cryptography in low-resource contexts, such as memory, battery power, connectivity, and/or bandwidth-constrained use cases. These devices may require unique considerations so as not to compromise device functionality or useful life.

One of the most common ways the authors have seen engineering staff at MDMs incorrectly implement security into their embedded medical device is to implement risk controls that were originally designed for high-resource CPUs with unlimited memory, standard network connectivity, and no restrictions upon power consumption. In some medical devices, such as a device utilizing a mains-powered, PC-grade CPU, these big solutions can be successfully implemented. This includes leveraging PKI; however, this is generally not possible (or difficult and may require compromise) with low-resource medical devices employing a point-to-point style of communication.

Many of the medical devices being designed today are body-worn or implanted, which results in physically small devices with limited-capacity batteries and power-conserving microcontrollers (MCUs) being utilized in these designs (or MCUs being down-clocked to reduce power consumption). Ten years ago, this type of MCU would have very limited resources, including CPU throughput, data RAM space, program flash memory space, and poor support for low power modes of operation.

However, the current MCU design is a completely different animal. Newer MCUs are designed for extremely low-power modes of operation, have much larger internal RAM and program memory sizes, higher throughput, and, most importantly, dedicated security functionality implemented directly in the core of these MCUs. While still low-performing compared to their PC cousins, newer MCUs are far more capable than previous models in this class.

Security features that can be found in many new MCUs are:

- True random number generator (TRNG).
- Hardware-accelerated cryptographic engine (AES, ECDF, SHA, HMAC, CMAC, RSA, ECC).
- Secure boot with a hardware root of trust.
- ARM TrustZone virtualized CPU for trusted executables.
- Locking and unlocking of debug port.
- Differential power analysis mitigations.
- Anti-tamper mitigations.
- Physically unclonable function (PUF) unique for each MCU.
- Secure key management and storage.
- Support for secure firmware updates.

This new set of security features permits engineers to create stronger security controls with good design practices and still have minimal impact on the rest of the MCU's resources (e.g., power consumption) and on device performance, availability, and longevity. The following sections will give examples of how these tools can be used to secure a medical device system.

12.8.1 Implementing Key Exchange in Low-Resource Devices

When performing encryption (or any other cryptographic operation that requires a key), the weakest part of the process is the key. All aspects of a key's lifecycle are a source of vulnerabilities, including key creation, sharing, storage, modification, or deletion.

12.8.1.1 Key Creation

Medical devices do have an inherent advantage over commercial products, such as web browsers or mobile apps, which are nonunique in nature and therefore require that a key be exchanged dynamically, via methods such as ECDF [26], RSA [27], or one of the many variations of encrypted key exchange [28]. Medical devices have been controlled during manufacture, which allows the MDM to inject an initial seed key that can be used to derive other subkeys using a key derivation function (KDF) for instantiated uses (the same key should never be reused between any of the various possible cryptographic utilizations). This injected seed key should be unique to each device and stored securely on the device. During manufacturing, the injected key should be associated with (but not derived from) a publicly available device identifier, such as a serial number.

An even better approach is to utilize a PUF that is already present in the MCU and unique to each MCU, thus removing the need to inject a key during manufacturing. Another variation would be the generation of the key on the device itself as that would prevent key exposure during manufacturing, especially when third parties are used.

12.8.1.2 Key Sharing

The approach of keys being pre-shared at the time of manufacturing has already been discussed, but there are other ways that key material can be shared, each of which varies according to use case and strength of protection.

If the two devices have human interfaces that allow for displaying information and a method for the user to input information, then this can be utilized as an OOB method of sharing a key via numbers or even as a phrase of randomized text.

OOB gives us many solutions to share a key, including utilizing a devices' LED to blink out a serial stream of data to a camera on a smartphone [29] or the device producing a high-frequency (above human hearing) modulated data stream to be received by a smartphone's microphone [30]. Another OOB approach is for the device to have a 2-D barcode label that can impart an authentication key to a smartphone via its camera, which, in turn, can be used by the smartphone app to allow it to connect to the device via a wireless connection. All of these very specific use cases employing OOB solutions depend on physical proximity to the device to convey trust.

12.8.1.3 Key Storage

Storing a key can be problematic as there are physical attacks that could expose these keys. However, there are separate electronic components called secure elements that can be used to securely store key values. Some newer MCUs offer secure internal memory or a separate, virtualized security CPU, such as TrustZone, which can be used to store keys in the MCU's memory.

Even when utilizing old MCUs, if the key values are unique to a device (as they should be) or change rapidly (ephemeral), then exposure of one of these keys has a very limited impact due to its short life.

12.8.1.4 Key Deletion or Modification

The respective methods used for created or shared keys impacts end-of-key viability, as there is no certificate authority to delete a key. This presents a challenge unless key use is coordinated by a central authority, such as the MDM's key-to-serial number database.

Another approach to key use is to make the viable lifespan of any key utilized for encrypting communications very short (ephemeral), as in perfect forward secrecy. This results in constantly generating new keys for new messages being transmitted, so exposure of a current key can at most only expose a limited amount of data (e.g., the device's latest status message). There have been multiple schemes created to achieve this system, such as using a Diffie-Hellman exchange for each key change, but probably the best-known and most widely used is the Signal messaging app, which employs the double ratchet protocol invented by Moxie Marlinspike and Trevor Perrin [31].

At the most fundamental level, Signal creates a new key for each message being transmitted, and this is performed on both the sending and the receiving devices through the use of predictable one-way functions (in the case of Signal, a KDF). While Signal is an extremely secure messaging app that can stand up to nation-state level attacks [32], the basic concept of perfect forward secrecy can be implemented in a number of ways, including leveraging SHA, HMAC, or CMAC hardware accelerators to rapidly hash the existing key value to create a hashtag that will be the next key value. This easy-to-implement approach makes for a very secure system; however, the challenging part of such a system is recovering from a loss of key synchronization between the sender and the receiver in a manner that cannot be exploited. It should also be pointed out that the encryption key used for one direction of communications should not be the same key used for the return direction and both should be ephemeral.

The observant reader may have recognized that there are multiple possible variations upon this theme, such as expanding the PUF value (by the use of a KDF) into a set of function-specific root keys, so that each root key can also be ephemeral as shown in Figure 12.5.

While such approaches are typically used for data in transit, it can be extended to create a data at rest encryption scheme as well.

12.8.2 Implementing Encryption in Low-Resource Devices

As discussed before, encryption is a method to convert a data set (plaintext) into another data set (cipher text) that appears to be random data, and thus cannot expose the contents of the original plaintext. Performing this transformation requires an encryption key; decrypting the cipher text back to the original plaintext requires a decryption key.

There are many different types of encryption but the two major categories are symmetric encryption, where the encryption key and the decryption key are the same value, and asymmetric encryption, where the encryption key and the decryption key are different but associated values referred to as a key pair. Individually, these keys are referred to as the public key (for encryption) and the private key (for decryption).

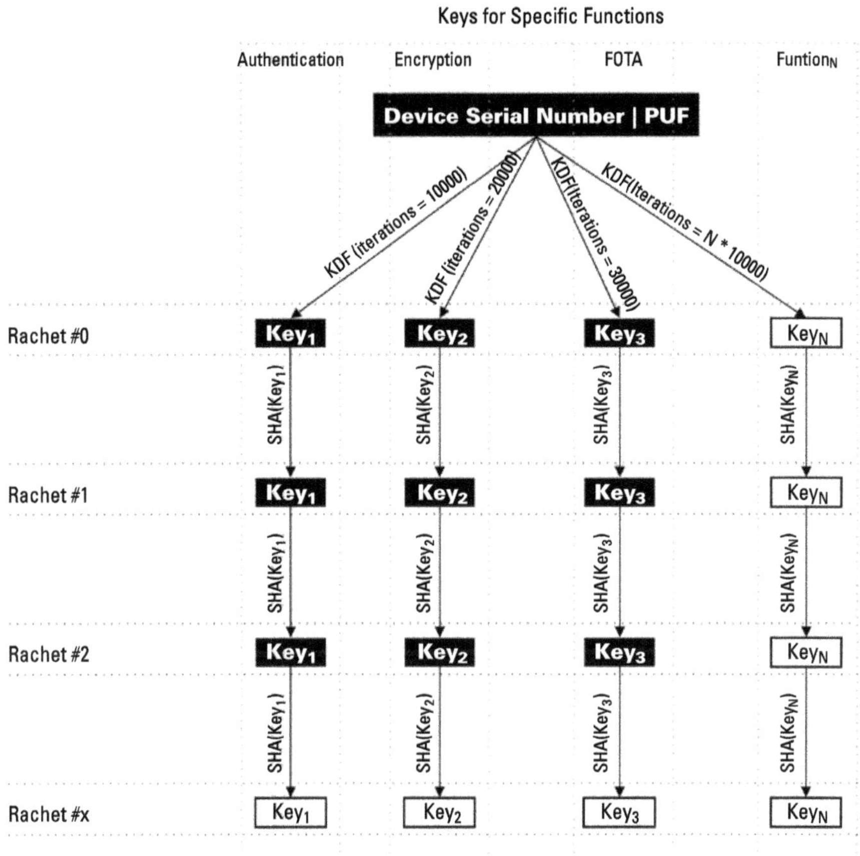

Figure 12.5 PUF value expansion using function-specific ephemeral root keys.

Asymmetric encryption is used on the internet due to the ease of key management; however, it is not a better or more effective encryption than symmetric. In many ways, asymmetric encryption has some specific disadvantages (e.g., asymmetric is mathematically thousands of times more complex and therefore slower to perform than symmetric and the length of key required to produce an equivalent strength of encryption is much larger with asymmetric encryption). NIST has calculated that a symmetric AES key of 128 bits in length as equivalent to an asymmetric RSA key of 3,072 bits in length, and a symmetric AES key of 256 bits in length to an asymmetric RSA key of 15,360 bits in length [33].

So, especially in resource-constrained devices there is a point to be made about the advantages of symmetric encryption. However, since all participants need to have knowledge of the same encryption key, protection of that key becomes the Achilles' heel of this approach. Due to that need for shared knowledge, key management and protection in symmetric encryption schemes requires great thought and effort across the entire lifecycle of the device (from provisioning to decommissioning) and across the entire infrastructure (from production to use environment). Yet even with the best planning and practices, symmetric keys will always remain at higher risk of compromise than asymmetric key pairs.

A possible solution to this dilemma, device resources and operating environment permitting, is to use Public Key Cryptography of Ephemeral Asymmetric Keys to create short-lived symmetric keys. Creating the temporary symmetric key is a relatively slow process as it uses mathematically more intensive asymmetric cryptography, but for the remainder of the communication a less resource-intensive approach via symmetric cryptography can be used. Again, this is a viable approach as long as the device resources and operating environment support this.

The Advanced Encryption Standard is the most likely form of symmetric encryption to be supported by MCU hardware acceleration. AES is frequently available in 128-bit and 256-bit lengths. Both were considered secure at the time of this writing; however, due to the expected long operating life of a medical device (up to 20 years) and the probable improvements to hardware that might be used against the device during its life, the highest available and executable level of encryption should be used.

When selecting an AES mode, hardware acceleration support may again be the deciding factor—not all modes are supported by these accelerators. Each mode has its advantages and disadvantages (shown in Table 12.1) that have to be weighed for each use case, but generally speaking, GCM mode is the mode that should be preferred. It has the fewest downsides and supports integral hashing of the ciphertext to provide not only confidentiality but integrity checking as well.

12.8.3 Implementing Authentication in Low-Resource Devices

One of the single most important security controls that can be implemented in a medical device system is authentication, both human-to-machine and machine-to-machine. Human-to-machine authentication has been in use for a much longer period of time and, as a result, the technical methods are more widely known, such as account credentials, passwords, biometrics, single factors, physical tokens, and salt values concatenated together and then hashed with SHA-256.

For machine-to-machine authentication, the methods are completely different. On the internet, we can leverage PKI (via x.509 certificates) to achieve authentication, but that requires the presence of a trusted third party, the CA.

Table 12.1
Supported Functionality of AES Modes

Mode	Name	Authenticating	None Required	Padding	Pseudostreamcipher	Storage Encryption	Self-Synchronizing	Chosen Plaintext Vulnerable
ECB	Electronic code book			✓				✓
CBC	Cipher block chaining			✓				✓
CTR	Counter				✓			
OFB	Output feedback				✓			✓
CFB	Cipher feedback				✓		✓	
XTS	XEX tweakable stealing		✓			✓		
CCM	Counter cipher message	✓						
OCB	Offset codebook	✓	✓					
GCM	Galois counter mode	✓						

In many medical devices, the connections to be authenticated are point-to-point connections without the possibility of a trusted third party. In these cases, a separate level of authentication needs to be performed between the two connecting devices and this needs to be performed in a manner that:

- Cannot be replayed (a replay attack is where a valid authentication is recorded by an attacker and then replayed back without understanding the values or algorithms being utilized).
- Does not expose any shared secrets to a man-in-the-middle attack against multiple iterations of authentication.
- Can be performed immediately following a connection being established before any other medical device functionality can be performed.
- Uses a unique value for each medical device so that a large-scale attack cannot be derived from an attack on one device.

The usual solution to this is a challenge-response process that is tailored to the limited resources of an MCU. Specifically, it leverages the one-way function of a keyed cryptographic hash (i.e., an HMAC) to demonstrate knowledge of a shared secret (in this case, the key). In the sequence diagram in Figure 12.6, Alice contributes half of the random value to Bob to prevent the possibility of a replay attack if Bob supplied all of the random value. The key value should be unique for each Alice.

12.8.4 Implementing Integrity in Low-Resource Devices

In the domain of medical device security, integrity can take on several meanings, such as "integrity of the intended function" or "integrity of data at rest/in motion," or specific use cases such as ensuring the integrity of a boot image or software update (with the software code being a specific type of data to be protected via code signing, discussed in more detail in Chapter 8). For the purposes of this section, we are only addressing data integrity.

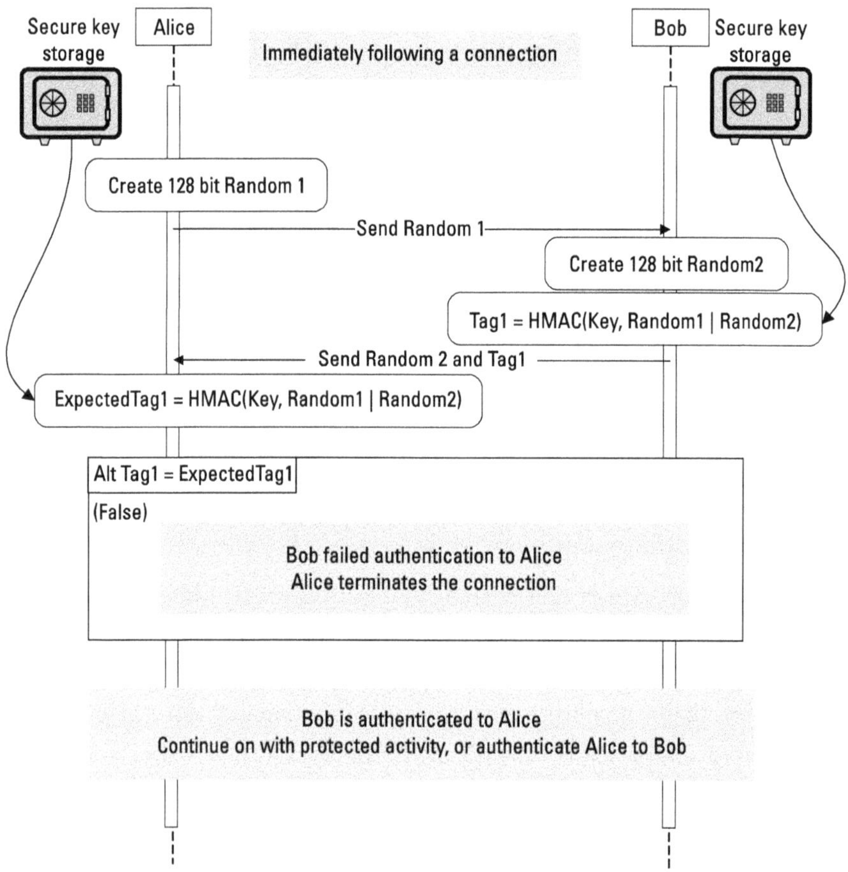

Figure 12.6 Challenge-response authentication.

Traditional data integrity tools consisted of algorithms such as a cyclic redundancy check (CRC), checksums, and parity bits. However, these are not cryptographically significant, so they need to be replaced with cryptographic hashes.

There are three fundamental ways to protect data: (1) by signing it, (2) by adding an HMAC/CMAC to the data, or (3) by utilizing a mode of encryption that conveys integrity as well as confidentiality (such as AES-GCM mode).

Signing is typically utilized in a system with asymmetric encryption. This is where the data set is hashed with a nonkeyed hash, such as SHA-256. The process is as follows:

- The data set is encrypted with the sender's private key.
- The cipher text is then hashed with a SHA-256 to create a hashtag value.
- The hashtag value is appended to the data set and then encrypted with the public key of the recipient.
- The recipient then uses their private key to decrypt the ciphertext.
- The recipient validates the SHA hashtag value.
- If the integrity is intact, then the recipient decrypts the data set back into plaintext using the sender's public key.

One way to think of this signing process is that the authenticity of the sender is conferred by the asymmetric encryption activity, while the integrity is ensured by the hashtag value. This is usually performed in the context of asymmetric encryption, but it does not have to be: the data set can be concatenated to the hashtag value and then encrypted with symmetric encryption. This is not as flexible or secure as when implemented in an asymmetric context, but it may still be a valid approach in some cases.

Directly utilizing an HMAC or a CMAC (both keyed hashes) is also a valid method to ensure the integrity of the data set. This is a symmetric approach to integrity, where both the sender and the receiver have previously established a shared secret in the form of a hash key. The data set is subjected to an HMAC function with the previously shared key and the resultant hashtag value is appended to the plaintext of the data set being transmitted. The receiver vets the integrity of the plaintext by performing the same HMAC operation and comparing it to the received hashtag value. Usually, after the sender appends the hashtag value to the data set, the entire structure is then encrypted (asymmetric or symmetric) before being sent to the recipient.

12.9 Further Medical Device Considerations

Most commercial cybersecurity solutions on the market today are reactive in nature. They are not necessarily designed to ensure that medical devices maintain their essential clinical performance in their deployment environment. These solutions are designed to detect issues as they arise and monitor things such as network traffic or system events from the sidelines. While firewalls and network IDS are absolutely necessary for securing hospital networks and medical devices, this is not a complete solution. Devices need to be secure by design such that they can still operate securely in a potentially hostile network environment [34] and even a physically insecure environment.

The common data security tools on the market today that are more proactive, and can potentially provide a trust hierarchy to drive data security in the device, are deeply rooted in the World Wide Web SSL/TLS certificate paradigm where a client, such as a web browser, is meant to talk to a vast number of server endpoints, or websites, to request information to download and display for the user. Many medical devices attempt to use these same security tools to secure the data in transit between their medical devices and hosted servers.

While a medical device connecting to a hosted server, in the cloud or a datacenter, is similar to the scenario where a web browser is connecting to a website, there are important differences. The first and most obvious one is that the medical device, as a security best practice, should only trust a specific subset of the servers on the internet and not just any server from any organization. The subset of servers with which the medical device wants to communicate is likely limited to just a few server endpoints that belong to the MDM or HDO. This type of restriction may be achieved by the MDM acquiring SSL/TLS certificates that are signed by an intermediate certificate for their organization and configuring the medical device to only trust servers that chain up to their organization's intermediate certificate. However, there are some difficulties in implementing this solution with certain medical devices. Among other issues, the PKI vendor issuing the certificates may not support an x509 format the device can handle, they may not support a certificate chain hierarchy that gives the medical device vendor confidence in the integrity of the root of trust, or they may not support the desired key algorithms and signing algorithms that would yield the appropriate or workable solution. Moreover, the SSL/TLS configuration likely requires special attention to ensure that the device does not inadvertently negotiate down to an insecure algorithm or trust a certificate chain that is outside of its organization.

One of the more difficult aspects of implementing a PKI solution to establish a root of trust is the information workflow and protocol required to manage keys and establish certificates. The current certificate signing request workflows and APIs in the market, as well as the way the outstanding certificate information is stored and tracked, can be problematic for medical device systems that represent different use cases from traditional PKI applications.

These are the issues we encounter with the simplest connected medical device use case. Many medical devices are not this simple. Medical devices range from a single device with a simple client-server configuration, to a configuration with many computers on a local subnetwork that connect to the internet through a gateway machine. In order to properly secure the range of medical device configurations we need tools that are specifically designed to address these use cases. For example, traditional SSL/TLS certificates and their provisioning processes are not appropriate for medical devices that talk to components on a local network where traditional, internet-centric, top-level domains and DNS resolution do not apply.

12.10 Cryptography—Conclusion

Cryptography can provide for very strong protections against many types of attacks, exploits, or compromise but requires careful selection of methodology, process, hardware, implementation, and supporting infrastructure, which requires identifying the right algorithm, key lifecycle scheme, and correct MCU, early on in the lifecycle of the device. Cryptography can be used effectively to protect data confidentiality (encryption), integrity (hashing), authenticity (signing), or any combination thereof. Considering the resource restraints of many medical devices and infrastructure/operational limitations in some use cases, not all are required at all times; therefore, why spend computation cycles on encryption if only integrity or authenticity is required?

However, technology is always on the move, and so are cyber adversaries. As hardware improves, asymmetric encryption solutions may become possible even in ultra-low-resource contexts, as both built-in and stand-alone hardware-accelerated implementations of ECC and other cryptographic technologies become more common and begin to close what is, at the time of this writing, an unacceptable performance gap for most embedded medical device use cases.

There are many more ways cryptographic primitives can be used to protect a medical device than we can address in this book, and many sources of additional information about how to apply them in a design. Foremost among these is Ferguson, Schneier, and Kohno's excellent book *Cryptography Engineering: Design Principles and Practical Applications* [1].

References

[1] Ferguson, N., B. Schneier, and T. Kohno, *Cryptography Engineering: Design Principles and Practical Applications*, New York: John Wiley & Sons, 2011, https://www.schneier.com/books/cryptography_engineering/.

[2] Moghimi, D., et al., "TPM-FAIL: TPM Meets Timing and Lattice Attacks," https://tpm.fail/tpmfail.pdf.

[3] Kelly, S., "Security Implications of Using the Data Encryption Standard (DES)," The IETF Trust, 2006, https://tools.ietf.org/html/rfc4772.

[4] MITRE, "CWE-338: Use of Cryptographically Weak Pseudo-Random Number Generator (PRNG)," July 19, 2006, https://cwe.mitre.org/data/definitions/338.html.

[5] Genkin, D., A. Shamir, and E. Tromer, "RSA Key Extraction Via Low-Bandwidth Acoustic Cryptanalysis," *Cryptology ePrint Archive*, December 18, 2013, https://eprint.iacr.org/2013/857.

[6] Bernstein, D., J. Buchmann, and E. Dahmen, (eds.), *Post-Quantum Cryptography*, Berlin: Springer-Verlag, 2009.

[7] NIST Computer Security Research Center, *Glossary*, https://csrc.nist.gov/glossary.

[8] Schaad, J., "Use of the Advanced Encryption Standard (AES) Encryption Algorithm in Cryptographic Message Syntax (CMS)," July 2003, https://tools.ietf.org/html/rfc3565.

[9] Gueron, S., "Intel Advanced Encryption Standard (AES) New Instructions Set," May 2010, https://www.intel.com/content/dam/doc/white-paper/advanced-encryption-standard-new-instructions-set-paper.pdf.

[10] Nir., Y., and Langley A., "ChaCha20 and Poly1305 for IETF Protocols," May 2015, https://tools.ietf.org/html/rfc7539.

[11] Moriarty, K., et al., "PKCS #1: RSA Cryptography Specifications Version 2.2," November 2016, https://tools.ietf.org/html/rfc8017.

[12] McGrew, D., K. Igoe, and M. Salter, "Fundamental Elliptic Curve Cryptography Algorithms," February 2011, https://tools.ietf.org/html/rfc6090.

[13] Arkko, J., et al., "Perfect-Forward Secrecy for the Extensible Authentication Protocol Method for Authentication and Key Agreement," July 25, 2019, https://tools.ietf.org/html/draft-ietf-emu-aka-pfs-00.

[14] International Telecommunication Union, "X.509: Information Technology - Open Systems Interconnection - The Directory: Public-Key and Attribute Certificate Frameworks," October 14, 2019, https://www.itu.int/rec/T-REC-X.509-201910-I/en.

[15] Barker, E., "Guideline for Using Cryptographic Standards in the Federal Government: Cryptographic Mechanisms," NIST, March 2020, https://doi.org/10.6028/NIST.SP.800-175Br1.

[16] Cooper, D., et al., "Internet X.509 Public Key Infrastructure Certificate and Certificate Revocation List (CRL) Profile," Network Working Group, May 2008, https://datatracker.ietf.org/doc/html/rfc5280.

[17] CA/Browser Forum, "Guidelines for the Issuance and Management of Extended Validation Certificates," November 30, 2022, https://cabforum.org/wp-content/uploads/CA-Browser-Forum-EV-Guidelines-1.8.0.pdf.

[18] Barker, E., and W. C. Barker, "SP 800-57 Part 2 Revision 1 'Recommendation for Key Management: Part 2 – Best Practices for Key Management Organizations'," May 2019. https://doi.org/10.6028/NIST.SP.800-57pt2r1.

[19] Cooper, D., et al., "Internet X.509 Public Key Infrastructure Certificate and Certificate Revocation List (CRL) Profile," May 2008, https://tools.ietf.org/html/rfc5280.

[20] NIST, "FIPS 140-3 'Security Requirements for Cryptographic Modules'," March 2019, https://csrc.nist.gov/pubs/fips/140-3/final.

[21] Chokhani, S., et al., "Internet X.509 Public Key Infrastructure Certificate Policy and Certification Practices Framework," November 2003, https://www.ietf.org/rfc/rfc3647.txt.

[22] Bernstein, D., et al., "How to Manipulate Curve Standards: A White Paper for the Black Hat," July 22, 2014, https://bada55.cr.yp.to/bada55-20140722.pdf.

[23] Bernstein, D., "Curve25519: New Diffie-Hellman Speed Records," September 2, 2006, https://cr.yp.to/ecdh/curve25519-20060209.pdf.

[24] FDA, "Cybersecurity in Medical Devices: Quality System Considerations and Content of Premarket Submissions," September 2023, https://www.fda.gov/regulatory-information/search-fda-guidance-documents/cybersecurity-medical-devices-quality-system-considerations-and-content-premarket-submissions.

[25] Mirsky, Y., et al., "CT-GAN: Malicious Tampering of 3D Medical Imagery Using Deep Learning," *USENIX Security 2019*, June 6, 2019, https://arxiv.org/pdf/1901.03597.pdf.

[26] du Preez, P. P., "Understanding EC Diffie-Hellman," *Medium*, October 2, 2020, https://medium.com/swlh/understanding-ec-diffie-hellman-9c07be338d4a.

[27] Harris, B., "RSA Key Exchange for the Secure Shell (SSH) Transport Layer Protocol," March 2006, https://tools.ietf.org/html/rfc4432.

[28] Sheffer, Y., et al., "An EAP Authentication Method Based on the Encrypted Key Exchange (EKE) Protocol," February 2011, https://tools.ietf.org/html/rfc6124.

[29] Duque, A., et al., "Decoding Methods in LED-to-Smartphone Bidirectional Communication for the IoT," *Global LiFi Congress*, Paris, France, February 2018, pp. 1–6, https://hal.inria.fr/hal-01683629/document.

[30] Waddell, K., "Your Phone Is Listening—Literally Listening—to Your TV: All in the Name of Serving You More Targeted Ads," *The Atlantic*, November 19, 2015, https://www.theatlantic.com/technology/archive/2015/11/your-phone-is-literally-listening-to-your-tv/416712/.

[31] Perrin, T., and M. Marlinspike, "The Double Ratchet Algorithm," November 20, 2016, https://signal.org/docs/specifications/doubleratchet/.

[32] Grauer, Y., "WikiLeaks Says the CIA Can 'Bypass' Secure Messaging Apps Like Signal. What Does That Mean?" *Slate*, March 8, 2017, https://slate.com/technology/2017/03/wikileaks-says-the-cia-can-bypass-signal-what-does-that-mean.html.

[33] Barker, E., "Recommendation for Key Management," *NIST*, January 2016, https://nvlpubs.nist.gov/nistpubs/SpecialPublications/NIST.SP.800-57pt1r4.pdf.

[34] NextGov, "FDA Official: Medical Devices Should Be Designed for Hostile Environments," October 2016, https://www.nextgov.com/digital-government/2016/10/fda-official-cyber-risks-internet-things/132360/.

13

Cybersecurity Failures, Myths, and Excuses

13.1 Notable Failures and Lessons Learned

Failures and lessons learned span the spectrum from technical failures and poor engineering practices, to misguided management decisions, to adverse cultural environments. Regardless of the nature of the failure or the person at fault, the outcomes can be consequential. These consequences can include regulatory rejection, market entry delay, impacts on care delivery or patient safety, as well as impacts on business sales, valuation, market share, and reputation.

13.1.1 The Classics

Granted, the medical device industry has made huge strides in security practices and culture. Looking back at the anecdotal reports provided by pioneering security researchers from 10 to 15 years ago can be eye-opening, but also encouraging, as we can use these examples to appreciate the long way we have come. These early researchers—people like Kevin Fu (who mentions these early experiences in his Foreword to this book), Jay Radcliffe, Billy Rios, and others—all reported similar experiences. Analyzing a device and discovering a security flaw were only part of the hard work. Next was the challenge to find somebody at the respective manufacturer who was actually willing to listen, often resulting in off-the-record meetings in secret locations. Equally common was pushback from manufacturers, ranging from disputing the existence of the vulnerability to challenging the researcher's rights to probe the device, up to and including legal threats.

In the early days of medical device security, most manufacturers' approach was one of more or less aggressive denial: "The vulnerability does not exist," or "if it exists, the risk would be low." In all fairness, there are also examples of security researchers having thought that they struck proverbial gold with a security discovery and having a "you dare to call my child ugly" reaction when the manufacturer (or others in the industry) plausibly laid out that this newly discovered vulnerability actually was not as serious as it may have initially appeared. Strong emotions can exist on either side. Yet we have come a long way, with many manufacturers now having a coordinated vulnerability disclosure (CVD) policy in place and/or participating in an ISAO. Security researchers know how to contact manufacturers securely and confidentially and can trust that there will be a person on the receiving end that takes his or her submission seriously and will respond based on a well-defined process, keeping the researcher in the loop, and, if warranted, notifying customers and providing an update or mitigation. We have come a long way, indeed.

13.1.2 Failed Implementations

While the quality of an implemented medical device and the security of a medical device are two separate processes, using different tools and methods, it cannot be denied that there is a link between the two. "Low-quality" medical devices are almost always vulnerable medical devices. The reciprocal statement is not as likely to be true, but when reviewing the security of a given medical device, researchers are often forced to question the manufacturer's commitment to quality.

By far the largest class of failed implementations is when the manufacturer uses an existing medical device as the starting point for a new medical device. The existing device usually dates from a time when security was not even a consideration made by any manufacturer. The manufacturer's goal is often to save time and money in the new development with the attitude that "It passed premarket approval the last time, why wouldn't it pass now?" There is so much wrong with this attitude that it is hard to itemize the mistakes being made:

- Complete disregard for the advances that have been made worldwide for medical device cybersecurity, sometimes due to a lack of awareness, but all too many times a result of simply not wanting to know that total lifecycle security needs to be designed into every new device.
- Prioritization of short-term profits over quality, safety, and efficacy.
- Ignorance of the fact that the old components called for by that old medical device design are much more expensive, offer fewer capabili-

ties, often have known vulnerabilities with widely disseminated exploit methods and tools, and will be unavailable sooner (or are already unavailable) compared to modern replacements.

- Failure to account for the time required for engineers to study and understand the old design and how to implement it. Often, this can be much greater than the time needed to utilize modern tools and components for the first time.
- Failure to consider that these old tools and components are not being updated (or even tested) for any security issues that should be patched.

What does this look like in an engineering team?

- The reuse of CPUs/MCUs that are more than 10 years old with no security features.
- There is only one person who can touch "that code" ("Go ask Ted how that works, no one else knows").
- "Tribal knowledge" is the primary form of documenting the old device.
- There is a "special PC" that is running some old version of Windows 95 or NT because it is the only one that can run the tools (in one specific case, this was taken to such an extreme that it was mandated that the "special PC" could never be turned off, as they feared that it would not boot back up and thus they could not build the code base).
- The management insists that all devices originate from "the code base" (said to the accompaniment of kettle drums). Upon closer inspection, this code base is revealed to be a huge amount of code, with thousands of conditional compile statements and hundreds of macro definitions, all in place to build a specific product and of course, no one (except for "that one guy") knows how it functions. No automated tools can digest this "conditional compile hell" (and in one specific case, this lexicon of macro definitions to product builds was only documented on the software engineering manager's 8-foot whiteboard).

However, all of that is just the "whys" of failed implementations, not the juicy examples of failures that we can all read about to feel smugly superior to the organizations that created these messes. Here is a partial list of cybersecurity failures that we have directly observed:

- An implanted device that had no authentication/encryption/integrity checking over a wireless interface.

- A cardiac device (both implanted and external to the body) that was written in C and had more "CERT Secure C Rule" violations than total lines of code.
- A life support medical device, originally designed to be a stand-alone device, that later had an ethernet interface bolted onto it—which caused the device to cease functioning due to the frequency of packet reception.
- An external medical device that used components that were over 30 years old and thus could not implement any form of security.
- An external patient-worn device with 44 manufacturing-related unauthenticated wireless commands that should have only been utilized during manufacturing, not when in clinical use and connected to a patient.

The sad truth is that we could fill this entire book with nothing but bad examples, largely due to the reuse of old code bases and hardware. The good news is that this is changing. We are seeing more concern being placed on the quality and security of medical devices, and for this, we can thank the FDA and other regulatory agencies that have followed its example. Also, if you are worried about the cybersecurity failures listed above, do not be. These have all been fixed or withdrawn from the market and replaced with new secure alternatives.

13.1.3 Cryptography Gone Wrong

In a previous chapter, we discussed cryptography implementations, and associated challenges, in great detail. In light of the complexities and required experience, we advised "don't roll your own crypto." There are plenty of examples of failures in cryptography, ranging from flawed implementations [1] to poorly written standards [2].

Examples of crypto implementation failures in medical devices include:

- Foregoing the use of cryptography altogether and using plaintext communication without ensuring confidentiality or integrity.
- Perceived design or device resource constraints may lead to the use of symmetric key cryptography with insufficient thought given to the protection of the symmetric key. Or even worse, try to implement your own way of scrambling data that is perceived to be (1) "secure enough" and/or (2) supportable by the device's limited resources (including, and we are not making this up, use of the 2,000-year old Caesar's Cypher) [3].
- Failure to account for the risks resulting from the need to protect secret materials (e.g., exposing cryptographic keys during provisioning or while in storage).

- Use of a single hard-coded key.
- The belief that your selected block cipher also provides integrity protection (some do, most do not).
- Compromises in key management processes to accommodate operational priorities (e.g., sharing of secret key material across organizational entities).
- The reuse of keys across multiple devices to address economical or logistical constraints, leading to the risk of larger quantities of devices being exposed if a single key is compromised.
- Vulnerable transport of keys (e.g., on insecure data carriers such as thumb drives).
- Use of a single key to support multiple security functions with different levels of risk, criticality, or lifecycle needs.
- Failure to manage key lifecycle and provide management features to support secure key generation, provisioning, storage, distribution, use, and destruction.
- Failures of proper certificate practices, such as the use of self-signed certificates; and use of excessively long (e.g., 30-year) certificate periods.
- Failure to anticipate the need for future crypto algorithm and library updates to account for predictable algorithm depreciation.
- Use of cryptographic concepts and architectures from the nondevice space (e.g., web browser-server communication or general IT use cases that do not account for unique device requirements).

13.1.4 Failures in Dealing with Regulators

As our industry navigates its current multiyear upgrade of regulators' cybersecurity expectations, transitioning from providing basic guidance to enforcement of defined standards, manufacturers need to adapt their expectations and demonstrate cybersecurity capabilities as part of their submissions. Failing to anticipate these requirements will result in rejection (as enforced via the FDA's eStar premarket submission process) [4] and delay in market entry.

Regulatory submissions need to demonstrate that a device meets a security baseline and can be maintained and kept secure. Failure to include supporting documentation will impact approval if the following inclusions are deemed missing:

- Documentation that demonstrates that:

    - The device does not contain any high-risk vulnerabilities.
    - Security requirements have been defined, controls have been implemented, and traceability to final testing has been established.
- Demonstration that the manufacturer is including cybersecurity considerations in their quality system and is following secure software development lifecycle practices, for example, evident through:
    - Security-specific lifecycle procedures.
    - Defined risk management processes.
    - Security engineering best practices such as threat modeling, secure coding practices, code review.
    - Supporting security documentation such as the SBOM and security architecture views.
- Documentation for the end-user or operator on how to ensure and maintain a device's security posture including general security and configuration instructions, network diagrams, SBOM, update instructions, event response and forensic evidence capture, backup and restore, authorized security tools (e.g., anti-malware), end-of-life and end-of-support information, and security communication processes and contacts.
- Plans to outline how the manufacturer intends to maintain the devices' security posture after release (i.e., postmarket), including supply chain risk management, coordinated vulnerability disclosure, vulnerability and update communication, and incident response plans.

Assuming that "I did not need this in my last application and it passed just fine" would be folly now that regulators have raised the bar through new laws and regulations and are approaching cybersecurity with newly clarified authority for enforcement.

13.2 Common Cybersecurity Myths and Excuses

After many decades of combined experience working with and for MDMs as engineers and security specialists, this book's contributors have heard just about every excuse for why the MDM did not secure their medical device. Ironically, it sometimes seems like more time and money were spent on excuses than it would have taken to secure the device.

Although we have seen improvement, the capacity of some management (from mid-level to senior) to willfully ignore security has often been nothing short of astonishing. The level of self-deception entailed sometimes rivals the proverbial ostrich's "head-in-the-sand" approach to threats. (Note that ostriches do not put their head in the sand in response to threats. Any species that behaved in this manner would rapidly go extinct—a realization that in and of itself should serve as a lesson to business leaders.) This is why it is such a pleasure to work with MDMs who address security as an essential aspect of their products and business and thus help to improve the overall industry. This is, in the end, the fruits of alert executive leadership, the right business priorities, and an organization's cybersecurity culture. We need more leaders like these.

We have also seen good people at some MDMs reach a point of such frustration with their organizations that they leave or give up trying to improve the organization's approach to security. This section of the book is for those people: wherever you are, at whatever organization you work for, take comfort in knowing that your situation is not unique. Ten years ago, you would have been isolated and fighting a lonesome battle, but now you are in good company with most of the leaders of the medical device industry and the worldwide regulatory agencies working to improve the security of medical devices. So—keep up the spirit and continue to fight the good fight! This section is dedicated to you and your efforts.

A decade ago, "management ostriches" could be explained away by the general lack of regulatory enforcement and market demand, as well as the lack of security awareness and training that existed for engineers and the MDM community in general. However, that has not been the case for several years now, and it is becoming obvious that this was never about ignorance or not knowing better, but always about "not wanting to fund the extra effort necessary to secure medical devices." This does not speak well for where our industry has come from. As we see, many members of the MDM community fully embracing security in medical devices, it makes the contrast with those who continue to reject it even starker.

Excuses that we have seen (and, sadly, still continue to see on a regular basis) include:

- *"We do not need to secure our medical device!"* This speaks to such a vast level of incompetence that it is always disheartening when we hear it. This speaker is revealing that he or she has spent years ignoring all U.S. and international regulations about medical device security, ignoring widely reported security incidents relating to medical devices in the news, and does not care if their company's business is damaged or if their patients are harmed by their medical devices as long as short-term profitability is achieved. It is very difficult to make headway with someone like this, as they will not acknowledge moral or even long-term financial concern for the consequences their actions may create.

  This is a good place to point out that if you do not have the cooperation of at least some members of the senior management team (see

Chapter 2), your efforts are guaranteed to fail. If you cannot get the "management ostrich" in question overruled by someone in a position of greater authority, the best approach is to leave the employment of the company or drop them as a client. That manager has hired you as a token nod toward security and will not hesitate to paint any negative security events that later befall as your fault if he or she finds it expedient to do so.

- *"Our engineers only write secure code!"* We love this one. It always comes from a pompous senior executive who does not understand security (or software engineering). Trying to keep from laughing when we hear it is a challenge. What this person is saying is some variant of, "I'm a senior executive and I know more about this than you do and my department is handling it. Even though we have no processes in place to track or measure it, trust us, we're doing a perfect job." It also reveals the executive's false assumption that all vulnerabilities are software-based.

  Once you start tunneling into specifics, these managers start getting quite evasive and push back about not knowing all the details, but they are confident that their engineers are doing this thing called "security," even though they cannot point to what processes or artifacts are necessary to design, develop, and manufacture a secure medical device system. When we then suggest that we get someone from their engineering group to ask, just to clarify their activities, surprisingly, in 100% of occurrences, that person or group cannot be reached for questions.

  Much like the executives who believe that they do not need to secure their medical device, this company is very difficult to make headway with and typically has a lot more issues than just security. If you ever get to see their source code, you will almost certainly be shocked at its lack of quality.
- *"Once the source code is compiled, nobody can read it!"* This is one of those odd myths that seem to permeate the software developer industry. How otherwise skilled software engineers can proclaim such utter hogwash is beyond us. All languages have cheap, readily available disassemblers and decompilers that can be used to quickly and easily expose the intellectual property, algorithms, and security mitigations and methods in the software. Just compiling your code (native or bytecode) in no way protects it from exposure or corruption.
- *"Nobody can hack our proprietary [fill-in-the-blank]!"* We hear this statement on something close to a monthly basis, stated as a fact by an MDM that their proprietary interface is secure, simply because it is proprietary. Also, some engineers may even equate "proprietary" with "encryption,"

a myth we have hopefully dispelled in Chapter 12. Please, dear reader, understand that security professionals reverse-engineer proprietary technology all the time. Your prized communications protocol is not that unique. Yes, there have been examples of security researchers who were stopped dead in their attack on a protocol by such simple technical details as data whitening, where long sequences of "1"s will have a "0" bit inserted into the long sequence, and likewise, long sequences of "0"s will have a "1" bit inserted into the long sequence [5]. This can be confusing to nonengineers who are just sniffing the RF signal and trying to make sense of it. However, anyone with any RF technical knowledge will immediately filter out such whitening to restore the original bitstream.

With the introduction of cheap software-defined radios (SDRs) (some as low as $20), and a whole host of free programs to control the SDR [6, 7], the process of reverse-engineering an RF protocol can be measured in minutes [8–10]. We can reverse-engineer any frequency (from kilohertz to gigahertz) and any modulation scheme and increase the range of any low-power RF to the point where it could be weaponized to attack patients in a "brush-by" attack, where the target device is only in range for a couple of seconds. None of these supposed mitigating controls are going to stop a determined, competent attacker. This is equally true for wired communications. Sniffers exist for every type of electrical connection you can find. Even less mainstream communication mediums, such as the inductive communications common to certain implantable medical device systems, while being wildly diverse and unstandardized, can be reverse-engineered with the same tools used for RF reverse engineering plus a few additional tools.

All proprietary interfaces share a common trait: There is absolutely no real cryptographic security being utilized—no authentication, no integrity checking, no encryption. Not unless the MDM layers these security features atop the proprietary interface, as it should be doing.

One final point worth noting is that if you submit any documentation to the FDA that claims that you are secure because of your "proprietary interface," we guarantee you will be questioned in detail and ultimately rejected.

- *"Our data is anonymized!"* Proper anonymization or deidentification of data is extremely difficult to do, and there are many examples of failed attempts to do so [11–14]. Encrypting the data, in contrast, is far easier and much more likely to be both successful and compliant with regulations. Do not be afraid of encryption.
  - Seek out assistance from trained people and organizations.

- There are many good cryptographic libraries or software tools available.
- If you're using hardware-accelerated cryptography, it does not take long to perform.
- Even purely software-based solutions are accomplishable in milliseconds (depending on your hardware).
- There are good external components, such as the Microchip ATECC608A, that can perform the operations in hardware if your selected CPU cannot support it.

The hardest part of encryption is the key exchange and storage. So why, then, will so many MDMs argue endlessly that the data that they are collecting about a patient is not PHI, instead of just encrypting the data? This is a rhetorical question: We have no sensible answer. The argument usually centers around not including a Social Security number with the data or an address or phone number. They attempt to make the argument that they have "deidentified" the data via the "HIPAA Breach Notification Safe Harbor" rule (further discussed in Chapter 6) [15]. Fortunately, with the increase in HDO contracts that call for "encryption of all data" [16], this argument will not survive for much longer.

The takeaway for this section is that you simply need to encrypt the data, PHI or not. It is not that difficult.

- *"The advantages of our device outweigh the insecurities of the device."* Do they really? Considering that security mitigations exist to protect your device's essential functionality (integrity and availability) and to protect patient and business-critical information (confidentiality), what advantages can your device possibly provide once it has been compromised? Moreover, would not your competitor's properly secured medical device deliver the same or similar advantages to the market, without the risks posed to patients by your unsecured device?
- *"We didn't have to secure the last device we made."* This is a valid point. Given the pace of recent changes in security regulations relative to the time needed to develop a new device, it is possible that your company did not have to secure its previous medical device to get premarket approval. However, knowing what you now know about what is going on in the world with cybersecurity and associated attacks, you have to ask yourself:
    - "Do I feel comfortable knowing that my insecure device is out in the market? Would I want a loved one to use one of these devices, or would I want to use it myself?"

- "Is there intellectual property on that device that could be extracted by a competitor?"
- "Is there a consumable product that is utilized as a part of your medical device that a competitor could clone?"

Yes, the regulatory environment has changed, and it is good for your business as well as for the health and safety of your patients. Most recently, the Consolidated Appropriations Act of 2023 amended the Federal Food, Drug, and Cosmetic Act to include new and significant medical device security provisions. Its impact on medical device cybersecurity in general and specifically on FDA's cybersecurity guidance was still playing out at the time of this writing. Explicit cybersecurity requirements are now enforceable under FDA's statutory authority [17].

- *"The FDA just doesn't understand our particular situation/device."* A less common form of the hubris of the aforementioned "senior executive"-type can be found in companies that just will not accept that the FDA has the right to regulate their medical device and thus enforce patient safety over the company's profitability and/or continued existence. While it is less common, it is not as rare as most of us would suppose. Now with the addition of cybersecurity to the FD&C Act (Section 524B), the FDA is mandated to define and enforce medical device cybersecurity, so this is going to be a very difficult position for any executive to adopt.

  This manager believes that they (or the company's lawyers) will successfully argue for a special exemption from the law due to the uniqueness of their device or company. What he or she does not understand is that (a) this simply is not true, and (b) security activities are now part of the "refuse to accept" list, or prescreening criteria, for all premarket submission applications to the FDA. What this means is that the company that refuses to take cybersecurity requirements seriously will never get to make its case at all.
- *"Our legal risk is lower if we don't take responsibility for security."* This argument was presented to one of us by an MDM's legal department. They laid out their rationale that if they implemented security processes and technology, they would be taking on responsibility for security. Therefore, even if it would be much less likely that a damaging security incident would occur, the MDM would now be legally liable if it did, since they are responsible for including the security features. Hence, the MDM would be better off not implementing security to reduce their legal liability.

  What this twisted thinking fails to take into account is current regulations in most markets worldwide, which declare that the MDM already has the responsibility to secure their device. Lack of security in no way limits or avoids that responsibility; it simply ensures that the MDM will be denied regulatory approval and market entry.

This has been further clarified through recent legal and regulatory developments, for example, NIS2 in Europe or in the United States through the National Cybersecurity Strategy (NCS, March 2023 [18]) and National Cybersecurity Strategy Implementation Plan (NCSIP, July 2023 [19]). Collectively, these are pushing cybersecurity responsibility from the user and operator towards the manufacturer of devices and software by introducing a "secure by design" and "secure by default" approach to product security. Specifically of note is the Consolidated Appropriations Act of 2023 (H.R.2617), which requires manufacturers' FDA submissions to include plans for postmarket monitoring, postmarket management (updates and patches), and SBOM, and, in general, demonstrate compliance with FDA cybersecurity requirements, such as following a secure development lifecycle and taking a risk-based approach to security controls and testing [20]. Under the Omnibus Act, the FDA's implicit authority has been elevated to an explicit authority and oversight on medical device cybersecurity, a shift that is nothing short of significant.

- *"We agree with what you are telling us about cybersecurity, but we still don't see the business case."* Again, a response by an MDM to one of us, stubbornly continuing in denial of the regulatory environment, threat landscape, and market opportunity that we have outlined in this book. The denial may be baseless, but in some places, it persists in the face of all reason. However, we do hope that growing customer demand for security and increasing regulatory and reputational risks will eventually make this argument go away.

13.3 In Closing

Our hope is that you can use some of these examples to help improve the product security posture at your organization, but if that does not happen, we hope they at least cheer you up! Although we are ending this chapter and book on a lighthearted note, these examples show that we are still far from cybersecurity for medical devices being generally accepted as a business priority and a must-have deliverable that requires support across the entire value chain. Only a mature, well thought-through, and carefully implemented secure total lifecycle process can accomplish this.

We and our cocontributors have spent decades of our careers on this topic and have seen regulators, device manufacturers, security researchers, and healthcare delivery organizations take significant steps forward, but, as you were always told as a kid on a long car ride, "we are not there yet." Not by a long way. However, we do need to recognize that the industry at large has made significant progress between the publication of the first edition of this book in 2020 and now. We hope that this second edition will help all of us to stay on track and keep up the good work. Our aim and desire is not to contribute to the headline-driven "The sky is falling!" view of the problem. Even though some news outlets and certain security researchers try to convey that we are in a world-ending crisis, we do not believe that we should panic. At the same time, the scope of the problem and risk to patient safety and national health systems dictate that we do need to proceed with a sense of urgency and take a systematic and deliberate approach to address this challenge, as outlined in this book.

One of the many things that have changed since the first edition of this book is that we now know with undeniable certainty of deaths attributable to cyberattacks [21]. In addition to these relatively straightforward and notable incidents, Josh Corman and his research team at CISA led an excellent, but sobering, study that correlates delays caused by cyberattacks (such as ransomware) to increased death rates among patients [22]. While we have long suspected this to be true, we now have definitive proof [23–25].

As we were finalizing the first edition of this book and the COVID pandemic was just getting started, we anticipated that "the world faces a health crisis such as we have not seen in a long time. The resulting changes to our healthcare system will accelerate the digitalization of care delivery and will require an even stronger focus on cybersecurity." Now, with the COVID pandemic phase behind and as we have entered its endemic phase, our past prediction appears well-validated. So we will add to it and say here that hopefully all of us have learned a valuable lesson with that pandemic and are taking appropriate steps to prepare for the next calamity. We have reached yet another critical juncture in human history. We have the knowledge and ability to prevent human suffering on a wide scale. Will we do the right thing and make the future a better place for our children and their children? Or will we continue to argue, deflect, and distract to obscure the self-destructive path that so many would have us follow in the name of short-term objectives?

We hope that, as a global people, we find our better natures and secure our future. Through this book, we are contributing what we can (little as it may be), so we would like to end this edition with the same phrase that ended the first edition:

We hope that this book will help clarify the right course and further our progress on the path to more secure devices.

References

[1] Zetter, K., "Code Kept Secret for Years Reveals Its Flaw—a Backdoor," *Wired*, July 24, 2023, https://www.wired.com/story/tetra-radio-encryption-backdoor/.

[2] Loshin, P., "Wired Equivalent Privacy (WEP)," *TechTarget*, August 2021, https://www.techtarget.com/searchsecurity/definition/Wired-Equivalent-Privacy.

[3] GeeksforGeeks, "Caesar Cipher in Cryptography," May 11, 2023, https://www.geeksforgeeks.org/caesar-cipher-in-cryptography/.

[4] https://www.fda.gov/medical-devices/how-study-and-market-your-device/estar-program.

[5] Freescale Semiconductor, "Implementing Data Whitening and CRC Verification in Software in Kinetis KW01 Microcontrollers," July 2015. https://www.nxp.com/docs/en/application-note/AN5070.pdf.

[6] Benchoff, B., "Shmoocon 2017: A Simple Tool for Reverse Engineering RF," *Hackaday*, January 15, 2017 https://hackaday.com/2017/01/15/shmoocon-2017-a-simple-tool-for-reverse-engineering-rf/.

[7] https://www.gnuradio.org/.

[8] Pasham, N., "RF Reverse Engineering Has Become Trivial — Thanks to the 'Opensource SDR' Movement," *Medium*, October 10, 2018, https://medium.com/@nihal.pasham/rf-reverse-engineering-has-become-trivial-thanks-to-the-opensource-sdr-movement-d1f9216f2f04.

[9] Tania, "Reverse Engineering a Wireless Home Alarm," *PandwaRF*, August 15, 2019, https://pandwarf.com/news/reverse-engineering-a-wireless-home-alarm/.

[10] "Reverse Engineering Digital RF Signals the Easy Way with DSpectrum," September 22, 2016, https://www.rtl-sdr.com/reverse-engineering-digital-rf-signals-the-easy-way-with-dspectrum/.

[11] Bode, K., "Researchers Find 'Anonymized' Data Is Even Less Anonymous Than We Thought," *Vice*, February 3, 2020, https://www.vice.com/en_us/article/dygy8k/researchers-find-anonymized-data-is-even-less-anonymous-than-we-thought.

[12] Ehrenkranz, M., "Researchers Reveal That Anonymized Data Is Easy to Reverse Engineer," *Gizmodo*, July 23, 2019, https://gizmodo.com/researchers-reveal-that-anonymized-data-is-easy-to-reve-1836629166.

[13] Campbell-Dollaghan, K., "Sorry, Your Data Can Still Be Identified Even If It's Anonymized," *Fast Company*, December 10, 2018, https://www.fastcompany.com/90278465/sorry-your-data-can-still-be-identified-even-its-anonymized.

[14] Thompson, S., and C. Warzel, "How to Track President Trump," *The New York Times*, December 20, 2019, https://www.nytimes.com/interactive/2019/12/20/opinion/location-data-national-security.html.

[15] U.S. Department of Health & Human Services, "Guidance Regarding Methods for De-identification of Protected Health Information in Accordance with the Health Insurance Portability and Accountability Act (HIPAA) Privacy Rule," November 26, 2012, https://www.hhs.gov/hipaa/for-professionals/privacy/special-topics/de-identification/index.html#standard.

[16] Gates, C., "HSCC's Model Contract Language Template Represents a Win-Win Cybersecurity Solution for HDOs and MDMs," *AAMI Array*, April 13, 2022, https://array.aami.org/content/news/hscc-s-model-contract-language-template-represents-win-win-cybersecurity-solution-hdos.

[17] McKeon, J., "Key Ways to Prepare for Revamped Medical Device Security Requirements," *Health IT Security*, June 1, 2023, https://healthitsecurity.com/features/key-ways-to-prepare-for-revamped-medical-device-security-requirements.

[18] The White House, "National Cybersecurity Strategy," March 1, 2023, https://www.whitehouse.gov/wp-content/uploads/2023/03/National-Cybersecurity-Strategy-2023.pdf.

[19] The White House, "National Cybersecurity Implementation Plan," July 2023, https://www.whitehouse.gov/wp-content/uploads/2023/07/National-Cybersecurity-Strategy-Implementation-Plan-WH.gov_.pdf.

[20] Connolly, G. E., "H.R.2617 - Consolidated Appropriations Act, 2023," *117th Congress*, December 29, 2022, https://www.congress.gov/bill/117th-congress/house-bill/2617/text.

[21] Miller, M., "The Mounting Death Toll of Hospital Cyberattacks," *Politico*, December 28, 2022, https://www.politico.com/news/2022/12/28/cyberattacks-u-s-hospitals-00075638.

[22] Corman, J., "Statement of Joshua Corman," Senate Health, Education, Labor & Pensions Committee (HELP) Full Committee Hearing, "Cybersecurity in the Health and Education Sectors," May 18, 2022, https://www.help.senate.gov/imo/media/doc/Corman.pdf.

[23] McGlace, C., H. Neprash, and S. Nikpay, "Hacked to Pieces? The Effects of Ransomware Attacks on Hospitals and Patients," *SSRN*, October 4, 2023, https://papers.ssrn.com/sol3/papers.cfm?abstract_id=4579292.

[24] Pham, T., et al., "Ransomware Cyberattack Associated with Cardiac Arrest Incidence and Outcomes at Untargeted, Adjacent Hospitals," NIH National Library of Medicine, April 10, 2024, https://pubmed.ncbi.nlm.nih.gov/38605720.

[25] Choi, S., E. Johnson, and C. Lehmann, "Data Breach Remediation Efforts and Their Implications for Hospital Quality," NIH National Library of Medicine, October 2019, https://pubmed.ncbi.nlm.nih.gov/31506956.

Afterword to the Second Edition

When I first began working on the intersection of cybersecurity and medical devices in 2014, I would regularly be asked, "Who would ever hack a medical device? Is that even possible?"

I was far from the first person concerned about this possibility who began looking at technical opportunities for improvement in this space. People like Kevin Fu and Ken Hoyme had begun their work years prior. However, even as recently as 2014, the notion that devices that treat and diagnose disease should be resilient to cybersecurity attacks was still a novel one for most in our industry. To see a book of this quality, with the number of distinguished collaborators that Axel, Christopher, and Jason have assembled, is proof that the time for this topic has arrived.

The healthcare industry has a reputation for being a decade or more behind other fields when it comes to technology adoption. However, due in part to the FDA's inclusion of cybersecurity in medical device regulations, healthcare has been propelled to the forefront of the "Shift Left" approach to cybersecurity. It is not a sustainable model to expect community hospitals to dedicate precious resources securing drug infusion pumps instead of actually treating patients. Instead, we need the products sold into hospitals to be secure by design, ensuring that these devices are safe and effective for years to come. Reversing the normal state of affairs, industries such as automotive, industrial control systems, and the Internet of Things will be watching and learning from the healthcare industry as we shift a portion of the burden of cybersecurity from our users to the manufacturers.

In the years since the first edition of this book was written, countless hospitals have been targeted in ransomware attacks, and there are probably only a few medical device manufacturers that have not received at least a few questions about the cybersecurity posture of their products from both hospitals and regulators. Device manufacturers now have directors of product security and have formalized processes to ensure that devices that they release are as secure as possible. We have even seen device manufacturers highlight their product's cybersecurity features prominently in marketing materials meant for their customers.

This market now needs thousands of cybersecurity experts who understand the need to balance security with usability and clinical functionality. These new professionals will come from mixed backgrounds in computer science, IT and IT security, biomedical engineering, regulatory, and even medical backgrounds. I encourage those readers who are early in their careers to map out their path forward and consider what they can contribute to this critical field.

I thank the authors of this book for providing a framework for our industry to use in evaluating the future of medical device cybersecurity, the clinicians at hospitals who stress the importance of security to their vendors, the academics studying the realm of security possibilities, the lawmakers and regulators who have stepped in where the free market has fallen short, and the medical device manufacturers for putting patient safety first.

The progress our industry has made in the past 10 years in securing the life-saving technologies on which we and our loved ones rely on is nothing short of staggering. I am looking forward to seeing all of the ways our community will continue to innovate in the years to come, and I am confident that this book will be a significant contributor to this progress.

Mike Kijewski
Founder and CEO, Medcrypt
October 2024

Additional Resources

General

US FDA Cybersecurity Landing Page, https://www.fda.gov/medical-devices/digital-health-center-excellence/cybersecurity.

CISA, "Cyber Resource Hub," https://www.cisa.gov/cyber-resource-hub.

CISA, "Services Catalog," https://www.cisa.gov/resources-tools/services.

CyBoK, "The Cyber Security Body of Knowledge v1.1," https://www.cybok.org/media/downloads/CyBOK_v1.1.0.pdf.

CyBOK, "Knowledge Base," https://www.cybok.org/knowledgebase1_1/.

IEEE Computer Society, "Software Engineering Body of Knowledge (SWEBOK)," https://www.computer.org/education/bodies-of-knowledge/software-engineering.

SEBoK Editorial Board, "The Guide to the Systems Engineering Body of Knowledge (SEBoK)," www.sebokwiki.org.

DHS, "Making the Business Case for Security," https://www.cisa.gov/sites/default/files/2023-03/isc_making_a_business_case_for_security_2023_edition_508c.pdf.

CISA, "Free Cybersecurity Services and Tools," https://www.cisa.gov/resources-tools/resources/free-cybersecurity-services-and-tools.

DoD, "Cybersecurity Resource and Reference Guide," February 2022, https://dodcio.defense.gov/Portals/0/Documents/Library/CSResourceReference-Guide.pdf.

HPH Sector Coordinating Council (HSCC) Joint Cyber WG (JCWG), "Health Care and Public Health Sector Cybersecurity Framework Implementation Guide, V2," March 2023, https://aspr.hhs.gov/cip/hph-cybersecurity-framework-implementation-guide/Pages/Acknowledgements.aspx.

Eidam, B., and Global Cybersecurity Forum, "The 116 Best Cybersecurity Tools and Tactics (2021)," 2021, https://www.linkedin.com/posts/julienprovenzano_116-best-cybersecurity-tools-tactics-activity-6828246738214748160-PuQ9/ or https://drive.google.com/file/d/1LD5d2G7-aifRcF2UON7z0oSi7u6gj6cX/view?trk=public_post_comment-text.

Ray, A., *Cybersecurity for Connected Medical Devices*, November 2021, https://shop.elsevier.com/books/cybersecurity-for-connected-medical-devices/ray/978-0-12-818262-8. (The second book on medical device cybersecurity.)

Education and Enablement

CISA, "Cybersecurity Workforce Training Guide," August 2021, https://www.cisa.gov/resources-tools/resources/cybersecurity-workforce-training-guide.

CISA, "CISA Tabletop Exercise Packages," https://www.cisa.gov/resources-tools/services/cisa-tabletop-exercise-packages.

NIST SP 800-181, "National Initiative for Cybersecurity Education (NICE) Cybersecurity Workforce Framework," November 2020, https://csrc.nist.gov/pubs/sp/800/181/r1/final.

ENISA, "European Cybersecurity Skills Framework (ECSF)," https://www.enisa.europa.eu/topics/education/european-cybersecurity-skills-framework.

Industry Organizations and Public-Private Partnerships

Health-ISAC, https://h-isac.org/.

MedISAO, https://www.medisao.com/.

Healthcare and Public Health Sector Coordinating Council (HSCC), https://healthsectorcouncil.org/.

Archimedes Center for Medical Device Security, https://www.secure-medicine.org/.

Association for the Advancement of Medical Instrumentation (AAMI) Cybersecurity, https://www.aami.org/HTM/htm-cybersecurity-page.

AdvaMed, https://www.advamed.org/.

Infragard Cyber Health Working Group, https://www.infragardncr.org/cyberhealthworkinggroup.

NIST National Cybersecurity Center of Excellence (NCCoE), https://www.nccoe.nist.gov/healthcare.

I Am the Cavalry, https://iamthecavalry.org/.

Frameworks

Health Sector Coordinating Council, "Medical Device and Health IT Joint Security Plan version 2 (JSP2)," March 2024, https://healthsectorcouncil.org/jsp2/.

NIST, "Framework for Improving Critical Infrastructure Cybersecurity, V2," February 2024, https://www.nist.gov/news-events/news/2024/02/nist-releases-version-20-landmark-cybersecurity-framework.

NIST, "Privacy Framework: A Tool for Improving Privacy Through Enterprise Risk Management," January 2020, https://www.nist.gov/privacy-framework.

NIST SP 800-12, "An Introduction to Information Security," June 2017, https://nvlpubs.nist.gov/nistpubs/SpecialPublications/NIST.SP.800-12r1.pdf.

NIST SP 800-37, "Risk Management Framework for Information Systems and Organizations: A System Life Cycle Approach for Security and Privacy," December 2018, https://csrc.nist.gov/pubs/sp/800/37/r2/final.

NIST SP 1500-201, "Framework for Cyber-Physical Systems: Volume 1, Overview," June 2017, https://www.nist.gov/publications/framework-cyber-physical-systems-volume-1-overview.

UK National Cyber Security Centre, "Cyber Assessment Framework V3.1," April 2022, https://www.ncsc.gov.uk/files/NCSC-CAF-v3-1.pdf.

IEEE, "2933 - Standard for Clinical Internet of Things (IoT) Data and Device Interoperability with TIPPSS – Trust, Identity, Privacy, Protection, Safety, Security," https://standards.ieee.org/ieee/2933/7592/.

CISA, "Secure by Design," https://www.cisa.gov/securebydesign.

Lifecycle Practices

IEC 81001-5-1:2021, "Health Software and Health IT Systems Safety, Effectiveness and Security, Part 5-1: Security Activities in the Product Life Cycle," 2021, https://www.iso.org/standard/76097.html.

H-ISAC, "Medical Device Cybersecurity Lifecycle Management," October 2020, https://h-isac.org/medical-device-cybersecurity-lifecycle-management/.

CISA, "Shifting the Balance of Cybersecurity Risk: Security-by-Design and Default Principles," April 2023, https://www.cisa.gov/news-events/alerts/2023/04/13/shifting-balance-cybersecurity-risk-security-design-and-default-principles.

CISA, "Zero Trust Maturity Model," https://www.cisa.gov/zero-trust-maturity-model.

CISA, "Cloud Security Technical Reference Architecture," https://www.cisa.gov/resources-tools/resources/cloud-security-technical-reference-architecture.

NIST SP 1800-35, "Implementing a Zero Trust Architecture," December 2022, https://www.nccoe.nist.gov/projects/implementing-zero-trust-architecture.

NIST SP 800-82, "Guide to Operational Technology (OT) Security," April 2022, https://csrc.nist.gov/pubs/sp/800/82/r3/ipd.

NIST SP 800-207, "Zero Trust Architecture," August 2020, https://csrc.nist.gov/pubs/sp/800/207/final.

NIST SP 800-218, "Secure Software Development Framework (SSDF)," February 2022, https://csrc.nist.gov/pubs/sp/800/218/final.

NIST SP 1500-201, "Framework for Cyber-Physical Systems, Vol 1, Overview," June 2017, https://www.nist.gov/publications/framework-cyber-physical-systems-volume-1-overview.

NIST SP 1800-36, "Trusted Internet of Things (IoT) Device Network-Layer Onboarding and Lifecycle Management," December 2022, https://www.nccoe.nist.gov/projects/trusted-iot-device-network-layer-onboarding-and-lifecycle-management.

NIST SP 800-53, "Security and Privacy Controls for Information Systems and Organizations," September 2020, https://csrc.nist.gov/pubs/sp/800/53/r5/upd1/final.

NIST SP 800-160, "Engineering Trustworthy Secure Systems," November 2022, https://csrc.nist.gov/pubs/sp/800/160/v1/r1/final.

NIST SP 800-210, "General Access Control Guidances for Cloud Systems," July 2020, https://csrc.nist.gov/pubs/sp/800/210/final.

NIST TIR 8259 Series Providing Guidance for Manufacturers of IoT Devices, https://www.nist.gov/itl/applied-cybersecurity/nist-cybersecurity-iot-program/nistir-8259-series.

NIST TIR 8228, "Considerations for Managing Internet of Things (IoT) Cybersecurity and Privacy Risks," June 2019, https://csrc.nist.gov/pubs/ir/8228/final.

NIST IR 8183, "Cybersecurity Framework Version 1.1 – Manufacturing Profile," October 2020, https://csrc.nist.gov/pubs/ir/8183/r1/final.

NIST IR 8320, "Hardware-Enabled Security," May 2022, https://csrc.nist.gov/pubs/ir/8320/final.

NIST IR 8425, "Profile of the IoT Core Baseline for Consumer IoT Products," September 2022, https://csrc.nist.gov/pubs/ir/8425/final.

NIST Cybersecurity White Paper, "Establishing Confidence in IoT Device Security – How Do We Get There?" (Draft), May 2021, https://csrc.nist.gov/pubs/cswp/18/establishing-confidence-in-iot-device-security/ipd.

Shostack, A., *Threat Modeling Designed for Security*, February 2014, https://shostack.org/books/threat-modeling-book.

MDIC, "Playbook for Medical Device Cybersecurity Threat Modeling," October 2021, https://mdic.org/project/medical-device-cybersecurity-threat-modeling/.

Procurement

NEMA, "Manufacturer Disclosure Statement for Medical Device Security (MDS^2)," https://www.nema.org/standards/view/manufacturer-disclosure-statement-for-medical-device-security.

HSCC, "Model Contract-Language for Medtech Cybersecurity (MC2)," March 2022, https://healthsectorcouncil.org/model-contract-language-for-medtech-cybersecurity-mc2/.

ENISA, "Procurement Guidelines for Cybersecurity in Hospitals," February 2020, https://www.enisa.europa.eu/publications/good-practices-for-the-security-of-healthcare-services.

CISA, "Internet of Things Security Acquisition Guidance," https://www.cisa.gov/sites/default/files/publications/20_0204_cisa_sed_internet_of_things_acquisition_guidance_final_508.pdf.

CISA, "Software Bill of Materials (SBOM) Sharing Lifecyle Report," April 2023, https://www.cisa.gov/resources-tools/resources/software-bill-materials-sbom-sharing-lifecycle-report.

CISA, "Securing the Software Supply Chain," (various reports), https://www.cisa.gov/topics/information-communications-technology-supply-chain-security.

HSCC, "Health Industry Cybersecurity-Managing Legacy Technology Security (HIC-MaLTS)," March 2023, https://healthsectorcouncil.org/legacy-tech-security/.

Supply Chain Risk Management

US FDA, "Cybersecurity for Networked Medical Devices Containing Off-the-Shelf (OTS) Software Guidance for Industry," January 2005, https://www.fda.gov/regulatory-information/search-fda-guidance-documents/cybersecurity-networked-medical-devices-containing-shelf-ots-software.

US FDA, "Discussion Paper: Strengthening Cybersecurity Practices Associated with Servicing of Medical Devices: Challenges and Opportunities," June 2021, https://www.fda.gov/medical-devices/quality-and-compliance-medical-devices/discussion-paper-strengthening-cybersecurity-practices-associated-servicing-medical-devices.

Hughes, C., T. Turner, and S. Springett, "Software Transparency: Supply Chain Security in an Era of a Software-Driven Society," June 2023, https://www.amazon.com/Software-Transparency-Security-Software-Driven-Society/dp/1394158483.

CISA, "Information and Communications Technology Supply Chain Security," https://www.cisa.gov/topics/information-communications-technology-supply-chain-security.

CISA, "Software Bill of Materials (SBOM) Initiative," https://www.cisa.gov/sbom.

CISA, "Software Bill of Materials (SBOM) Sharing Lifecycle Report," April 2023, https://www.cisa.gov/resources-tools/resources/software-bill-materials-sbom-sharing-lifecycle-report.

NIST SP 800-161, "Cybersecurity Supply Chain Risk Management Practices for Systems and Organizations," May 2022, https://csrc.nist.gov/pubs/sp/800/161/r1/final.

NIST TIR 8276, "Key Practices in Cyber Supply Chain Risk Management," February 2021, https://csrc.nist.gov/News/2021/nistir-8276-key-practices-in-c-scrm.

NIST, "Case Studies in Cyber Supply Chain Risk Management – Summary of Findings," February 2020, https://nvlpubs.nist.gov/nistpubs/CSWP/NIST.CSWP.02042020-1.pdf.

NTIA SBOM page, https://www.ntia.gov/page/software-bill-materials.

NTIA, "The Minimum Elements For a Software Bill of Materials (SBOM)," July 2021, https://www.ntia.gov/report/2021/minimum-elements-software-bill-materials-sbom.

ENISA, "Good Practices for Supply Chain Management," June 2023, https://www.enisa.europa.eu/publications/good-practices-for-supply-chain-cybersecurity.

University of Minnesota, Center for Medical Device Cybersecurity, "Uncovering Systemic Cybersecurity Challenges along the Medical Device Supply Chain," 2023, https://cse.umn.edu/cmdc/uncovering-systemic-cybersecurity-challenges-along-medical-device-supply-chain.

CISA, "Secure Software Development Attestation Form," https://www.cisa.gov/resources-tools/resources/secure-software-development-attestation-form.

Risk and Vulnerability Management

NIST, "National Vulnerability Database," https://nvd.nist.gov/vuln.

FIRST, "Vulnerability Database Catalog," https://www.first.org/global/sigs/vrdx/vdb-catalog.

ENISA, "Coordinated Vulnerability Disclosure Policies in the EU," https://www.enisa.europa.eu/publications/coordinated-vulnerability-disclosure-policies-in-the-eu.

Roytman, M., and E. Bellis, *Modern Vulnerability Management*, Norwood, MA: Artech House, 2022, https://us.artechhouse.com/Modern-Vulnerability-Management-Predictive-Cybersecurity-P2278.aspx.

MITRE, "Rubric for Applying CVSS to Medical Devices," October 2020, https://www.mitre.org/news-insights/publication/rubric-applying-cvss-medical-devices.

NIST SP 800-216, "Recommendations for Federal Vulnerability Disclosure Guidelines," May 2023, https://csrc.nist.gov/pubs/sp/800/216/final.

NIST IR 8286, "Integrating Cybersecurity and Enterprise Risk Management (ERM)," October 2020, https://csrc.nist.gov/pubs/ir/8286/final.

NIST TIR 7946, "CVSS Implementation Guidance," April 2014, https://csrc.nist.gov/News/2014/NIST-Interagency-Report-7946,-CVSS-Implementation.

NTIA, "Improving Cybersecurity Through Enhanced Vulnerability Disclosure," https://www.ntia.gov/blog/improving-cybersecurity-through-enhanced-vulnerability-disclosure.

CERT, "Guide to Coordinated Vulnerability Disclosure," August 2017, https://resources.sei.cmu.edu/asset_files/SpecialReport/2017_003_001_503340.pdf.

IoT Security Foundation, "Vulnerability Disclosure Best Practices Guide, Release 2.0," September 2021, https://www.iotsecurityfoundation.org/wp-content/uploads/2021/09/IoTSF-Vulnerability-Disclosure-Best-Practice-Guidelines-Release-2.0.pdf.

ISO/IEC 29147:2018, "Information Technology – Security Techniques, Vulnerability Disclosure," 2018, https://www.iso.org/standard/72311.html.

ISO/IEC 30111:2019, "Information Technology – Security Techniques, Vulnerability Handling Processes," 2019, https://www.iso.org/standard/69725.html.

Incident Management and Forensics

MITRE, "Medical Device Cybersecurity Regional Incident Preparedness and Response Playbook," November 2022, https://www.mitre.org/news-insights/publication/medical-device-cybersecurity-regional-incident-preparedness-and-response.

CISA, "Cybersecurity Incident & Vulnerability Response Playbooks," https://www.cisa.gov/sites/default/files/publications/Federal_Government_Cybersecurity_Incident_and_Vulnerability_Response_Playbooks_508C.pdf.

NIST SP 800-61, "Computer Security Incident Handling Guide," August 2012, https://csrc.nist.gov/pubs/sp/800/61/r2/final.

NIST NCCoE, "Responding to and Recovering from a Cyber Attack – Cybersecurity for the Manufacturing Sector," December 2022, https://www.nccoe.nist.gov/manufacturing/responding-and-recovering-cyber-attack.

NISTTIR 8428, "Digital Forensics and Incident Response (DFIR) Framework for Operational Technology (OT)," June 2022, https://www.nist.gov/publications/digital-forensics-and-incident-response-dfir-framework-operational-technology-ot.

FEMA, "Planning Considerations for Cyber Incidents: Guidance for Emergency Managers," November 2023, https://www.fema.gov/sites/default/files/documents/fema_planning-considerations-cyber-incidents_2023.pdf.

Patching and OTA

Gates, C., "Updating Medical Device Software in the Field," June 2023, https://www.mpo-mag.com/contents/view_columns/2023-06-06/updating-medical-device-software-in-the-field/.

NIST SP 800-40, "Guide to Enterprise Patch Management Planning: Preventive Maintenance for Technology," April 2022, https://csrc.nist.gov/pubs/sp/800/40/r4/final.

NIST SP 1800-31, "Improving Enterprise Patching for General IT Systems: Utilizing Existing Tools and Performing Processes in Better Ways," April 2022, https://csrc.nist.gov/pubs/sp/1800/31/final.

NIST Cybersecurity White Paper, "Security Considerations for Code Signing," January 2018, https://csrc.nist.gov/pubs/cswp/5/security-considerations-for-code-signing/final.

NTIA, "Voluntary Framework for Enhancing Update Process Security," https://www.ntia.gov/files/ntia/publications/ntia_iot_security_update_framework.pdf.

Cryptography

NIST SP 800-108, "Recommendation for Key Derivation Using Pseudorandom Functions," August 2022, https://csrc.nist.gov/pubs/sp/800/108/r1/final.

NIST SP 800-130, "A Framework for Designing Cryptographic Key Management Systems," August 2013, https://csrc.nist.gov/pubs/sp/800/130/final.

NIST SP 1800-38 (draft), "Migration to Post-Quantum Cryptography: Preparation for Considering the Implementation and Adoption of Quantum Safe Cryptography," April 2023, https://csrc.nist.gov/pubs/sp/1800/38/iprd.

Artificial Intelligence and Machine Learning

NIST AI 100-1, "Artificial Intelligence Risk Management Framework (AI RMF 1.0)," January 2023, https://www.nist.gov/publications/artificial-intelligence-risk-management-framework-ai-rmf-10.

NIST AI 100-2, "Adversarial Machine Learning," March 2023, https://csrc.nist.gov/pubs/ai/100/2/e2023/ipd.

Google, "Google's Secure AI Framework (SAIF)," https://safety.google/cybersecurity-advancements/saif/.

FDA, "Artificial Intelligence Program: Research on AI/ML – Based Medical Devices," https://www.fda.gov/medical-devices/medical-device-regulatory-science-research-programs-conducted-osel/artificial-intelligence-program-research-aiml-based-medical-devices.

Testing

Chantzis F., et al., "Practical IoT Hacking: The Definitive Guide to Attacking the Internet of Things," April 2021, https://www.amazon.com/dp/1718500904?ref_=cm_sw_r_cp_ud_dp_B673PC76EABS8NF75C9Z.

NIST SP 1800-34, "Validating the Integrity of Computing Devices," December 2022, https://csrc.nist.gov/pubs/sp/1800/34/final.

NIST, "Guidelines on Minimum Standards for Developer Verification of Software," July 2021, https://www.nist.gov/publications/guidelines-minimum-standards-developer-verification-software.

Selected Topics

U.S. FDA, "Best Practices for Communicating Cybersecurity Vulnerabilities to Patients," October 2021, https://www.fda.gov/about-fda/cdrh-patient-science-and-engagement-program/best-practices-communicating-cybersecurity-vulnerabilities-patients.

IMDRF, "Principles and Practices for the Cybersecurity of Legacy Medical Devices," IMDRF/CYBER WG/N70FINAL:2023, https://www.imdrf.org/documents/principles-and-practices-cybersecurity-legacy-medical-devices.

NIST SP 800-121, "Guide to Bluetooth Security," January 2022, https://csrc.nist.gov/pubs/sp/800/121/r2/upd1/final.

NIST SP 800-55, "Performance Measurement Guide for Information Security," November 2022, https://csrc.nist.gov/pubs/sp/800/55/r2/iwd.

NIST SP 800-88, "Guidelines for Media Sanitization," December 2014, https://csrc.nist.gov/pubs/sp/800/88/r1/final.

Mission Secure, "A Comprehensive Guide to Operational Technology (OT) Cybersecurity," https://www.missionsecure.com/resources/comprehensive-guide-to-operational-technology-security-ebook.

ETSI, "Consumer IoT Security," https://www.etsi.org/technologies/consumer-iot-security.

U.S. Department of Energy, "Cybersecurity Capability Maturity Moel (C2M2)," June 2022, https://www.energy.gov/ceser/cybersecurity-capability-maturity-model-c2m2.

SALT, "API Security Best Practices," June 2022, https://salt.security/blog/api-security-best-practices.

MDM Cybersecurity Website Examples

Philips Product Security, https://www.usa.philips.com/healthcare/about/customer-support/product-security.

Becton Dickinson Cybersecurity Trust Center, https://www.bd.com/en-us/about-bd/cybersecurity.

Medtronic, https://global.medtronic.com/xg-en/product-security.html.

Siemens Healthineers, https://www.siemens-healthineers.com/en-us/support-documentation/cybersecurity.

Glossary

AAMI Association for the Advancement of Medical Instrumentation, https://www.aami.org/.

Artifact "A piece of evidence, such as text or a reference to a resource, that is submitted to support a response to a question." [1] Artifacts are generated for and maintained in the device's Design History File (DHF) and are used to support such activities as regulatory submissions, postmarket surveillance, vulnerability management, and incident response. Artifacts may be physical or electronic in nature.

ASL Approved Supplier List

Attack Surface "The set of ways in which an adversary can [gain access] to a system and/or execute a threat." [2] "Physical and functional interfaces of a system that can be accessed and, therefore, potentially exploited by an attacker." [3]

Attack Vector Also referred to as an Attack Path. "The steps that an adversary takes or may take to plan, prepare for, and execute an attack." [2]

Authentication The process of "verifying the identity of a user, process, or device." [1]

Authorization "The act of verifying the identity of a user, process, or device as a prerequisite to allowing access to the device, its data, information, or systems, or provision of assurance that a claimed characteristic of an entity is correct." [4]

Availability "The property of data, information, and information systems to be accessible and usable on a timely basis in the expected manner (i.e., the assurance that information will be available when needed)." [4]

BOM Bill of materials

Business risk Encompasses potential negative outcomes such as exposure of IP or trade secrets, loss of reputation or customer relationship and market share, legal repercussions, and financial losses. Excludes risks involving potential harm to device users. *See also* Risk.

CAPA Corrective and Preventive Action

CERT Computer Emergency Response Team, a division of the Carnegie Mellon Software Engineering Institute. CERT is a registered trademark of the SEI. Other CERTs exist (in agreement or cooperation with CERT), for example, US-CERT is the national computer security incident response team (CSIRT) for the United States of America.

CERT/CC CERT Coordination Center

CISA Cybersecurity & Infrastructure Security Agency, a division of the United States Department of Homeland Security.

Compensating Control "A safeguard or countermeasure deployed, in lieu of, or in the absence of controls designed in by a device manufacturer. These controls are external to the device design, configurable in the field, employed by a user, and provide supplementary or comparable cyber protection for a medical device." [4]

Confidentiality "The property of data, information, or system structures to be accessible only to authorized persons and entities and are processed at authorized times and in the authorized manner, thereby helping ensure data and system security. Confidentiality provides the assurance that no unauthorized users (i.e., only trusted users) have access to the data, information, or system structures." [4]

COTS Commercial off-the-shelf

CSRC Computer Security Resource Center, a division of NIST.

CVD Coordinated vulnerability disclosure. *See also* Disclosure, Coordinated.

DAST Dynamic Application Security Testing

Defense-in-Depth "The application of multiple countermeasures in a layered or stepwise manner to achieve security objectives. The methodology involves layering heterogeneous security technologies in the common attack vectors to ensure that attacks missed by one technology are caught by another." [1] "Approach to defend the system against any particular attack using several independent Methods... DEFENSE-IN-DEPTH implies layers of SECURITY and detection, even on single systems, and provides the following features:

- Is based on the idea that any one layer of protection, can and probably will be defeated;
- Attackers are faced with breaking through or bypassing each layer without being detected;
- A flaw in one layer can be mitigated by capabilities in other layers;
- System SECURITY becomes a set of layers within the overall network SECURITY;
- Each layer is autonomous and not rely on the same functionality nor have the same failure modes as the other layers." [3]

DFU Device firmware update. *See* Update

DHF Design history file

DHS United States Department of Homeland Security

Disclosure, coordinated Disclosure scenario where a reporter (e.g., a security researcher) informs an MDM either directly or via a coordinator (a trusted individual or an organization such as an ISAO) about a newly discovered vulnerability in a software component or finished product. The key to the process is a cooperative approach between reporter and MDM and coordinated disclosure (typically delayed to allow development of a patch) to the public. *See also* ISO/IEC 29147:2018 "Information Technology — Security Techniques — Vulnerability Disclosure" (note that responsible vulnerability disclosure and coordinated vulnerability disclosure are typically synonyms but the latter is the more current use).

Disclosure, full Disclosure scenario where the reporter publishes the analysis of the vulnerability as quickly as possible, making the analysis information accessible to everyone without delay or restriction.

Disclosure, non- Disclosure scenario where the reporter does not share the discovered cybersecurity information, except (a) only with the MDM under contract, bug bounty, or NDA; or (b) when shared with the intelligence community as a secret vulnerability for use in a future attack.

Disclosure, responsible Disclosure scenario where the reporter informs the MDM of the vulnerability first, sharing the analysis with nobody besides the MDM until either: (a) the MDM releases remediation to address the vulnerability or initiates a voluntary recall (if warranted); or (b) an agreed-upon period of time has elapsed. (Note that responsible vulnerability disclosure and coordinated vulnerability disclosure are typically synonyms but the latter is the more current use.)

Efficacy The device's ability to provide clinically significant results when used as designed. *See also* 21 CFR Section 860.7(e)(1)

ENISA The European Union Agency for Cybersecurity (originally European Network and Information Security Agency)

EOGS End of Guaranteed Support

EOL End of Life (typically from the MDM perspective, that is, meaning end of production and sale)

EOS End of support/end of service

ETSI European Telecommunication Standards Institute

EU European Union

Event An observable occurrence in an information system or network or attempt to "change the security state of the system (e.g., change access controls, change the security level of a user, change a user password). Also, any [occurrence] that attempts to violate the security policy of the system (e.g., too many logon attempts)." [1]

Exploit "Defined way to breach the security of information systems through some vulnerability." [3]

Failsafe Mode A mode of operation with reduced functionality that maintains the essential performance of the medical device.

FDA US Food and Drug Administration

Fielded A fielded device or device "in the field" and that has received regulatory approval for use by entities external to the MDM (e.g., for trials, demos, or active clinical or home health care use).

Forensics "The practice of gathering, retaining, and analyzing computer-related data for investigative purposes in a manner that maintains the integrity of the data." [1]

FOTA Firmware over-the-air. *See* Update

Guidance Recommended approaches and methodologies for complying with preexisting law, often used in the context of providing additional interpretation and clarification of existing laws or regulations.

H-ISAC Health Information Sharing and Analysis Center

Hardening "A process intended to eliminate a means of attack by patching vulnerabilities and turning off nonessential services. Hardening, when applied to computing, is the practice of reducing a system's vulnerability by reducing its attack surface. Hardening may involve a reduction in attack vectors by culling the pathways or vectors attackers would use (e.g., patching vulnerabilities and turning off nonessential services)." [4]

HDO Healthcare Delivery Organization

HIMSS Health Information Management and Systems Society, https://www.himss.org

HSCC Health Sector Coordinating Council, https://healthsectorcouncil.org/

IDS Intrusion Detection System. Can be appended with H (Host IDS) or N (Network IDS).

IETF Internet Engineering Task Force

IMDRF International Medical Device Regulators Forum, https://www.imdrf.org/

Incident "An occurrence that actually or potentially jeopardizes the confidentiality, integrity, or availability of a system or the information the system processes, stores, or transmits or that constitutes a violation or imminent threat of violation of security policies, security procedures, or acceptable use policies." [1]

Integrity "The property of data, information and software to be accurate and complete and have not been improperly or maliciously modified." [4]

IPS Intrusion Prevention System. Can be appended with H (Host IPS) or N (Network IPS).

ISAC Information Sharing and Analysis Center

ISAO Information Sharing and Analysis Organization

ISO International Standards Organization

Legacy medical device "Medical devices that cannot be reasonably protected against current cybersecurity threats." [5]

Lifecycle "All phases in the life of a medical device, from initial conception to final decommissioning and disposal." [4]

Likelihood "A weighted factor based on a subjective analysis of the probability that a given threat is capable of exploiting a given vulnerability." [1]

MC2 Model Contract Language for Medtech Cybersecurity, an HSCC publication

MDCG Medical Device Coordination Group

MDIC Medical Device Innovation Consortium

MDM Medical Device Manufacturer

MDR Medical Device Regulation (European Union)

MDS^2 Manufacturer Disclosure Statement for Medical Device Security

MFA Multifactor authentication. *See* Authentication

Mitigation Preventive measures intended to reduce the ease of exploitation and/or impact of a risk. Generally, the application of one or more measures to increase the difficulty of an unwanted occurrence and/or lessen its consequences. Implementing appropriate risk-reduction controls is based on risk management priorities and analysis of alternatives.

NIST National Institute of Standards and Technology, a division of the United States Department of Commerce.

NTIA The National Telecommunications and Information Administration, a division of the United States Department of Commerce.

OS Operating system

OTS Off-the-shelf. *See also* COTS

Patch "A 'repair job' for a piece of programming; also known as a 'fix'. A patch is the immediate solution to an identified problem that is provided to users. The patch is not necessarily the best solution for the problem, and the product developers often find a better solution to provide when they package the product for its next release. A patch is usually developed and distributed as a replacement for or an insertion in compiled code (that is, in a binary file or object module). In many operating systems, a special program is provided to manage and track the installation of patches." [4]

PHI Protected health information

PII Personally identifiable information

Postmarket The period of the device lifecycle that begins when the MDM receives regulatory approval (if required) and places its device on the market, and ends on the MDM's declared end-of-support date.

Postmarket surveillance Monitoring activities for possible threats, vulnerabilities, and incidents that potentially impact the device during its post-market period.

Premarket The period of the device lifecycle that encompasses all activities up until the date the MDM has obtained market approval and places its device on the market.

PUF Physically Unclonable Function

Regulation/Regulatory Referring to the legal context, including laws, policies, and overseeing bodies, that apply to a given medical device market.

Remediation "The act of mitigating a vulnerability or a threat." [1]

Reporter "An individual or organization that notifies a vendor or coordinator of a potential vulnerability." [6]

Risk In cybersecurity, the potential for an unwanted or adverse outcome resulting from an incident, event, or occurrence, as determined by the level of difficulty for a particular threat to exploit a particular vulnerability and weighted using the exploit's associated consequences (i.e., severity).

Risk management "The process of identifying, analyzing, assessing, and communicating risk and accepting, avoiding, transferring or controlling it to an acceptable level considering associated costs and benefits of any actions taken. Includes: 1) conducting a risk assessment; 2) implementing strategies to mitigate risks; 3) continuous monitoring of risk over time; and 4) documenting the overall risk management program." [2] "Systematic application of management policies, procedures and practices to the tasks of analysing, evaluating, controlling and monitoring risk." [3]

Root cause The vulnerability or vulnerabilities leveraged in a cybersecurity incident.

Safety "Freedom from unacceptable risk." [3]

SAST Static Application Security Testing

SBOM Software Bill of Materials, "A formal inventory of software components and dependencies, information about those components, and their hierarchical relationships." [4]

Security systems engineering "Systems security engineering is a specialty engineering field strongly related to systems engineering. It applies scientific, engineering, and information assurance principles to deliver trustworthy systems that satisfy stakeholder requirements within their established risk tolerance." [1]

Severity A vulnerability scoring metric intended to represent and rank the total potential impact of a given vulnerability being exploited by a threat.

SOUP Software of Unknown Provenance

Standard "A published statement on a topic specifying the characteristics, usually measurable, that must be satisfied or achieved to comply with the standard." [1]

Supplier "An organization or individual that enters into an agreement with the acquirer or integrator for the supply of a product or service. This includes all suppliers in the supply chain. Includes (i) developers or manufacturers of information systems, system components, or information system services; (ii) vendors; and (iii) product resellers." [1] For purposes of this book, "agreement" as used in this definition includes general public licenses and open-source licenses as employed by TPSCs.

Testing, dynamic application A "black box" testing approach that analyzes an application in its running state and observes its responses to inputs, including inputs that simulate an attack.

Testing, fuzz A type of security testing in which "invalid data is input into the application via the environment, or input by one process into another process. Fuzz testing is implemented by tools called fuzzers, which are programs or script that submit some combination of inputs to the test target to reveal how it responds." [1]

Testing, penetration A type of "security testing in which evaluators mimic real-world attacks in an attempt to identify ways to circumvent the security features of an application, system, or network. Penetration testing often involves issuing real attacks on real systems and data, using the same tools and techniques used by actual attackers. Most penetration tests involve looking for combinations of vulnerabilities on a single system or multiple systems that can be used to gain more access than could be achieved through a single vulnerability." [1]

Testing, Red Team A Red Team is "a group of people authorized and organized to emulate a potential adversary's attack or exploitation capabilities against an enterprise's security posture. The Red Team's objective is to improve enterprise cybersecurity by demonstrating the impacts of successful attacks and by demonstrating what works for the defenders (i.e., the Blue Team) in an operational environment." [1]

Testing, static application A type of security testing that "detects software vulnerabilities by examining the app source code and binary and attempting to reason over all possible behaviors that might arise at runtime." [1]

Threat "Any circumstance or event with the potential to adversely impact the device, organizational operations (including mission, functions, image, or reputation), organizational assets, individuals, or other organizations through an information system via unauthorized access, destruction, disclosure, modification of information, and/or denial of service. Threats exercise vulnerabilities, which may impact the safety or effectiveness of the device." [4]

Threat modeling "A methodology for optimizing system, product, network, application, and connection security by identifying objectives and vulnerabilities, and then defining countermeasures to prevent, or mitigate the effects of, threats to the system." [4]

TPSC Third-party software components

Update "Corrective, preventative, adaptive, or perfective modifications made to software of a medical device." [4]

V&V Verification and validation

Vulnerability "A weakness in an information system, system security procedure(s), internal control(s), human behavior, or implementation that could be exploited." [4] "Flaw or WEAKNESS in a system's design, implementation, or operation and management that could be exploited to violate the system's SECURITY policy." [3]

Vulnerability assessment "Systematic examination of an information system or product to determine the adequacy of security measures, identify security deficiencies, provide data from which to predict the effectiveness of proposed security measures, and confirm the adequacy of such measures after implementation." [1]

Workaround *See* Compensating Control

References

[1] NIST CSRC Glossary, https://csrc.nist.gov/glossary.

[2] U.S. CERT NICCS Glossary, https://niccs.us-cert.gov/about-niccs/glossary.

[3] IEC 81001-5-1, "Health Software and Health IT Systems Safety, Effectiveness and Security – Part 5-1: Security – Activities in the Product Life Cycle," 2021, https://www.iso.org/standard/76097.html.

[4] U.S. FDA, "Cybersecurity in Medical Devices: Quality System Considerations and Content of Premarket Submissions - Guidance for Industry and Food and Drug Administration Staff," September 2023, https://www.fda.gov/regulatory-information/search-fda-guidance-documents/cybersecurity-medical-devices-quality-system-considerations-and-content-premarket-submissions.

[5] IMDRF, "Principles and Practices for the Cybersecurity of Legacy Medical Devices," April 11, 2023, https://www.imdrf.org/documents/principles-and-practices-cybersecurity-legacy-medical-devices.

[6] ISO/IEC 29147:2018 "Information Technology—Security Techniques—Vulnerability Disclosure," October 2018, https://www.iso.org/standard/72311.html.

About the Authors

Axel Wirth is the chief security strategist at Medcrypt. In this role, he helps guide the company and its clients in critical security strategy decisions and assists in the adoption of leading security technology by the healthcare industry. As an advocate for compliance, privacy, and security, and ultimately patient safety, in healthcare, Wirth draws from over 40 years of international experience in the industry. He is an active participant in industry organizations, serves on boards and committees, and is a frequent speaker at conferences, forums, and webcasts on subjects such as healthcare cybersecurity and privacy, medical device security, regulatory compliance, and related healthcare-specific topics. In recognition of his accomplishments, he was awarded the 2018 ACCE/HIMSS Excellence in Clinical Engineering & IT Synergies Award and the ACCE 2019 Clinical Engineering Advocacy Award; in addition, he has been recognized as a Fellow by the Association for the Advancement of Medical Instrumentation (AAMI) and the Healthcare Information and Management Systems Society (HIMSS). His extensive background in the healthcare IT and medical device industries includes engineering leadership as well as strategic business development and marketing roles with Siemens Medical, Analogic, Mitra, Agfa Healthcare, and Symantec. His education includes a BS in electrical engineering (BSEE) from the University of Applied Sciences, Düsseldorf (Germany) and an MS in engineering management (MSEM) from The Gordon Institute of Tufts University.

Christopher Gates is the director of product security at Velentium. He has over 50 years of experience developing and securing medical devices for numerous industry-leading device manufacturers and frequently collaborates with regulatory and standard bodies including CISA, HSCC, OWASP, Bluetooth SIG, IEEE, and the SBOM Forum to present, define, and codify tools, techniques, and processes that enable the creation of secure medical devices. Mr. Gates is both a highly experienced medical device development engineer and a cybersecurity subject matter expert, which gives him unique perspectives on the tools and techniques utilized to secure medical devices.

Mr. Gates has also created a student-paced certificate training program to instruct medical device engineers with the goal of elevating them to be experts in all aspects of medical device cybersecurity, including threat modeling, regulations, practical use of cryptography, secure coding, key security metrics, and side channel mitigations. The training program and this book are intended to help to alleviate the shortfall of security-capable engineers in the medical device industry.

Everywhere he goes, Mr. Gates promotes the secure total product lifecycle, an approach that eases the burden on developers and ensures high-quality products that work as intended to save and improve lives without putting patients at risk. His motivation became tangible when his first grandchild was born prematurely, before his lungs could fully close. When Mr. Gates visited him in the NICU, he realized that he had helped to design some of the machines that were allowing his grandson to breathe.

Jason Smith is the director of client services at Mission Triangle, a nonprofit organization that convenes, networks, trains, and equips other nonprofits and for-cause organizations with the tools they need to become proficient in the core competencies required to thrive as healthy, sustainable, effective organizations. He previously served as Velentium's senior strategist for marketing and communications, where he helped hundreds of medical device companies from startups to Fortune 500s engage the expertise they needed to change lives for a better world. Mr. Smith has been working as a professional writer and communicator in various fields and capacities for almost 25 years and was first approached about applying his skill set to the medical device field in 2016. In addition to *Medical Device Cybersecurity for Engineers and Manufacturers,* Jason is the coauthor of *28 Days to Save the World: Crafting Your Culture to be Ready for Anything,* a bestselling book that tells the story of "Project V" (2020's emergency ventilator manufacturing scale-up) and lays out the practical steps to build team cultures that survive and thrive amid crisis.

Contributors

The authors gratefully acknowledge the contributions of medical device cybersecurity leaders from both public and private sectors who lent their expertise on key subjects to this book.

Kevin Fu, Foreword to the Second Edition. Dr. Fu is the director of the Archimedes Center for Health Care and Medical Device Cybersecurity at Northeastern University, a professor in the Department of Electrical & Computer Engineering in the Khoury College of Computer Sciences and the Department of Bioengineering in the Kostas Research Institute for Homeland Security at Northeastern University, and the former Acting Director of Medical Device Security at the US Food and Drug Administration (FDA). Dr. Fu is an ACM Fellow, IEEE Fellow, AAAS Fellow, and a Dr. Dwight E. Harken Memorial Lecturer.

Greg Garcia, Foreword to the First Edition. Mr. Garcia is the executive director for the Health Sector Coordinating Council's Cybersecurity Working Group and the nation's first Assistant Secretary of Cybersecurity and Communications for the US Department of Homeland Security from 2006 to 2009.

Shannon Lantzy for her discussion of communicating cyberneeds, costs, and risks to senior leadership in Section 2.3. Dr. Lantzy is an expert in structured and quantitative benefit-risk assessment, human decision-making, and regulatory innovation, and product cybersecurity strategy. She is currently focused on getting medical technology innovation to market faster, better, an d cheaper. Dr. Lantzy has spent more than 20 years consulting in research and development for NASA, the FDA, and various medical device manufacturers. Her PhD research and academic publications center around the economics of information asymmetry, behavioral economics, and consumer decision-making under uncertainty.

Eric Henry for providing Chapter 3 on global regulations, standards, and guidances. Mr Henry is a senior quality systems and compliance advisor in the FDA and Life Sciences practice of King & Spalding. In this role, Mr. Henry advises King & Spalding clients on a variety of regulatory compliance matters, including quality system establishment and remediation, establishment of quality organizations, audit and inspection management and response, compliance-related due diligence reviews for mergers and acquisitions (M&A), training, and serving as expert witness during litigation. He is a 30-year industry veteran who has led and coached global organizations through a wide variety of quality and compliance challenges. Complementing his leadership experience is an extensive, hands-on skillset specializing in QMS (see Section 3.1) process development, software design controls (including regulatory frameworks for cybersecurity and machine learning systems), risk management, and Corrective and Preventative Action (CA/PA).

David Finn for his contribution of the discussion of healthcare facility types and their respective security needs in Section 4.2. Mr. Finn is the executive vice president (EVP) of governance, risk, and compliance at First Health Advisory. Most recently he served as VP at the College of Healthcare Information Management Executives (CHIME), responsible for providing programs, services, and educational opportunities focused on the professional growth and development of leaders within the healthcare security, technology, and data/application focus areas. Previously he served as EVP at CynergisTek, a top-ranked security provider in healthcare; health IT officer, Symantec; vice president and CIO/privacy and information security officer, Texas Children's Hospital, and integrated delivery system. David has also made significant contributions to the industry, both through his 2017 appointment to the Health Care Industry Cybersecurity Task Force, and as a prolific writer and speaker.

William Harding for his contribution to Section 4.3, Home Care Environments. Dr. Harding is a biomedical engineer and a distinguished technical fellow in medical devices, missile technology, and drug interdiction systems. He is the lead author of several publications and has designed and built the infrastructure and solutions for several medical device manufacturing sites. He has a doctorate in technology integration, holds 23 patents and trade secrets, and is considered a subject matter expert (SME) advisor to investors and executives for manufacturing, medical devices, data, and applications for surgical and clinical environments over the product lifecycle. Dr. Harding and co-contributor Sherri Douville are IEEE colleagues who have worked on 32 chapters, 3 books, and the industry's Clinical IoT P2933 Trust, Identity, Privacy, Safety, and Security technical standard together.

Sherri Douville for her contribution to Section 4.3, Home Care Environments. Ms. Douville is the CEO at Medigram, the mobile medicine company. She is also series editor for Trustworthy Technology & Innovation in Healthcare from Taylor & Francis, for which she leads a team of teams with domain expertise across medicine, hospital administration, IT, cybersecurity, and informatics that are collectively recognized as a top ecosystem of the best of the best across their respective disciplines, comprising in total 83 coauthors and editorial reviewers. Ms. Douville also helps to lead the industry's IEEE technical standard and serves on the healthcare CIO leadership College of Healthcare Information Management (CHIME) certification board for cybersecurity. She has given dozens of talks, panels, keynotes, university lectures, and podcasts, including for professional associations, governments, and prestigious universities internationally. She teaches corporate board education for cybersecurity at Santa Clara University.

Dennis Seymour contributed Section 4.3, Government and Military Agencies. Mr. Seymour has worked in healthcare cybersecurity since 1995, when he started as information security officer for the Department of Veterans Affairs after 10 years as a US Army Military Policeman and 8 years in federal law enforcement. He was appointed the director of the VA Center for Healthcare Information Security in 2003, where he initiated such projects as Medical Device Isolation Architecture and Medical Device Security Assessments. He served as the chairman of the HIMSS Privacy and Security Committee, HIMSS Risk Assessment Task Force, and the Medical Device Security Taskforce for HIMSS, leading the team that developed the Manufacturers Disclosure Statement for Medical Device Security (MDS^2). Mr. Seymour later served as the senior Program Manager of the VA Security Control Assessment (SCA) team responsible for the assessments of all 163 VA Medical Center IT systems in accordance with NIST SP 800-53. After retiring with 31 years of federal service, he continued in this role and others as a contractor supporting both the VA and Defense Health Agency programs. He is currently supporting a VA project Community Image Exchange Service as senior security engineer, a project that is working to allow veteran medical records to include magnetic resonance imaging (MRI), computed tomography (CT), and other images available from community care providers.

Allan Friedman for his contributions to the SBOM topic in Chapter 7. Dr. Friedman is the senior advisor and strategist at the Cybersecurity and Infrastructure Security Agency. He coordinates the global cross-sector community efforts around the software bill of materials (SBOM). Dr. Friedman was previously the director of cybersecurity initiatives at NTIA, where he led pioneering work on vulnerability disclosure, SBOM, and other security topics. Prior to joining the federal government, he spent over a decade as a noted information security and technology policy scholar at Harvard University's Computer Science Department, the Brookings Institution, and George Washington University's Engineering School. Dr. Friedman is the coauthor of the popular text *Cybersecurity and Cyberwar: What Everyone Needs to Know.*

Ken Hoyme for his contributions to Chapter 8, Postmarket Lifecycle. Mr. Hoyme is the CEO of Dark Star Consulting. He has nearly 40 years of experience in the design of regulated safety-critical secure systems. He has been active in many cross-industry initiatives, including H-ISAC's Medical Device Security Information Sharing Council (MDSISC), AAMI's Device Security

Working Group, HSCC JCWG working groups including the development of the Joint Security Plan, Medical Device Innovation Consortium (MDIC), and the University of Minnesota's Center for Medical Device Cybersecurity. He is a frequent speaker at medical device security conferences as well as FDA workshops on medical device security. Mr. Hoyme was formerly the senior Fellow of Global Product Cybersecurity at Boston Scientific and previously senior Fellow at their Cardiac Rhythm Management division. Before that, as a Distinguished Scientist at Adventium Labs, he performed government-funded research on the intersection of safety and security for cyber-physical systems. Earlier, Mr. Hoyme worked at Honeywell's Corporate Research for 18 years where he was a senior Fellow. He was awarded Honeywell's highest technical recognition for his work on the Boeing 777. He holds 40 US and 9 international patents.

Eric Pancoast for his contributions to the cryptogarphy section (Section 10.2). Mr. Pancoast is a cofounder and CTO at Medcrypt, a company focused on building proactive security solutions for medical devices. He was the cofounder of Gamma Basics, a radiation oncology-focused software startup that was acquired by Varian Medical Systems in 2013. In an earlier role at Mimeya Technology, he served as a lead architect and a software engineer.

Mike Kijewski for providing the Afterword. Mr Kijewski is the founder and CEO of Medcrypt. Prior to starting MedCrypt, he was the founder of Gamma Basics, a radiation oncology-focused software startup that was acquired by Varian Medical Systems in 2013. Mr. Kijweski holds an MBA from the Wharton School and a Master of Medical Physics from the University of Pennsylvania.

Index

www.ingramcontent.com/pod-product-compliance
Ingram Content Group UK Ltd.
Pitfield, Milton Keynes, MK11 3LW, UK
UKHW020944180725
460909UK00005B/30

9 781630 819910